Michael Eckert · Michael Kauffeld · Volker Siegismund

Natural Refrigerants: Applications and Practical Guidelines

AF894117

Michael Eckert · Michael Kauffeld · Volker Siegismund

Natural Refrigerants: Applications and Practical Guidelines

VDE VERLAG GMBH

ATMOsphere

Disclaimer

VDE VERLAG, shecco (as ATMOsphere), the editors, the authors, and the companies and organisations that employ them do not endorse the performance, worker safety, or environmental acceptability of any of the technical options discussed. Every industrial operation requires consideration of worker safety and proper disposal of contaminants and waste products. Moreover, as work continues – including additional toxicity evaluation – more information on health, environmental and safety effects of alternatives and replacements will become available for use in selecting among the options discussed in this document.

VDE VERLAG, shecco (as ATMOsphere), the editors and the authors, in furnishing or distributing this information, do not make any warranty or representation, either express or implied, with respect to the accuracy, completeness, or utility; nor do they assume any liability of any kind whatsoever resulting from the use or reliance upon any information, material, or procedure contained herein, including but not limited to any claims regarding health, safety, environmental effect or fate, efficacy, or performance, made by the source of information.

Mention of any company, association, or product in this document is for information purposes only and does not constitute a recommendation of any such company, association, or product, either express or implied by VDE VERLAG, shecco (as ATMOsphere), the editors, the authors, and the companies and organisations that employ them.

Title photo: Michele Cornelius/123RF.com

Bibliographic information published by the Deutsche Nationalbibliothek
The Deutsche Nationalbibliothek lists this publication in the Deutsche Nationalbibliografie; detailed bibliographic data are available on the internet at http://www.dnb.de.

ISBN 978-3-8007-5330-7 (Print)
ISBN 978-3-8007-5331-4 (E-Book)

ISBN 978-2-8052-0611-5 (Print)
ISBN 978-2-9602-8050-0 (E-Book)

VDE VERLAG GMBH · Berlin · Offenbach
Bismarckstr. 33, 10625 Berlin/Germany

shecco SPRL (ATMOsphere)
Avenue de la Chasse 133–135, Brussels 1040/Belgium

Printed by Plump Druck + Medien GmbH, Rheinbreitbach
Produced in Germany

2022-02

This textbook is supported by:

Advertorial

Norwegian Apartment Blocks Choose Advansor's CO_2 Heat Pump for Sustainable Heating Needs

The green apartment project Vestre Billingstad has been using an Advansor CO_2 heat pump successfully for two heating seasons already to provide a sustainable solution with high energy efficiency.

Expanding into the heating sector

Advansor is known for manufacturing sustainable refrigeration racks that use natural refrigerant CO_2. However, this technology can also be used to produce heat and based on the experience from building more than 8,000 refrigeration racks, Advansor's heating solution was developed several years ago already. This solution is now gaining traction as demand for more sustainable products in the heating market increases too.

About the project

Over a period of 10 years, Vestre Billingstad will be the home of almost 4,000 people living in 1,650 apartments supplied from 120 energy wells in the ground. The ambition is renewable energy coverage of around 90%.

When the project started, the end user was looking for a heat pump with environmentally- and climate-friendly refrigerants that could be expanded over 10 years. This need for expansion of several stages made Advansor's CO_2 heat pump a perfect solution thanks to its low part-load performance and the fact that the three-exchanger system was specially adapted for tap water and space heating, resulting in the lowest possible energy consumption.

Finding the best solution

The Advansor SteelXL 5x0 was selected to provide an environmentally- and climate-friendly solution for the first stage of the project. It is an 800kW heat pump complete with heat recovery and chiller module. Everything was built on the rack, which reduced the work time and installation cost on site.

Advansor did more than just build the equipment. With its expert knowledge, the team also helped with calculations for many different scenarios to ensure the best possible result. The rack is linked to 22 sub-centres run by four different contractors with slightly different heating systems that required different calculations. This is the first Advansor rack equipped with three heat exchangers as to improve the efficiency.

The customer, Winns AS, was very happy with the end result delivered by Advansor.

For more information, visit: https://www.advansor.com/

ADVANSOR
a DOVER company

The art of achieving the perfect temperature
For more than ten years, Beijer Ref has been offering the market refrigeration units and heat pumps based on natural refrigerants
BEIJER REF
Provides global expertise in temperature control products and solutions
www.beijerref.com

Do you have a CO_2OL Power?
COOL
The transcritical CO_2 solution for high power applications
Power CO_2OL
Logistics warehouses
Processes
Sport venues
Heat pumps
©2020 Carrier. All Rights Reserved.
Featuring: CO_2OLtec Evo
▪ 15,000+ natural refrigerant CO_2 mechanical systems in Europe
▪ Patented modulating vapour ejector assisted by a CO_2 pump
▪ Up to 30%* energy savings
▪ High efficiency in all climates
▪ Simplicity in use
* versus a standard transcritical system
Carrier
Learn more on our solutions:
www.carrier-refrigeration.com

COPPER TUBES
TALOS®
XS
X-TRA STRONG
COPPER ALLOY TUBES CO2
TALOS® XS tubes are manufactured using a high strength copper-alloy (CuFe2P) material, in order to satisfy the demands of today's high-pressure CO2 systems in refrigeration, as well as, other HVAC&R applications. TALOS® XS tubes are designed to withstand high operating pressures of up to 120/130bar and provide a reliable and safe solution in a complete range of sizes from 3/8" to 2 5/8".
Halcor specializes in the production, processing and marketing of copper and copper alloy products with dynamic commercial presence in the European and global markets. For more than 80 years, Halcor has been offering innovative and added-value solutions that meet contemporary client demands in fields, such as HVAC&R, plumbing, renewable energy, architecture, engineering and industrial production.
HALCOR
COPPER TUBES DIVISION of:
ELVALHALCOR
www.halcor.com

395 YEARS OF EXPERIENCE
IN VENTILATED HEAT EXCHANGER TECHNOLOGY
Fincoil 1950
CONTARDO® 1928
Helpman 1924

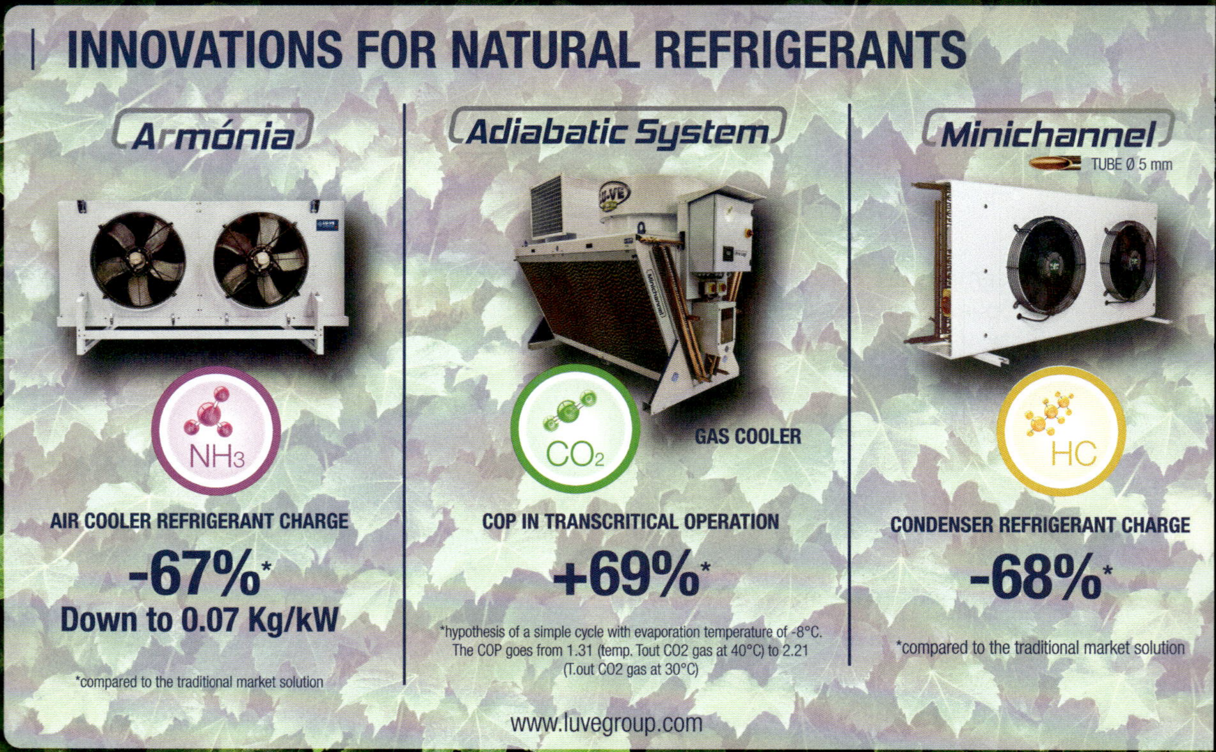
INNOVATIONS FOR NATURAL REFRIGERANTS
Armónia
NH_3
AIR COOLER REFRIGERANT CHARGE
-67%*
Down to 0.07 Kg/kW
*compared to the traditional market solution
Adiabatic System
CO_2
GAS COOLER
COP IN TRANSCRITICAL OPERATION
+69%*
*hypothesis of a simple cycle with evaporation temperature of -8°C. The COP goes from 1.31 (temp. Tout CO2 gas at 40°C) to 2.21 (T.out CO2 gas at 30°C)
Minichannel
TUBE Ø 5 mm
HC
CONDENSER REFRIGERANT CHARGE
-68%*
*compared to the traditional market solution
www.luvegroup.com

More than cold.
LR Kältetechni
ENERGY-EFFICIENT REFRIGERATION TECHNOLOGY WITH NATURAL REFRIGERANTS
Future-orientated - Efficient - Environmentally friendly
Refrigeration systems with propane, propene, ethane, CO_2, ammonia, in accordance with ATEX.
L&R Kältetechnik GmbH & Co.KG
Hachener Strasse 90a-c • 59846 Sundern-Hachen
T +49 2935 9652 0 • info@lr-kaelte.de • www.lr-kaelte.de

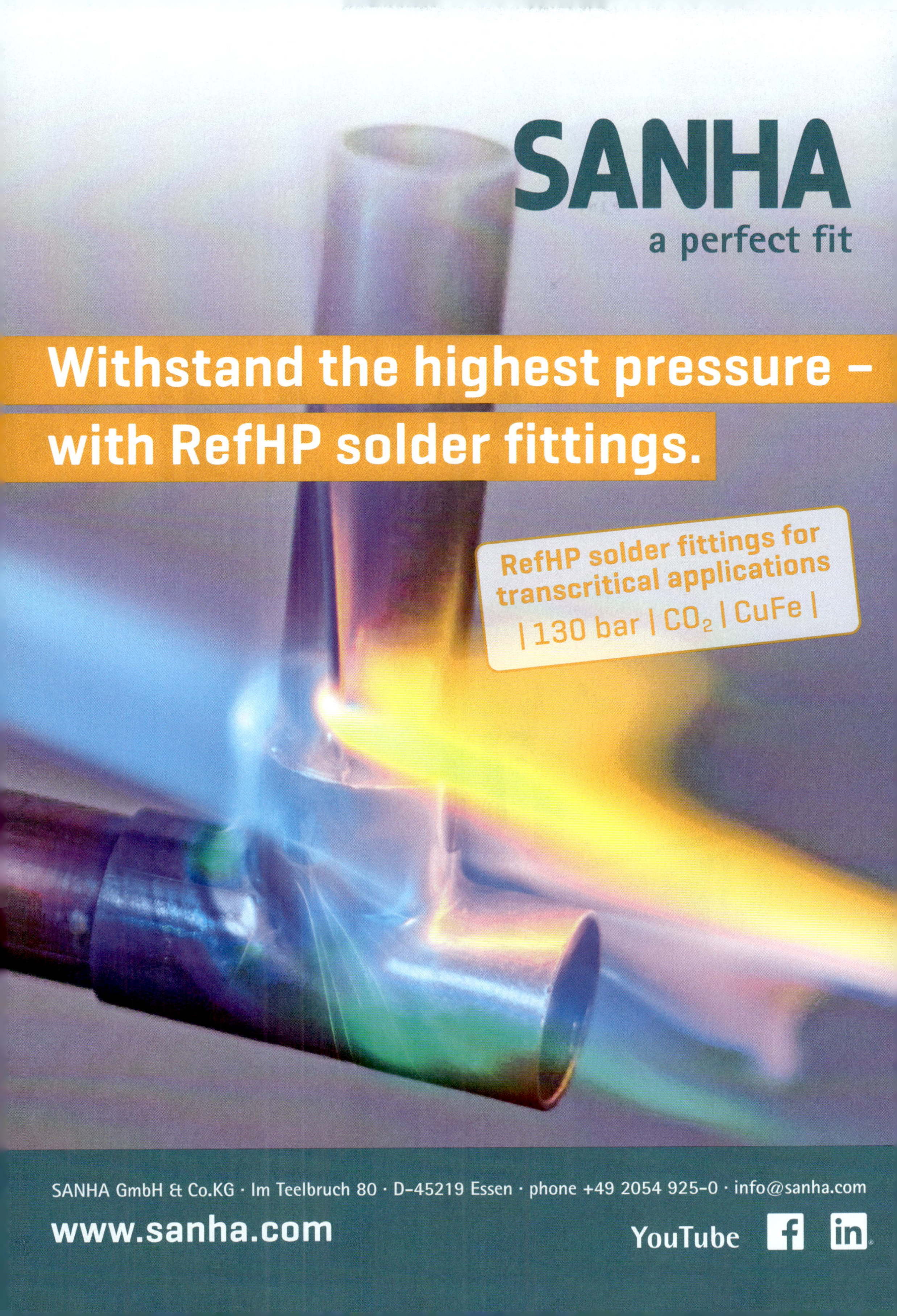
SANHA
a perfect fit
Withstand the highest pressure –
with RefHP solder fittings.
RefHP solder fittings for
transcritical applications
| 130 bar | CO_2 | CuFe |
SANHA GmbH & Co.KG · Im Teelbruch 80 · D-45219 Essen · phone +49 2054 925-0 · info@sanha.com
www.sanha.com
YouTube

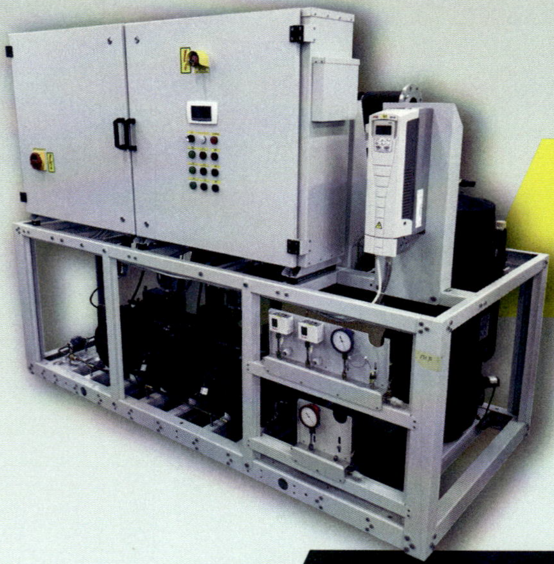

VULKAN
LOKRING
www.vulkan.com
BRAZE-FREE
TUBE SOLUTIONS
LOKRING® STOPPER
SEALING OF REFRIGERANT CIRCUITS
Advantages
One pneumatic tool for crimping and sealing
No heat generation during assembly
Lower failure rate than brazing and ultrasonic welding
Suitable for HC, HFC and HFO
Nearly maintenance free
LOKRING® TUBE CONNECTIONS
FOR VRV/VRF INSTALLATIONS
Key facts
More than 40 years of experience
2 billion LOKRING connections have successfully been made
Suitable for HC, HFC, HFO and CO_2 (subcritical)
UL, TÜV and ISO approved
German technology

WITT
future-proof
refrigeration technology
Your Expert in CO_2 and NH_3
th-witt.com
For your successful F-Gas Exit.

We dedicate this book to Professor Johannes Reichelt, an advocate of natural refrigerants since the 1990s. We acknowledge his lifelong teaching and training efforts in the field of refrigeration and heat pump technology, especially with natural refrigerants. The editors and authors donate their revenues to the Valerius Füner Foundation, founded by Johannes Reichelt in 2001 and named after the first professor, who introduced refrigeration technology at Karlsruhe University of Applied Sciences in 1948. The Foundation supports students of refrigeration technology at Karlsruhe University of Applied Sciences, where several of the authors of this book were inspired to work in our industry.

Preface by the editors

The Norwegian refrigeration technology professor Gustav Lorentzen, † 1995, stated at an IIR meeting in Gent in 1993: "*In the present situation, when the CFCs and in a little longer perspective the HCFCs are being banned by international agreement, it does not seem very logical to try to replace them by another family of related halocarbons, the HFCs, equally foreign to nature.*"

Does it then make sense to replace hydrofluorocarbons (HFC), which are on a downward trend in many countries, with another family of related halocarbons? Certainly not! Unsaturated hydrofluorocarbons (uHFC) – marketed as hydrofluoroolefins (HFO) – are not likely to be a long-term solution either due to the influence of their atmospheric breakdown products on the regional environment.

Alternative solutions have been available for over 100 years, namely:

- Ammonia,
- Carbon dioxide,
- Hydrocarbons,
- Water and
- Air, or possibly N_2O mixed with CO_2 for low-temperature applications (< −50 °C / −58 °F).

They have all been found in nature for millions of years. Therefore, their emissions do not cause any unknown changes in nature, such as CFCs depleting the ozone layer or CFCs and HFCs contributing substantially to global warming or HFOs resulting in large amounts of trifluoroacetic acid (TFA).

It took over 40 years from the invention of chlorofluorocarbons (CFC) and hydrochlorofluorocarbons (HCFC) to the discovery of these synthetically produced substances' negative effect on the ozone layer and another 20 years before they were banned under the Montreal protocol. The Kigali Amendment to the Montreal Protocol signed on October 16, 2016, eventually will phase-out their replacement fluids, namely HFCs. Let us not wait such a long time to replace HFOs due to TFA formation in the atmosphere. Let us go back to the early beginning of refrigeration technology and re-use our old friends: natural refrigerants. Find out how to do it in this book!

The main reason for the decision to "phase down" HFCs under the Kigali Amendment is their high global warming potential (GWP). However, the direct contributions to global warming due to refrigerant emissions from refrigeration or air conditioning systems are often lower than the indirect contribution from CO_2 emissions from generating the electricity to operate the systems. Regardless of which refrigerant you want to use, the system must be more efficient in the future than its predecessor! Fortunately, many installations with natural refrigerants are very energy efficient due to the good thermodynamic properties of these refrigerants, as detailed in this book.

Yet another environmental aspect is the CO_2-equivalent emissions during production of the refrigerants. Natural refrigerants score by far better than HFCs and especially HFOs, resulting in one tenth or less of CO_2-equivalent emissions during their manufacturing process.

Each of the natural refrigerant solutions has pros and cons, and therefore different applications. Accordingly, in this book they are all covered in a separate chapter in which their preferred applications are described. For each natural refrigerant, we were able to enlist respective experts as authors. We thank all of them for their tremendous work writing this comprehensive book on natural refrigerants. A special thank you goes to Andy Pearson who extensively reviewed Chapters 1 to 4 with special emphasis on the ammonia chapter, which to some extent was based on his IIR Guide "Ammonia as Refrigerant" published by the International Institute of Refrigeration in 2008. We also thank Bernd Hansemann of VDE Verlag and Ilana Koegelenberg of ATMOsphere for their patience with us and their countless hours of work in making this book possible. Thank you to ATMOsphere and the whole team supporting this English version. As non-English native speakers we apologise for any inconvenience because of possible language errors.

A final thank you goes to you, the reader of our book. May you find some guidance in applying what we love most – NATURAL REFRIGERANTS.

Karlsruhe, Germany	Ludwigsburg, Germany	Assamstadt, Germany
Michael Kauffeld	Michael Eckert	Volker Siegismund

Introduction by ATMOsphere

The future is natural

Time and time again, the lack of suitable training materials in the natural refrigerant space has been cited as an excuse for slowed progress towards more sustainable HVAC&R solutions.

As an independent global market accelerator for natural refrigerants, ATMOsphere has been striving to remove barriers for the uptake of climate-neutral cooling systems for nearly two decades. That is why, when we first came across VDE VERLAG's German natural refrigerant textbook in 2019, we immediately saw an opportunity to also bring this expert knowledge to the global stage and address the training barrier.

The "*Natürliche Kältemittel – Anwendungen und Praxiserfahrungen*" German textbook ticks all the boxes, offering a comprehensive technical look at how to work with all natural refrigerants. By simply looking at the authors' list, it was clear that the textbook was the resource we had been waiting for.

After interviewing Professor Michael Kauffeld, who initially came up with the idea for the original textbook, I became even more convinced that this was something that could benefit a much wider audience, not just the German-speaking industry. Professor Kauffeld was kind enough to put me in touch with the publisher of the book, Bernd Hansemann of VDE Verlag, who immediately saw the potential of an English translation as well.

It would be safe to say that none of the parties involved (and there were many) had any idea of the sheer complexity of the project we had taken on. It was a pre-COVID world and we were optimistic that things would progress quickly with the translation and republishing in English. But things were a bit more complex than that. The volume of the book and the highly technical nature of the material meant that a lot of revision was required – all while managing the new pandemic reality and full-time jobs.

Finally seeing the completed work, I am so excited about what this textbook can do for raising the bar for natural refrigeration installations around the world – not only from a safety point of view but also in terms of efficiency. However, it is not just about improving these systems, it is also about driving the move away from harmful, climate-heating and water-polluting refrigerants.

I firmly believe that by making this knowledge more accessible, we can greatly accelerate the adoption of natural, clean cooling technologies around the world – something that is more urgent than ever.

Happy learning!

A word of thanks

Allow me a few lines to thank the incredible team who made this journey to an English natural refrigerant textbook possible. Without everyone's hard work and patience, "*Natural Refrigerants: Applications and Practical Guidelines*" would not exist.

First and foremost, I would like to thank Bernd Hansemann, who always kept things moving and never lost hope that we would one day arrive at a complete product.

Thank you to Prof. Kauffeld, for his vision, his long hours, and trusting us to translate the book.

Thank you to all the authors, for taking the time to rework and improve the translated manuscript; also to the legend himself, Andy Pearson, for getting involved with the revisions; and everyone else who worked behind the scenes.

Of course, thank you also to all the companies that supported the financing of this book project by placing an advertisement of their specific natural refrigerant solution. Without your support we would not be able to offer this book at such a reasonable cost. It is exciting to see such a variety of companies offering components, systems, and knowhow for, and with, natural refrigerants in all applications. It's an ever growing community we are very happy about!

Last but definitely not least, I want to thank the persons I relied on the most – Jan Wisniewski, Maureen Garry, and Joseph Herzog – our English native-speaking proofreaders who ensured the best quality of the text.

I remain convinced that all this hard work will be worth it in the end.

Ilana Koegelenberg
ATMOsphere Co-Founder & Market Intelligence Manager

Table of Contents

Abbreviations and units

AB	Alkylbenzene
ATEL	Acute Toxicity Exposure Limit
CFC	Chlorofluorocarbon
COP	Coefficient of Performance
EER	Energy Efficiency Ratio
GWP	Global Warming Potential
HC	Hydrocarbon
HCFC	Hydrochlorofluorocarbon
HFC	Hydrofluorocarbon
HFO	Hydrofluoroolefin
HP	High Pressure
IDLH	Immediately Dangerous to Life and Health
IHX	Internal Heat Exchanger
LEL	Lower Explosive Limit
LFL	Low Flammability Limit
LP	Low Pressure
LT	Low (Temperature) Cooling
M	Mineral oil
MPRV	Medium Pressure Reduction Valve
MT	Medium (Temperature) Cooling
NF	Nonflammable
ODL	Oxygen Deprivation Limit
ODP	Ozone Depletion Potential
PAG	Polyalkylene Glycol
PAO	Polyalphaolefin
PE	Primary Energy
PED	Pressure Equipment Directive
PL	Practical Limit
POCP	Photochemical Ozone Creation Potential
POE	Polyolester
RSW	Refrigerated Sea Water
TEWI	Total Equivalent Warming Impact
UEL	Upper Explosion Limit
UL	Upper limit

VOC	Volatile Organic Compounds
R50	Methane CH_4
R170	Ethane C_2H_6
RC270	Cyclopropane C_3H_6
R290	Propane C_3H_8
R717	Ammonia NH_3
R718	Water H_2O
R744	Carbon Dioxide CO_2
R744A	Nitrous Monoxide N_2O
R600	Butane C_4H_{10}
R600a	Isobutane C_4H_{10}
R1150	Ethene C_2H_4
R1270	Propene C_3H_6

Units – Please note:

SI units and "neutral" numbers as percent (%), ratios, factors etc. are written in *SI standard* with comma (,) as decimal separator/radix character. Solely for pure *imperial units*, the Anglo-American standard with dot (.) as decimal separator is used.

1 Introduction

Michael Kauffeld

In Egypt, more than 2000 years ago, drinks were cooled in open-pore clay jugs. Water evaporated on the outside of these vessels, thus extracting heat from the amphora. In India, a few thousand years ago, open flat clay bowls were also placed under the starry sky at night – many hundreds of such bowls each formed a thin layer of ice overnight. The harvested ice was then used for a variety of cooling purposes. In Europe, it was not until 1755 that the Scotsman William Cullen in Glasgow made ice by evaporating water. Whether Egypt, India or Scotland – water was the refrigerant of choice for the first simple attempts to reach temperatures below the ambient temperature. It was Jacob Perkins (1834) and Ferdinand Carré (1859) who first introduced ethyl ether and ammonia. Since both substances have a higher vapour pressure than water, the corresponding chillers could be made to be more compact. Thaddeus Lowe (1867) patented even more compact ice machines by using carbon dioxide.

Most of these substances share a common characteristic: they occur in large quantities in nature. This was also the only available choice for the early refrigeration pioneers, because there was no chemical industry that could have produced other substances in the 19th century. Since the first natural refrigerants all have at least one disadvantage – ammonia is toxic, ethyl ether burns, carbon dioxide requires high operating pressures and water, on the other hand, leads to very low system pressures – for decades refrigeration systems using these natural refrigerants have been restricted to use in industrial applications. The first household refrigerators with methyl chloride or sulfur dioxide regularly led to accidents. Therefore, propane was promoted as the odourless safety refrigerant in the 1920s, as it was not toxic, in contrast to methyl chloride or sulfur dioxide. The development of CFCs was led by Thomas Midgley Jr. because the opinion at that time – not only of Midgley's employer Frigidaire, a General Motors subsidiary – was that household refrigerators would only prevail if a non-toxic and non-flammable refrigerant could be found.

None of the developers of CFCs could have imagined that the successful adoption of these substances across many applications (refrigerants, propellants, foam blowing agents, fire extinguishers, aerosol, etc.) would damage the ozone layer that shields the earth from harmful UV light. The end of this universal CFC deployment came with the discovery of the ozone hole, caused by the splitting of the stratospheric ozone by the chlorine atoms in the CFCs. The manufacturers of CFCs were quick to find substitutes for the refrigeration and air-conditioning industry. This led to the development of hydrofluorocarbons (HFCs) – halogenated hydrocarbons that only use fluorine as the substitutional atom. When HFCs such as R134a were launched in 1992, the first cautionary voices were already raising the alarm concerning the high global warming potential of these fluids. Since the Paris Climate Agreement (2015) and in addition, the Montreal Protocol Amendment of Kigali (2016), it has become clear that HFCs have no commercial future. Europe, which is a pioneer in this area, has enacted regulations on fluorinated greenhouse gases (F-Gas Regulation), setting limits on their use.

Accustomed to the local safety of CFCs and HFCs, parts of the refrigeration and air-conditioning industry are now again looking to the chemical industry. Their answer to CFC's ozone depletion and the HFC's greenhouse effect is hydrofluoroolefins (HFOs). According to Mark McLinden, this is the last group of chemicals which chemistry can offer for vapour compression refrigeration cycles[1)] [1-1 and 1-2]. Olefins are hydrocarbons with one or more carbon-carbon double bonds. Due to the double bond between two carbon atoms, these molecules disintegrate in the atmosphere within less than 11 days. Their global warming potential is therefore low (around 1 to 4). However, the rapid atmospheric decay also means that a lot of trifluoroacetic acid (TFA) can be formed locally under certain weather conditions, namely lack of wind. R1234yf – the refrigerant used in vehicle air conditioning systems – yields significantly more TFA than as the HFC it replaced (R134a). Holland et al. quantify in their study from 2021 [1-3] the effects on atmospheric TFA of a full transition from HFC-134a to HFO-1234yf use in industry using an updated atmospheric fate model applied to the 2015 emissions of R134a and its regional distribution. A 33-fold increase in global atmospheric TFA burden is predicted following a complete transition with a maximum 250-fold increase over Central Europe as loss processes cannot compensate with the increase in emissions of TFA precursors [1-3]. This is more than previous simulations have predicted and will need verification by other researchers. Considering the five to eight times higher TFA formation rate from R1234yf compared to R134a, and assuming that the natural TFA loss processes have capacity limitations, the increases in TFA predicted by Holland et al. seem to be reasonable – and are frightening: TFA washes out of the atmosphere and can accumulate in closed water reservoirs and eventually may end up in ground water. Rainwater deposition measurements of TFA at eight different sampling sites of the German Meteorological Service (DWD) systematically distributed over Germany [1-4] reveal a four-fold higher TFA deposition during the actual measurement period from 2018 to 2020 compared to 1995/1996. Behringer et al. [1-4] conclude that “the trifluoroacetate concentrations in rainwater determined in the measurement program are significantly higher than previously reported in the relevant literature”; thereby supporting the findings of Holland et al. [1-3] that there will be a much higher increase in TFA deposition in central Europe than previously anticipated [1-5]. TFA is very difficult to remove from water and is now found in drinking water in small amounts. There is currently assumed to be no health risk for people. However, the situation is different for aquatic organisms.

Gustav Lorentzen's statement in 1994, that “it does not seem very logical to try to replace one chemical by another family of related halocarbons equally foreign to nature” is as relevant today as it was a quarter of a century ago.

To ensure sustainability, it makes sense to use refrigerants only that have been in the natural cycle for millions of years and therefore do not have any unforeseen negative environmental effects. Local safety risks, if using flammable or poisonous refrigerants, are in general much easier to deal with than global environmental problems caused by careless handling of large quantities of synthetic substances.

The refrigerants that were already in use over a hundred years ago remain as working materials for refrigeration systems with vapour compression even today.

These natural refrigerants are:

- ammonia,

1) Mark McLinden and his co-workers looked at 60 million different chemicals and their suitability as refrigerants in vapour compression cycles [1-1]. “The results of McLinden et al. indicate [...] that there are no fundamentally new classes of chemicals available for use in vaporcompression refrigeration systems. [...] The situation may be to return (recycle?) to earlier generations, such as the ammonia, CO_2, and hydrocarbons of the first generation.” [1-2]

- carbon dioxide,
- hydrocarbons (e.g. ethane, isobutane, propane, propene)
- water, and
- nitrous oxide (N_2O) with CO_2 as additive (for temperatures below –50 °C/–58 °F).

In addition to the discussion about refrigerants, one important aspect of refrigeration and air-conditioning systems has to be addressed: energy efficiency. Refrigeration and air-conditioning systems contribute, albeit indirectly, to the greenhouse effect through their electric energy consumption. The development of systems with higher refrigeration efficiency is at least as important as the actual use of low-GWP refrigerants. There is still a lot of room for improvement here. Much of the increase in efficiency is based on better system controls. However, changes in the system layout or system operation can also lead to reduced energy consumption. By cleverly combining some measures, over 50 % of the (electrical) drive energy can easily be saved [1-6]. Both the Association of German Mechanical Engineering Industry Association (VDMA) and the International Institute of Refrigeration (IIR) state that around 15 % of the total electricity consumption is used for refrigeration and air-conditioning systems. This share could therefore be reduced to 7 to 8 % relatively easily. The exceptional thermodynamic properties of natural refrigerants will make a significant contribution to the higher energy efficiency of refrigeration and air-conditioning systems.

To give an overview of these natural refrigerants, a basic description of them is given here. Subsequently, individual chapters deal in detail with the respective technologies.

1.1 Ammonia

The thermodynamic properties of ammonia (NH_3 or R717) are unusually beneficial compared to other refrigerants. It has the highest specific heat capacity (see Chapter 4) and enables a particularly high heat transfer coefficient during evaporation of liquid and condensation. Ammonia is the most frequently used refrigerant in the food industry's refrigeration systems. The National Aeronautics and Space Administration (NASA) relies on this fluid for its space expeditions, because the excellent heat transfer properties result in extremely compact heat exchangers. In general, on a like-for-like basis, ammonia is inherently more efficient than most other refrigeration fluids.

In addition to the high energy efficiency of the refrigeration systems, ammonia has another positive feature: it is priced much lower than any fluorocarbon. In addition, although at concentrations above 5000 ppm ammonia is a danger to life, its odour can be detected by humans at concentrations of around 5 ppm, which acts as an early warning system. The risk of inhaling dangerous amounts of ammonia is therefore very small.

Ammonia is an excellent refrigerant for water chillers, although the refrigeration system has to be placed outdoors or else the use of ammonia is restricted to a machine room. There are numerous examples of installations in department stores, hotels and airport buildings. Installation locations outside a building are preferred because the light ammonia gas rises quickly into the atmosphere. For installation locations within a building plant room ventilation is used to clear the air indoors in the event of an ammonia leak. The high affinity of ammonia for water can be used for improving safety. In the state parliament building in Düsseldorf, Germany, for example, the ammonia system is installed in the basement. In the event of ammonia gas leakage, water mist is sprayed into the ventilation system and the ammonia vapour is chemically bound before it can escape into the atmosphere. Alternatively, a scrubber can be used to perform the same job. Please note that it is not a good idea to spray water on liquid ammonia spills as this will result in even more ammonia transferred into gas phase.

Ammonia dissolves in water in the atmosphere and then acts as a nitrogen-rich fertiliser for the plants in the surrounding area. Its global warming potential is therefore virtually zero.

Depending on the application, ammonia can be used in direct evaporation, with a secondary liquid coolant circuit or as the upper stage of a cascade refrigeration system. Direct evaporation is the preferred option, where possible, for all applications without public traffic and with a small, trained group of people who are operationally on site, for example in breweries or cold stores.

Indirect refrigeration systems are preferred for applications with a larger workforce or when serving areas are accessible to the public. In this way, the high energy efficiency of ammonia need not be sacrificed. The higher inherent efficiency of the ammonia refrigeration circuit can generally compensate for the energy disadvantage of a secondary coolant circuit. Applications with a secondary coolant circuit are also found in the food industry and even in large supermarkets. For low-temperature applications (freezer), a cascade of ammonia and CO_2 is applicable – a technology that has been widely adopted in the food industry.

The chemical compatibility of ammonia with materials normally used in the construction of smaller refrigeration systems is not as good as for the other natural refrigerants. Contact with copper has to be avoided under all circumstances. As soon as there is even a small amount of water in the ammonia, the copper reacts very quickly. The same applies to zinc. In contrast, copper-free and zinc-free aluminium alloys are ammonia-resistant – at least as long as the water content is below 10 %.

As part of the worldwide exit from HFCs, ammonia is being used as a refrigerant in significantly more refrigeration applications than before, because ammonia refrigeration systems are characterised by high energy efficiency. The slight increase in capital costs for installing the systems is rapidly recouped through the lower cost of refrigerant compared with HFC and HFO blends, the lower operating costs, and fewer leak checks.

1.2 Carbon Dioxide

Carbon dioxide (CO_2 or R744) was first proposed and patented as a refrigerant by the American inventor Alexander Catlin Twining in 1850. After the first land-based CO_2 refrigeration systems from, among others, Thaddeus S. C. Lowe, Franz Windhausen, Everard Hesketh and Carl von Linde in the late 19th century, CO_2 became the refrigerant of choice at the beginning of the twentieth century, especially for ship refrigeration systems. CO_2 was later replaced by HCFC or HFC refrigerants because of the high pressures required in CO_2 systems and its comparatively low critical temperature of 31 °C/88 °F (at 74 bar/1073 psi). The main reason for the success of the early ship refrigeration systems was the low toxicity compared to sulfur dioxide and ammonia and the fact that CO_2 is not flammable. Both are still deciding arguments for favouring the renewed interest in this refrigerant.

CFCs and HFCs are only hazardous to health at relatively high concentrations due to oxygen displacement. However, the human body is adversely affected if there is approximately 4 % or 40000 ppm CO_2 in the air we breathe and from approximately 10 %, death is possible. The discussion about ozone depletion by CFCs and the high global warming potential of most halogenated hydrocarbons means that CO_2 refrigeration systems are now again being installed, but with appropriate safety technology.

From around 1998, the first small-scale, high-pressure compressors became available. The first supermarket refrigeration systems with carbon dioxide as the sole refrigerant were built as a cascade (CO_2/CO_2) system, because initially there were concerns about proper lubricating oil return to the individual compressors for medium and low temperature. During the last 15 years two-stage CO_2 systems have been developed for supermarket refrigeration systems, which increases their energy efficiency. All these systems operate transcritically in summer and rely on the regulation of the discharge (high side) pressure according to the outside air temperature. Up to now, this has been a disadvantage of CO_2 technology. With high outside air temperatures (above approximately 22 °C / 72 °F), simple transcritical CO_2 vapour compression plants require more energy than their HFC counterparts, although in lower ambient temperatures they are significantly more efficient. In Europe, there has been a CO_2 "equator" line at about the level of the Alps. South of it, the ambient temperature was too often above this temperature and hence inefficient transcritical operation of CO_2 systems took too many hours a year. If the ambient temperature is high these systems lose too much refrigeration capacity but the power input remains much the same, hence these systems could not be engineered to be both economic and energy efficient.

The situation is different for cascade systems, with CO_2 in the lower stage only and ammonia or hydrocarbons in the upper stage, thus achieving very good energy efficiency, even at high ambient temperatures.

Recent developments in transcritical CO_2 systems do provide for small external refrigeration systems to increase sub-cooling, for condensers/gas coolers moistened with water (adiabatic cooling), for additional parallel CO_2 compressors, and more recently for ejectors and flooded evaporators. Even expansion machines which recover expansion energy are also being assessed. If these measures are skillfully incorporated into the transcritical CO_2 plant, then CO_2 refrigeration systems can be as viable as HFC refrigeration systems even in areas with higher ambient temperatures. However, for smaller supermarkets (less than approximately 1000 m^2 / 10000 ft^2 of retail floor space), the investment costs are approximately 10–20 % higher than those of a comparable HFC refrigeration system. In larger supermarkets, the higher volumetric cooling capacity and the consequently more compact components can partially offset the higher costs with CO_2.

Other applications where CO_2 is clearly advantageous are hot water heat pumps and heat recovery from refrigeration systems. The reason for this is the high de-superheating heat or the continuous temperature change of CO_2 during supercritical cooling. This is an excellent way of heating water from the lowest possible temperature up to 70 °C (158 °F) or more. Many companies in Japan offer CO_2 hot water heat pumps and according to the Association of Electricity Suppliers in Japan, over five million such devices were installed as of 2016. These consume only a quarter of the electricity used by electric water boilers.

1.3 Hydrocarbons

Hydrocarbons are the base molecules of all CFCs, HCFCs, HFCs and HFOs – the "HC" in the respective abbreviations stands for hydrocarbon. The synthetic refrigerants all come from methane (e.g. R12 or R32), ethane (e.g. R134a), propane (e.g. R227ea, which is contained in R423A and 425A) or propene (e.g. R1234yf). The hydrocarbons are more reactive than fluorocarbons and therefore – with the exception of methane – broken down much more quickly in the atmosphere; usually within a few days. By contrast, methane has an atmospheric life of 12,4 years. This is still significantly shorter than most HFCs.

In addition to the low greenhouse effect, hydrocarbons do not have ozone depletion potential. Their outstanding thermodynamic properties result in higher efficiency than comparable systems with HFCs. Improvements of 10 to 15 % are not uncommon. The lower pressure ratios often lead to quieter systems. Since hydrocarbons are also non-corrosive and have excellent oil compatibility, they are actually close to the ideal refrigerants apart from their flammability.

In order for any fire to start, and also in the case of refrigerant leaks, three things are necessary: fuel, oxygen – both in the right ratio to each other – and the right ignition source. If one of them is missing, there will be no fire. It is therefore important to exclude at least one of the three conditions through appropriate design and construction measures. Standards EN 378 and IEC 60335-2-89 state no restrictions for refrigerant charge quantities up to 150 g (5.3 oz) in a single refrigeration circuit. With larger refrigerant charge quantities, EN 378 specifies special requirements for explosion protection. The pressure equipment directive 97/23/EC is also applicable for manufacturers of hermetic compressors. From a certain compressor size (pressure-volume product greater than 50 bar · litres), this requires special supervision during production, which is why the range of corresponding hermetic compressors approved for hydrocarbons for larger refrigeration capacity is still quite limited. However, air-conditioning systems from Chinese, Indian and Italian manufacturers as well as plug-in refrigeration units, chillers, heat pumps, prototypes of container cooling systems, and various other devices charged with hydrocarbons are already on the market. Safety analysis tests were carried out at a fire research institute in China. The safety tests for single split air-conditioning systems from this institute have shown that the use of propane or propene in such devices is less dangerous than the use of isobutane in household refrigerators, which has been used successfully and safely for 30 years. The density of propane and propene is lower than that of isobutane, the indoor units of the air-conditioning system usually hang on or close to the ceiling and a fan ensures that the air is mixed well, so that the risk of ignition is reduced. Even for vehicle air-conditioning systems, hydrocarbons would be the better and more energy efficient alternative than carbon dioxide. In Australia and China, hundreds of thousands of cars with hydrocarbons as refrigerants are said to be on the road. Compared to the 30 to 50 kg (66 to 110 lbs) of hydrocarbon compounds contained in the fuel tank of each car, the approximately 250 g (8.8 oz) propane in the air-conditioning system should be regarded as less dangerous.

The use of combustible HFCs such as R32 and R1234yf especially in air conditioning systems could change the attitude towards combustible refrigerants in the future. At the beginning of the 1990s, the management board of a large household appliance manufacturer decided, when choosing between R152a and R600a for their household refrigerators, that "If it is combustible, then it may as well be really combustible – but with a better environmental image." It seems probable, therefore, that hydrocarbons have a bright future in solving the current challenges in refrigeration and air-conditioning technology. Hydrocarbons reduce the direct contribution to the greenhouse effect by not releasing greenhouse gases, and the indirect contribution, due to their high energy efficiency.

1.4 Water as a refrigerant

From an environmental point of view, water is the optimal refrigerant: it is exceptionally efficient, it is non-toxic, it neither contributes to the greenhouse effect nor does it destroy the ozone layer, and it does not burn. It does not smell and is available in boundless quantities in most places of the world. These excellent credentials indicate that a closer look at water's properties as a refrigerant is required. A few companies and institutes have been doing this in recent years. But refrigeration systems with water as the refrigerant have been in use for many decades. In the case of thermally driven water chillers, water dominates the market in the form of lithium bromide water absorption chillers and steam jet refrigeration systems. The latter have had a certain market share in process engineering for many decades.

At 4,2 kJ/(kgK) or 1.0 Btu/(lb°F), water has one of the highest specific heat capacities of all naturally occurring substances – only ammonia is slightly better (approximately 5 kJ/(kgK)). Of all non-metallic liquids, water has the highest thermal conductivity: 0,6 W/(m·K) at 20 °C (0.3455 Btu/(h ft °F) at 68 °F). Of all the natural refrigerant substances, it has the highest specific evaporation enthalpy – approximately 2500 kJ/kg at 0 °C (1075 Btu/lbs at 32 °F). This is twice as much as ammonia and five times as much as propane or almost ten times more than CO_2. Since the physical size of a refrigeration system is rather volume-specific than mass-specific, the volumetric heat of evaporation for water is unfortunately less suitable than that of the other substances mentioned. The volumetric cooling capacity of water at 0 °C (32 °F) is only 12 kJ/m^3 (0.32 Btu/ft^3), that of CO_2 in contrast is 22500 kJ/m^3 (604 Btu/ft^3). The 0 °C/32 °F temperature limit is another sticking point when using water as a refrigerant: water freezes at 0 °C (32 °F.) This means that water-based refrigeration systems can only be operated down to this evaporation temperature.

The pressure ratio (the condensing pressure in relation to the evaporation pressure) is also an important parameter for the operation of a refrigeration system because it affects the compressor efficiency. In this respect water is at a disadvantage, with a pressure ratio that is four to five times higher compared to other refrigerants at the same temperature lift.

So, why deal with water as a refrigerant? If the appropriate – preferably oil-free – compressors for high volume flows delivering high-pressure ratio can be obtained, a designer will be rewarded with at least theoretically very high cooling efficiencies. To achieve this in practice, it is important to avoid any pressure losses in the piping and components. Because with evaporation pressures of only a few millibars, even the smallest pressure loss is a real problem and significantly reduces energy efficiency. Air in the system represents a further challenge for vacuum-operated refrigeration systems with water as the refrigerant, which is why most water vapour refrigeration systems are equipped with appropriate vacuum pumps to remove the air. The vapour pressure of water is only above the atmospheric air pressure above 100 °C (212 °F).

Because of the high-volume flows, most manufacturers use turbo machines as compressor for their refrigeration systems with water as the refrigerant[2]. These are usually designed as radial turbo-compressors. An Israeli company is still the manufacturer with the highest number of large vapour compression refrigeration systems using water as the refrigerant. Corresponding installations can be found in Denmark, Germany, Japan, Switzerland, and South Africa. Two German companies also have some systems in operation, for example, at the University of Essen, at an automobile manufacturer in Dresden, at Daimler in Sindelfingen, and at the Zwickau University of Applied Sciences.

Water has a great future as a refrigerant – especially for high-temperature heat pumps. The higher the application temperature, the higher the vapour pressure of the water, and the disadvantages mentioned above have less impact.

1.5 Literature

[1-1] McLinden, M.O., Brown, J.S., Brignoli, R., Kazakov, A.F., Domanski, P.A.: *Limited options for low-globalwarming potential refrigerants.* Nat. Comm. 8, 14476, 2017

[1-2] McLinden, M.O.: *Thermodynamics of the new refrigerants.* 25th IIR International Congress of Refrigeration, Montreal, 24.–30. August 2019

[1-3] Holland, R., Khan, M.A.H., Driscoll, I., Chantyal-Pun, R., Derwent, R.G., Taatjes, C.A., Orr-Ewing, A.J., Carl J., Percival, C.J., Shallcross, D.E.: I*nvestigation of the Production of Trifluoroacetic Acid from Two Halocarbons, HFC-134a and HFO-1234yf and Its Fates Using a Global Three-Dimensional Chemical Transport Model.* ACS Earth Space Chem, 2012

[1-4] Behringer, D., Heydel, F., Gschrey, B., Osterheld, S., Schwarz, W., Warncke, C., Feeling, F., Nödler, K., Henne, S., Reimann, S., Bleep, M., Jörß, W., Liu, R., Ludig, S., Rüdenauer, I., Gartiser, S.: *Report No. FB000452/ENG, Persistent degradation products of halogenated refrigerants and blowing agents in the environment: type, environmental concentrations, and fate with particular regard to new halogenated substitutes with low global warming potential,* German Environment Agency, May 2021

[1-5] Kauffeld, M., Dudita, M.: *Environmental impact of HFO refrigerants & alternatives for the future.* Open Access Government, June 11, 2021

[1-6] Kauffeld, M.: Current and future carbon-saving options for retail refrigeration. in Evans, J.A., Foster, A.M.: Sustainable retail refrigeration, Wiley Blackwell, 2015, ISBN 978-0-470-65940-3, p. 125–158

[2] Oil-free screw compressors derived from vacuum pumps are under development.

2 Thermodynamic properties of natural refrigerants

Michael Kauffeld

When selecting the refrigerant, several properties, sometimes contradictory, have to be taken into account. These can be divided into three main groups:

- thermodynamic properties for the design and operation of the plant and its energy efficiency;
- physical, chemical and physiological properties that guide the selection of materials and processes to achieve the safety of equipment and personnel;
- environmental properties like ozone depletion or global warming potential, contribution to ground smog and impact on water, flora and fauna.

In addition, local, national and international requirements, such as laws, regulations and standards, as well as specific customer requirements must be taken into account. Last but not least, the operating and investment costs are of paramount importance when choosing the system.

To assess the properties of a refrigerant, the following criteria among others must be examined and evaluated:

- vapour pressure level / saturation temperatures
- critical temperature and pressure
- molar mass, heat of vaporisation, specific heat capacity, viscosity
- volumetric cooling capacity

The size of compressor required depends on the volume of gas to be passed in order to generate the required cooling capacity. This depends on the suction-side saturation pressure and the latent heat of the refrigerant. Volume flow and suction pressure are roughly inversely proportional. Selecting a refrigerant with a higher suction pressure at the desired cooling temperature means that a lower volume flow is required, since the suction gas density is higher.

At the same time, the critical temperature of the refrigerant must be taken into account. For the conventional vapour compression cycle, the condensation temperature ought to be well below the critical temperature. This ensures that the latent heat of condensation on the high-pressure side almost corresponds to the latent heat of evaporation at suction pressure so that the proportion of flash gas at the expansion valve is reduced to a minimum.

The molar mass not only determines the gas density of the refrigerant and thus the displacement volume of the compressor, but also the speed of sound of the gas which limits the flow velocities that may not be exceeded in pipes, valves and at the compressor connections with reasonable pressure losses. Lightweight molecules like ammonia can be subjected to higher gas velocities without incurring substantial pressure drop.

In the following section, the thermodynamic properties of the natural refrigerants are compared.

Figure 2.1 compares the specific, mass-related latent heat of evaporation of different natural refrigerants. The enthalpy change of water is twice that of ammonia and six times that of propane. This reduces the mass flow required for a given cooling duty.

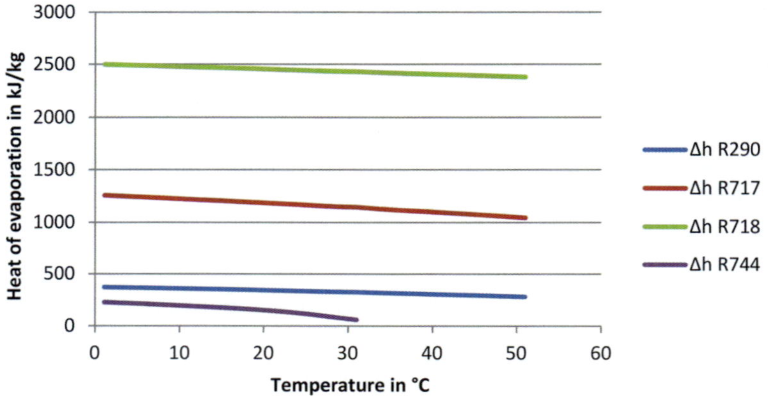

Figure 2.1 Latent heat of evaporation vs. temperature. Please note that the critical point of CO_2 is at 31 °C (87.8 °F) and that no phase change occurs and hence heat of evaporation does not exist for CO_2 above that temperature

Since the compressor size is more dependent on the volume flow rate of the refrigerant than the mass flow rate, the specific volume of the refrigerants at the compressor inlet, i.e. as first approximation the specific volume of the saturated vapour, is also important (see Figure 2.2).

Using the specific volume and the latent heat of evaporation, the specific volume flow per kW cooling capacity can be determined (see Figure 2.3).

The higher-than-average specific volume of water, compared to the other refrigerants, results from the low vapour pressure of water at the temperatures prevailing in refrigeration systems. This is significantly lower for water than for all other substances, see Figure 2.4. With similar low vapour pressures, the other refrigerants would also require correspondingly high-volume flows. However, these low pressures do not occur at the temperatures found in refrigeration systems with refrigerants other than water.

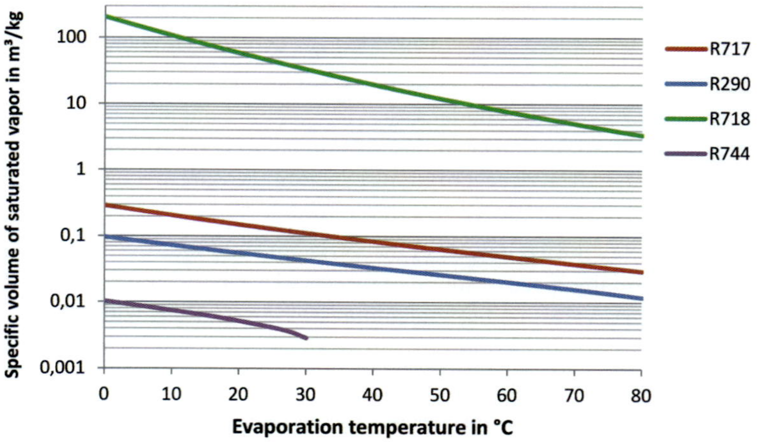

Figure 2.2 Specific volume of saturated vapour, i.e. on the dew point lines, for ammonia (R717), propane (R290), CO_2 (R744) and water (R718). Please note that the critical point of CO_2 is at 31 °C (87.8 °F) and no saturated CO_2 vapour exists above that temperature

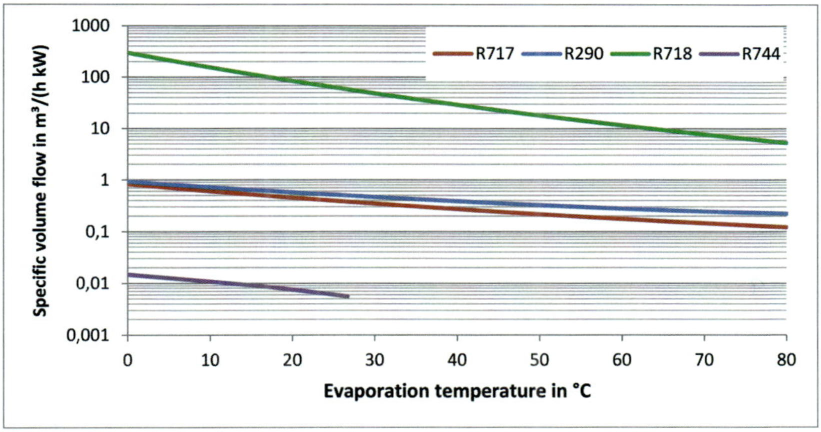

Figure 2.3
Specific volume flow per kW cooling capacity for ammonia (R717), propane (R290), CO_2 (R744) and water (R718). Please note that the critical point of CO_2 is at 31 °C (87.8 °F) and no evaporation of CO_2 occurs above that temperature.

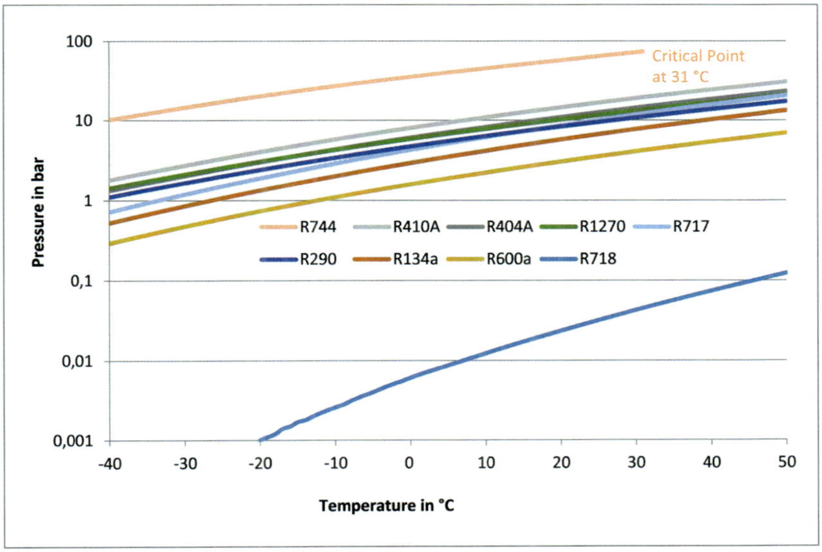

Figure 2.4
Pressure-Temperature relationship of various refrigerants

However, the decisive factor for any refrigeration application is not the pressure of the refrigerant, but the temperature required in the evaporator. Hence, for a given operating temperature, CO_2 with its comparatively high pressures leads to very compact refrigeration systems, while water with its very low pressures requires very large compressor, component and pipe dimensions.

For the design of evaporators, the pressure imposed on the refrigerant is important. This pressure may result from a column of liquid at the evaporator inlet, or a pump may produce it. Either way the result is that the boiling of the refrigerant is suppressed until either the pressure has dropped to the saturation point or the temperature has risen. This static head is, in effect, subcooling the liquid and making it more difficult to evaporate. Due to the low vapour pressures and the relatively high density of liquid water, the effects of a static head or liquid column on the boiling point of water as a refrigerant are significantly greater than with other refrigerants, as shown in Figure 2.5.

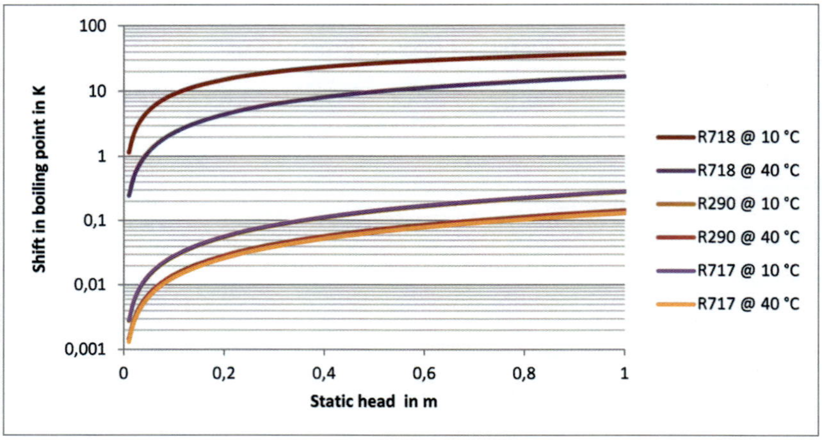

Figure 2.5 Shift in boiling point for ammonia (R717), propane (R290) and water (R718) dependent on static head pressure applied to the refrigerant

The specific heat capacity and the thermal conductivity of the liquid have a significant effect on the way in which the refrigerant behaves in the evaporator. Water (R718) has one of the highest specific heat capacities of all naturally occurring substances (see Figure 2.6). Only ammonia (R717) has a higher specific heat capacity, although liquids close to the critical point, for example CO_2 (R744) in Figure 2.6, show locally elevated values.

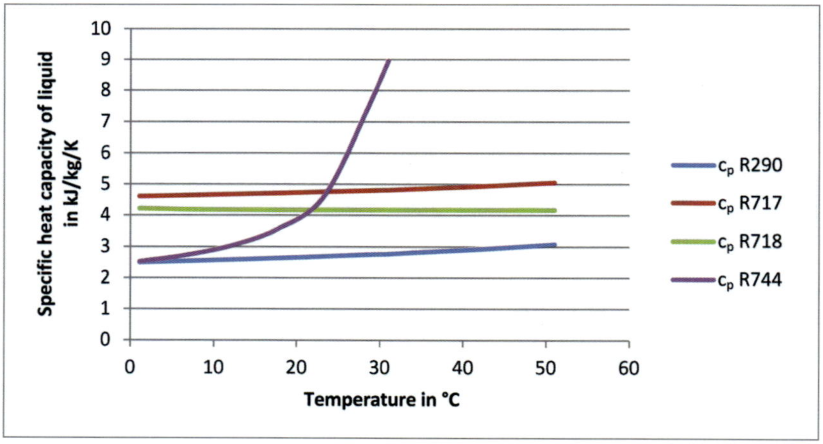

Figure 2.6 Specific heat capacity of liquid natural refrigerants for ammonia (R717), propane (R290), water (R718) and carbon dioxide (R744)

Figure 2.7 shows the thermal conductivity of various liquid natural refrigerants. The thermal conductivity of water is the highest, followed by ammonia. A high thermal conductivity ensures good heat conduction through the liquid in heat exchangers.

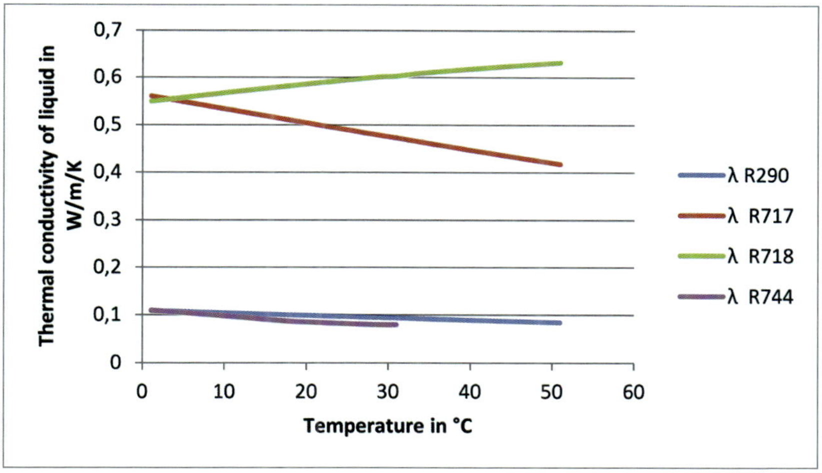

Figure 2.7 Thermal conductivity of natural refrigerants in liquid phase for ammonia (R717), propane (R290), water (R718) and carbon dioxide (R744)

The combination of high specific heat capacity and high thermal conductivity promises good heat transfer performance in evaporators especially for ammonia and water, but also propane and carbon dioxide have good properties – better than most HFC refrigerants.

3 Comparisons of natural refrigerants

Michael Kauffeld

McLinden and Huber (2020) comprehensively describe the century-long quest for the perfect refrigerant and offer a theoretical comparison of energy efficiency vs. volumetric cooling capacity for some of the current refrigerants using NIST's *"CYCLE_D-HX Vapour Compression Cycle Model accounting for Refrigerant Thermodynamic and Transport Properties"* [3-1]. The simulation was performed for an air conditioner employing typical forced-convection, air-to-refrigerant heat exchangers, which were optimised for a particular refrigerant. Evaporation and condensation temperatures were selected to be +10 °C and +40 °C respectively. Assuming an isentropic efficiency of the compressor around 0,7 as well as including pressure drop and heat-transfer limitations, it turned out that ammonia yields the highest COP followed by R32, propene and propane (Figure 3.1).

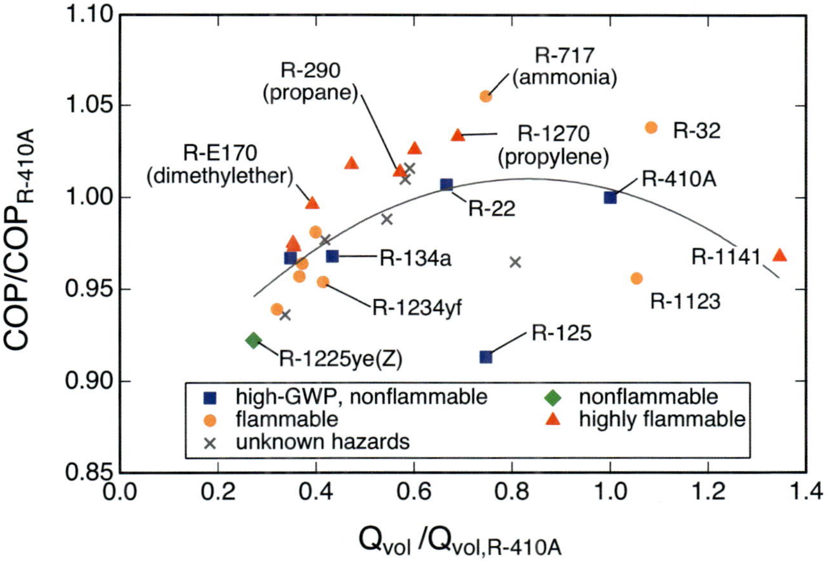

Figure 3.1
COP and Q_{vol} of selected low-GWP fluids relative to R-410A in the basic vapour-compression cycle including pressure drop and heat-transfer limitations; the curve indicates the general trend [3-1]

Not surprisingly, when applying a suction line heat exchanger propane and propene gain somewhat in efficiency and score highest, while ammonia is losing efficiency [3-2]. Due to the high isentropic exponent, i.e. the high specific heat ratio ($\kappa = c_p/c_v$), of ammonia vapour, any superheat is decreasing the efficiency of an ammonia system while the low specific heat ratio of propane allows for an efficiency increase when employing an internal heat exchanger. The efficiency gain is even larger with isobutane with an even lower heat capacity ratio, which is one of the reasons why most residential refrigerators with isobutane have an internal heat exchanger.

A comparison of natural vs. selected synthetic refrigerants is shown in Table 3.1. This table compares certain refrigerant parameters, with grey blue shading for the natural refrigerants. The theoretical COP calculations using coolpack as simulation tool, yield similar results as the simulations performed by McLinden et al. [3-2].

Table 3.1 Comparison of various refrigerants (SI units)

	Normal Boiling Point in °C	Critical Temperature in °C	Pressure in bar at boiling point of			Pressure ratio 0 °C / 40 °C	Theoretical COP 0 °C / 40 °C[1)]	Flammable	Toxic	GWP_{20} [UNEP RTOC 2018][4)]	Volumetric heat of evaporation at 0 °C in kJ/m^3
			-30 °C	0 °C	40 °C						
R32	–51,7	78,4	2,7	8,1	24,8	3,0	3,8	yes	no	2530	6970
R134a	–26,1	101,1	0,8	2,9	10,2	3,5	3,8	no	no	3810	2870
R404A	–46,5	72,1	2,1	6,1	18,2	3,0	3,4	no	no	6600	5070
R407C	–43,6	86,0	1,9	5,6	17,5	3,4	3,7	no	no	4100	4230
R410A	–51,4	72,5	2,7	8	24,3	2,9	3,6	no	no	4400	6780
R507A	–46,7	70,9	2,1	6,2	18,7	2,8	3,6	no	no	6700	5230
R1234yf	–29,4	94,7	0,99	3,16	10,18	3,2	3,6[2)]	yes	no	< 1	2882
R600a Isobutane	-11,7	143,7	0,5	1,6	5,3	3,4	3,9	yes	no	< 1	1510
R290 Propane	-42,2	96,7	1,7	4,7	13,7	2,9	3,8	yes	no	< 1	3880
R1270 Propene	-47,7	92,4	2,1	5,9	16,5	2,8	3,8	yes	no	< 1	4670
R717 Ammonia	-33,3	132,3	1,2	4,3	15,5	3,6	4,1	(yes)	yes	0	4360
R718 Water	100	390,0	(0,0004)[3)]	0,006	0,074	12	3,8	no	no	0	12
R744 Carbon Dioxide	(–78,4)[3)]	31,0	14,3	34,8	90–120	2,6–3,45	2,0	no	< 10 % no	1	22550

[1)] Calculated with coolpack1.50. Isentropic efficiency = 0,7. No sub cooling, no superheating, no internal heat exchanger

[2)] Calculated with NIST Cycle D-HC1.0. Isentropic efficiency =0,7. No sub cooling, no superheating, no internal heat exchanger

[3)] Sublimation

[4)] The GWP data in the UNEP RTOC 2018 report is mainly derived from the WMO 2014 and IPCC 2014 reports. GWP_{20} is chosen, because most HFC refrigerants have considerably shorter atmospheric lifetimes than 100 years (R134a for example remains 14 years in the atmosphere) and hence have a higher contribution to global warming when a time horizon fitting better to their atmospheric lifetime is chosen. In addition, we do not have 100 years time for reversing global warming.

[5)] Gas cooler pressure: 105 bar

Table 3.2 Comparison of various refrigerants (imperial units)

	Normal Boiling Point in °F	Critical Temperature in °F	Pressure in psi at boiling point of			Pressure ratio 32 °F / 104 °F	Theoretical COP 32 °F / 104 °F[1)]	Flammable	Toxic	GWP_{20} [UNEP RTOC 2018][4)]	Volumetric heat of evaporation at 32 °F in Btu/ft^3
			−22 °F	32 °F	104 °F						
R32	−61.1	173.1	39.16	117.48	359.69	3.0	3.8	yes	no	2530	187.14
R134a	−15.0	214.0	11.60	42.06	147.94	3.5	3.8	no	no	3810	77.06
R404A	−51.7	161.8	30.46	88.47	263.97	3.0	3.4	no	no	6600	136.13
R407C	−46.5	186.8	27.56	81.22	253.82	3.4	3.7	no	no	4100	113.58
R410A	−60.5	162.5	39.16	116.03	352.44	3.0	3.6	no	no	4400	182.04
R507A	−52.1	159.6	30.46	89.92	271.22	3.0	3.6	no	no	6700	140.43
R1234yf	−20.9	202.5	14.36	45.83	147.65	3.2	3.6[2)]	yes	no	< 1	77.38
R600a Isobutane	10.9	290.7	7.25	23.21	76.87	3.4	3.9	yes	no	< 1	40.54
R290 Propane	−44.0	206.1	24.66	68.17	198.70	2.9	3.8	yes	no	< 1	104.18
R1270 Propene	−53.9	198.3	30.46	85.57	239.31	2.8	3.8	yes	no	< 1	125.39
R717 Ammonia	−27.9	270.1	17.40	62.37	224.81	3.6	4.1	(yes)	yes	0	117.07
R718 Water	212.0	734.0	(0.01)[3)]	0.01	1.07	12.0	3.8	no	no	0	0.32
R744 Carbon Dioxide	(−109.1)[3)]	87.8	207.40	504.73	1305–1740	2.6–3.45	2.0[5)]	no	< 10 % no	1	605.47

[1)] Calculated with coolpack1.50. Isentropic efficiency = 0.7. No sub cooling, no superheating, no internal heat exchanger

[2)] Calculated with NIST Cycle D-HC1.0. Isentropic efficiency = 0.7. No sub cooling, no superheating, no internal heat exchanger

[3)] Sublimation

[4)] The GWP data in the UNEP RTOC 2018 report is mainly derived from the WMO 2014 and IPCC 2014 reports. GWP_{20} is chosen, because most HFC refrigerants have considerably shorter atmospheric lifetimes than 100 years (R134a for example remains 14 years in the atmosphere) and hence have a higher contribution to global warming when a time horizon fitting better to their atmospheric lifetime is chosen. In addition, we do not have 100 years time for reversing global warming.

[5)] Gas cooler pressure: 1523 psi

With regard to the global warming potential in particular, natural refrigerants have less effect than HFCs and are similar to R1234yf and other HFOs in this respect. The differences in the volumetric heat of vaporization of water and carbon dioxide compared to the other refrigerants are also very clear. Water has a very low and CO_2 a very high value. This leads to correspondingly large compressor volume flows for water and very small ones for CO_2. Looking at the pressure ratio of a refrigeration system operating between let's say 0 °C (32 °F) evaporation temperature and 40 °C (104 °F) condensing temperature, all of the natural refrigerants shown are at similar values except for water. The high compression ratio for water as a refrigerant requires special care when selecting the heat exchangers in order to keep the temperature lift and thus the pressure ratio as small as possible.

Based on the explanations in Chapters 1 and 2 and the information from Table 3.1 / Table 3.2, a rough evaluation of the refrigerants can be made, see Table 3.3. As can be seen, no refrigerant or refrigerant family has a universal positive rating, shown with a green background. So a compromise has to be found in all cases. Either one allows for the toxicity of ammonia, the high pressures with CO_2, the flammability of hydrocarbons (or HFO for that matter), or the large dimensions and low vacuum pressures with water as the refrigerant. These factors must be addressed individually when designing a refrigeration system in order to achieve a satisfactory energy and cost efficiency.

Table 3.3 Overview of refrigerant properties

Refrigerant	GWP	Flammable	Toxic	Comparative Costs		Swept Volume	Theoretical COP
				Refrigerant	Installation		
HFC	high	no	no	moderate	low	medium	good
HFO	low	moderate	not acute	high	moderate	medium	good
Hydrocarbons	low	yes	no	low	moderate	medium	very good[1])
Carbon Dioxide	low	no	above 10 %	low	moderate	small	moderate
Ammonia	none	yes	yes	low	moderate	medium	very good
Water	none	no	no	low	moderate	very large	good

[1]) With internal heat exchanger (suction gas heat exchanger)

The following chapters provide more detailed information about the optimal applications and system designs for respective natural refrigerants and their advantages and disadvantages when compared to synthetic refrigerants.

3.1 Literature

[3-1] Mark O. McLinden, M. O., Huber, M. L.: *(R)Evolution of Refrigerants.* Journal of Chemical & Engineering Data 2020 65 (9), 4176-4193, DOI: 10.1021/acs.jced.0c00338

[3-2] McLinden, M. O., Brown, J. S., Brignoli, R., Kazakov, A. F., Domanski, P. A.: *Limited options for low-global-warming-potential refrigerants.* Nat. Commun. 2017, 8, 14476

4 Ammonia as a refrigerant

Michael Eckert

This chapter is based on the publication by Dr. A.B. ("Andy") Pearson, "Ammonia as Refrigerant, International Institute of Refrigeration – IIR. 2008". It has been supplemented in some places with additional information. We thank Andy Pearson and the IIR for allowing us to use text and graphics. The final text was extensively reviewed by Andy Pearson and Michael Kauffeld.

4.1 Introduction

Ammonia (R717 or NH_3) has proven itself in industrial refrigeration for over 120 years. However, since the 1990s, there has been a clear change and expansion in the areas of application. Driven by a more conscious environmental policy and the pursuit of greater economic efficiency, traditional ammonia technology has been adapted for use in new applications with the help of experienced refrigeration engineers and system designers.

New products and production techniques have made it possible to design modern, compact ammonia machines with very small refrigerant charge and the technology has been developed further, particularly with regard to safety and tightness. This makes it possible to use ammonia in areas that were previously occupied by synthetic refrigerants. This includes applications such as the air conditioning of office buildings, airports or exhibition halls, as well as the cooling of data centres and process cooling in the industry.

In terms of maintenance, reliability and space requirements, the synthetic refrigerants were much easier to handle and more self-sufficient than the older ammonia technology. The new ammonia machines are very easy to maintain and operate. The operating concepts are self-explanatory and safe.

Ammonia refrigeration systems are often installed outdoors or on top floors. In contrast to the earlier chiller rooms, the machines are now often installed together with the general building technology.

With today's systems and safety technology, ammonia refrigeration systems can be very compact. Even the smallest leaks are immediately recognised by the very intense smell of ammonia and can be quickly localised and repaired. In addition, electronic ammonia sensors are available for automated leakage detection. The key points for the use and construction of ammonia plants are summarised in the following paragraphs.

4.1.1 General characteristics and manufacture of ammonia

Ammonia is a colourless gas with a pungent smell. This is noticeable from very low concentrations (3 ppm), far below the maximum workplace concentration, Table 4.1.

Table 4.1 Exposure levels at workplace

Peak exposure – short term	35 ppm	24 mg/m^3	3.99E–05 oz/ft^3
Average exposure – continuous	25 ppm	17 mg/m^3	1.99E–05 oz/ft^3

Ammonia can be liquefied at room temperature at approximately 6 bar (87 psi) pressure. Ammonia is very soluble in water. Ammonia solutions are quite alkaline. The density of ammonia gas at ambient temperature is lower than the density of air – ammonia vapours therefore rise. Below approximately −28 °C (−18 °F), saturated ammonia vapour will be heavier than air at ambient temperature. Ammonia vapour blowing off a low-temperature plant may therefore initially not rise. Ammonia-air mixtures are flammable in the range of 15 % to 28 % vol [4-2]. This corresponds to a concentration of 108000–197650 mg/m^3 (0.108 to 0.197 oz/ft^3). In practice, if the machine room ventilation is designed correctly, it is almost impossible for a concentration to reach this level.

NH_3 arises in natural, biological processes and is an important part of the natural biosphere nitrogen cycle (N-cycle). The amount of ammonia produced industrially by humans only corresponds to approximately 3 % of the total amount of ammonia found in nature.

Ammonia is created in an equilibrium reaction from the elements hydrogen and nitrogen, according to the equation:

$$\mathbf{N_2 + 3\,H_2 \rightleftharpoons 2\,NH_3}$$

The first practical catalytic method for ammonia production was patented by Fritz Haber in 1908. The Haber-Bosch process is a large-scale chemical process for the synthesis of ammonia. It is named after the German chemists Fritz Haber and Carl Bosch, who developed the process at the beginning of the 20th century. The central step of the process, the ammonia synthesis from atmospheric nitrogen and hydrogen, is carried out on an iron-containing catalyst at pressures of approximately 150 to 350 bar (2175 to 5076 psi) and at temperatures of approximately 400 to 500 °C (752 to 932 °F). The process has much lower CO_2-equivalent emissions related to it than the manufacturing of HFCs, e.g., 20 % of the CO_2-equivalent emissions associated with manufacturing R1234yf [4-4, 4-5]. As an important chemical process with a production output of more than 100 million metric tons per year, it covers the major part of the worldwide need for ammonia.

Ammonia has no effect on the ozone layer. Due to its short atmospheric lifetime of just a few days, ammonia has no direct influence on the greenhouse effect.

As can be seen in Table 4.2, ammonia (R717) has excellent thermodynamic properties. Compared to other refrigerants, ammonia refrigeration systems usually operate significantly more efficiently than refrigeration systems with other refrigerants. Initially, this leads to a lower electrical power consumption of ammonia chillers and consequently to lower CO_2 emissions if the electricity is generated using fossil fuels. This reduces both the direct and indirect greenhouse effect of chillers through the use of ammonia.

Table 4.2 Comparative COP for air-cooled refrigeration systems

Application	t_{evap}/t_{cond}	R717	R1234yf	R134a	R407A
Air conditioning	+2 / +40 °C 35.6 / 104 °F	4,8	4,2	4,2	3,9
MT	−8 / +40 °C 17.6 / 104 °F	3,4	2,8	3,2	2,8
LT	−28 / +40 °C −18.4 / 104 °F	1,8	(1,4)	(1,4)	1,6

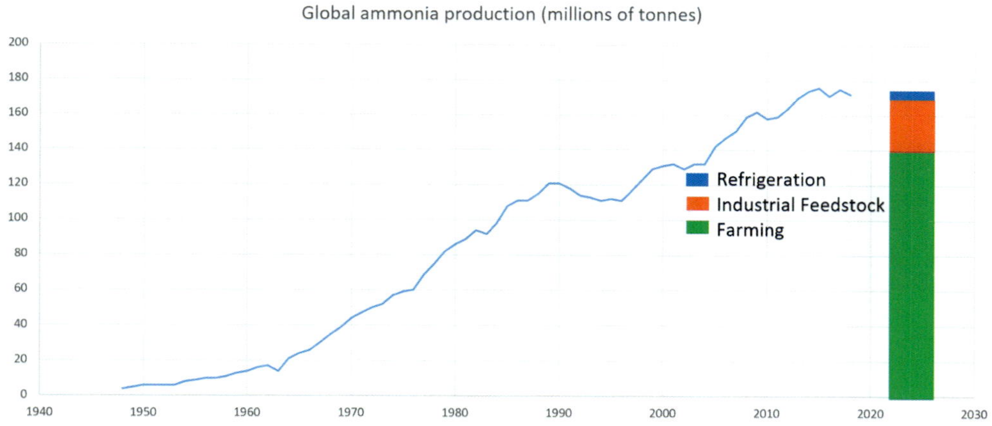

Figure 4.1 Annual ammonia production growth since 1950

4.1.2 Use of ammonia

As Figure 4.1 shows, ammonia production has increased steadily since the Second World War. Above all, the increasing use of fertilisers worldwide contributed to this. In the 1950s, ammonium sulphate and nitrate were used for the first time in grain and cotton cultivation in the US. As a result, the yield could be increased tenfold. Without the use of these fertilisers, today's global demand for food could no longer be sustained.

In 2018, almost 180 million metric tons of ammonia were produced worldwide. Over 80 % of today's production is used for the production of fertilisers. When used in agriculture, ammonia is returned to the natural nitrogen cycle and remains in nature.

The vast majority of the remaining ammonia production is applied in the chemical industry. The urea obtained from ammonia is used, for example, in the manufacture of synthetic textiles and fibres, in the pharmaceutical industry and in diesel engine technology as a urea solution for nitrogen reduction. In metallurgy, ammonia is applied to bright anneal steel and to extract zinc and nickel. Another area of application is the flue gas cleaning of power plant emissions. There, the nitrogen oxides and sulfur dioxide contained in the flue gas are catalytically cleaned using urea derived from ammonia, similar to diesel engine technology. Lately, research has started to also use ammonia produced by renewable energy as climate-friendly fuel in fuelcells and internal combustion engines. Ammonia is much easier to store and has a higher energy density than its base product hydrogen, produced by electrolysis from renewable energy sources. Around 2 % of the ammonia produced is used as a refrigerant in refrigeration technology. It should be taken into account that a much higher degree of purity is required for refrigeration applications than for other chemical applications; 99,5 % purity is sufficient for fertiliser production, but a purity of 99,95 % must be achieved for refrigeration. In refrigeration technology, one therefore speaks of anhydrous or dry ammonia.

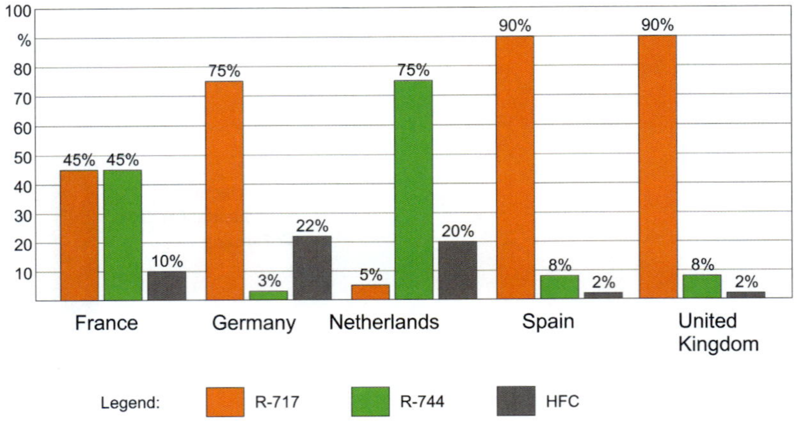

Figure 4.2 Refrigerant usage in European food industry [4-1]

4.1.3 Use and economic importance in refrigeration

Due to excellent thermodynamic properties and environmental neutrality, ammonia has the potential to play an important role in the future of refrigeration as new "old friend". Historically, ammonia has been used primarily in industrial refrigeration technology. More recently the area of application has expanded to applications that were previously reserved only for fluorinated refrigerants.

Figure 4.2 shows that the acceptance and use of ammonia in the individual European countries are very different. The reasons for this are different national laws and professional experience. However, due to Regulation No. 517/2014 on fluorinated greenhouse gases (the "F-Gas Regulation") that has been in force in the EU since January 2015, it can be assumed that the use of ammonia will increase across Europe.

4.2 Thermodynamics

To better classify the respective properties of ammonia, they will be compared with the properties of R1234yf, R134a and R407A in the subsequent sections.

4.2.1 Main features

In order to obtain an initial assessment of a refrigerant's performance, the volumetric cooling capacity can be used, calculated by the product of the specific enthalpy of vaporisation and saturated gas density.

Table 4.3 Physical properties of ammonia compared to three synthetic refrigerants

Property		R717	R1234yf	R134a	R407A
Normal boiling point – NBP (boiling temperature at atmospheric pressure)[1)]	°C	–33,3	–29,5	–26,1	–45,0
Triple point temperature	°C	–77,7	–53,2	–103,3	–
Critical temperature	°C	132,3	94,8	101,1	82,36
Critical pressure	bar abs.	113,3	33,6	40,6	45,3
Molar mass	g/mol	17,03	114	102,04	86,2
Specific evaporation enthalpy at –10 °C	kJ/kg	1297	169	206	216,06
Liquid density at –10 °C	kg/m^3	652	1212	1328	1247
Vapour density at –10 °C	kg/m^3	2,39	12,5	10,04	15,3
Temperature glide[2)]	K	0	0	0	6,6
Property		**R717**	**R1234yf**	**R134a**	**R407A**
Normal boiling point – NBP (boiling temperature at atmospheric pressure)[1)]	°F	–27.94	–21.28	–14.98	–49.00
Triple point temperature	°F	–107.86	–63.8	–153.94	–
Critical temperature	°F	270.14	202.64	–	180.24
Critical pressure	psi abs.	1643.28	487.32	588.85	657.02
Molar mass	lb/lb-mol	17.03	114	102.04	86.2
Specific evaporation enthalpy at 14 °F	Btu/lb	0.55	72.65	88.56	92.88
Liquid density at 14 °F	lb/ft^3	40.70	0.07	0.082	0.07
Vapour density at 14 °F	lb/ft^3	0.14	0.78	0.62	0.95
Temperature glide[2)]	K	0	0	0	6.6

[1)] Start of boiling for zeotropic refrigerant mixture R407A.

[2)] Total temperature glide from boiling point curve to dew point curve, i.e., from saturated liquid line to saturated vapour line.

Based on the values in Table 4.3, the following results are obtained for the volumetric cooling capacities (at t_{evap} = –10 °C / 14 °F):

R717	3100 kJ/m^3	83.2 Btu/ft^3
R1234yf	2112 kJ/m^3	56.7 Btu/ft^3
R134a	2080 kJ/m^3	55.8 Btu/ft^3
R407A	3306 kJ/m^3	88.7 Btu/ft^3

The values show that ammonia has a high volumetric cooling capacity compared to the synthetic refrigerants shown. R407A also has a high volumetric cooling capacity, but due to the relatively low critical temperature, this capacity cannot be fully utilised. In practice, this is shown by the high flash gas content during expansion.

As can be seen in Tables 4.4 and 4.5 and in Figures 4.3 and 4.4 (R134a was used as a reference refrigerant), ammonia has the following special properties in comparison:

- The specific heat capacity of liquid ammonia is exceptionally high – even compared to water = 4,180 kJ/(kg·K) (0.998 Btu/(lb·°F)).
- The ratio of isochoric and isobaric heat capacity (also known as isentropic exponent or coefficient for ideal gases or in refrigeration as index of compression) is very high compared to that of other refrigerants. This results in a very high hot gas temperature in the compressor discharge port. Due to the high isentropic exponent, superheat after the evaporator should be kept as low as possible, and an internal heat exchanger (suction gas heat exchanger) will not increase the COP of the refrigeration system. For every Kelvin superheat, the isentropic efficiency of an ammonia compressor reduces by approximately 0,25 %.
- The critical temperature of ammonia is significantly higher than that of other refrigerants, so the energy efficiency tends to be better, particularly in applications with high condensing temperatures like in air-cooled systems in summer.

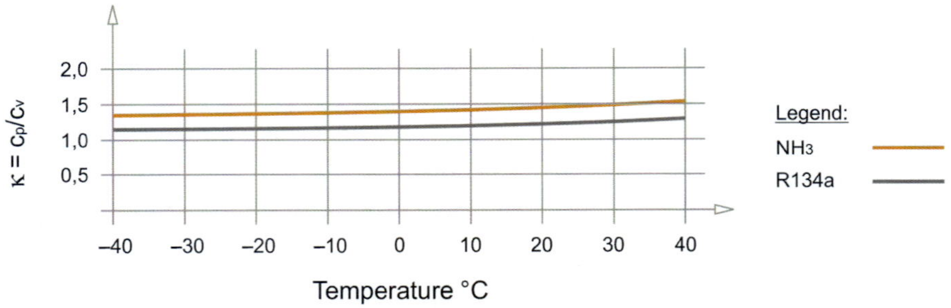

Figure 4.3 Comparison of isentropic exponent $\kappa = c_p/c_v$ for NH_3 and R134a

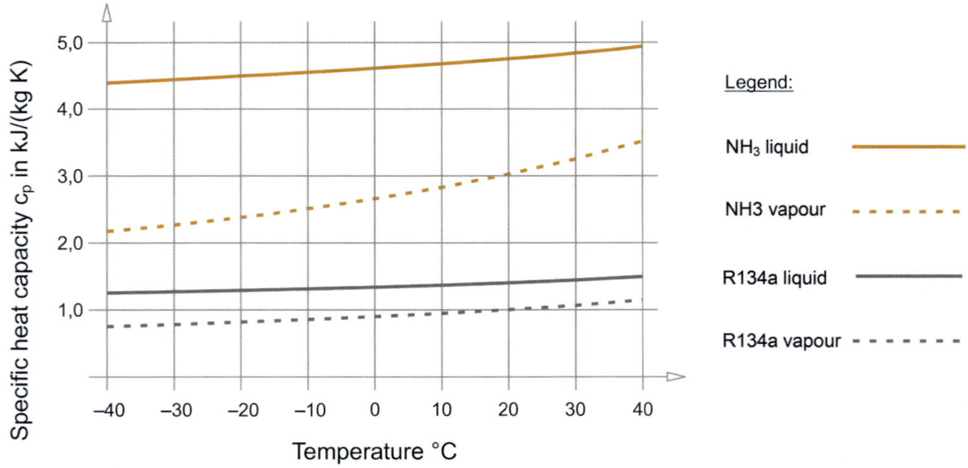

Figure 4.4 Specific heat capacity of NH_3 and R134a

Table 4.4 Ammonia properties

Temperature °C	Specific heat capacity c_p kJ/(kg·K)		Isentropic exponent c_p/c_v	Dynamic viscosity 10^{-6} Pa s		Thermal conductivity 10^{-3} W/(m·K)	
	liquid	vapour		liquid	vapour	liquid	vapour
−70		2,008	1,337	460,4	7,06		13,65
−60		2,047	1,341	391,8	7,33		14,68
−50		2,102	1,346	33,3,1	7,61		15,62
−40	4,396	2,175	1,352	287,0	7,90	601,4	16,79
−33,33	4,430	2,235	1,357	261,9	8,10	587,8	17,48
−30	4,448	2,268	1,360	250,7	8,19	581,0	17,83
−20	4,501	2,379	1,370	221,3	8,49	560,7	18,96
−10	4,556	2,510	1,383	196,8	8,79	540,5	20,29
0	4,617	2,660	1,400	175,8	9,09	520,2	21,84
+10	4,683	2,831	1,422	157,6	9,4	499,8	23,55
+20	4,758	3,027	1,451	141,6	9,71	479,2	25,38
+30	4,843	3,252	1,489	127,6	10,02	458,3	27,30
+40	4,943	3,516	1,538	115,2	10,35	437,1	29,34
+50	5,066	3,832	1,602	104,3	10,70	415,6	31,54
+60	5,225	4,221	1,687	94,5	11,08	393,6	34,00
+70	5,441	4,716	1,801	85,7	11,50	371,0	36,86
+90	6,211	6,302	2,192	70,0	12,56	323,1	44,57
Temperature °F	**Specific heat capacity c_p Btu/(lb·°F)**		**Isentropic exponent c_p/c_v**	**Dynamic viscosity 10^{-6} lb·s/ft²**		**Thermal conductivity 10^{-3} Btu/(h·ft·°F)**	
	liquid	vapour		liquid	vapour	liquid	vapour
−94		0.480	1.337	309.375	4.744		7.887
−76		0.489	1.341	263.277	4.926		8.482
−58		0.502	1.346	223.833	5.114		9.025
−40	1.050	0.519	1.352	192.855	5.309	347.482	9.701
−28	1.058	0.534	1.357	175.989	5.443	339.624	10.100
−22	1.062	0.542	1.360	168.463	5.503	335.696	10.302
−4	1.075	0.568	1.370	148.707	5.705	323.966	10.955
14	1.088	0.600	1.383	132.243	5.907	312.295	11.723
32	1.103	0.635	1.400	118.132	6.108	300.566	12.619
50	1.119	0.676	1.422	105.902	6.317	288.779	13.607
68	1.136	0.723	1.451	95.151	6.525	276.877	14.664
86	1.157	0.777	1.489	85.743	6.733	264.801	15.774
104	1.181	0.840	1.538	77.411	6.955	252.552	16.952
122	1.210	0.915	1.602	70.086	7.190	240.129	18.223
140	1.248	1.008	1.687	63.501	7.445	227.418	19.645
158	1.300	1.126	1.801	57.588	7.728	214.360	21.297
194	1.483	1.505	2.192	47.038	8.440	186.684	25.752

Table 4.5 R134a properties

Temperature °C	Specific heat capacity c_p kJ/(kg·K)		Isentropic exponent c_p/c_v	Dynamic viscosity 10^{-6} Pa s		Thermal conductivity 10^{-3} W/(m·K)	
	liquid	vapour		liquid	vapour	liquid	vapour
−70	1,210	0,6665	1,148				
−60	1,223	0,6924	1,146				
−50	1,238	0,7197	1,146				
−40	1,255	0,7490	1,148				
−30	1,273	0,7809	1,152	381,1	9,719	106,8	9,142
−20	1,293	0,8158	1,158	337,2	10,16	102,4	10,11
−10	1,316	0,8544	1,167	298,9	10,59	98,06	11,03
0	1,341	0,8972	1,179	265,3	11,02	93,67	11,96
10	1,370	0,9455	1,196	235,4	11,46	89,25	12,92
20	1,405	1,001	1,219	208,7	11,91	84,78	13,93
30	1,446	1,065	1,249	184,6	12,38	80,27	15,01
40	1,498	1,145	1,293	162,7	12,89	75,69	16,19
50	1,566	1,246	1,354	142,7	13,47	71,05	17,54
60	1,660	1,387	1,448	124,1	14,15	66,36	19,14
70	1,804	1,605	1,607	106,6	15,04	61,69	21,17
80	2,065	2,012	1,924	89,69	16,31	57,15	24,00
90	2,756	3,121	2,819	72,22	18,53	52,93	28,88

Temperature °F	Specific heat capacity Btu/(lb·°F)		Isentropic exponent c_p/c_v	Dynamic viscosity 10^{-6} lb·s/ft²		Thermal conductivity Btu/(h·ft·°F)	
	liquid	vapour		liquid	vapour	liquid	vapour
−94	0.289	0.159	1.148				
−76	0.292	0.165	1.146				
−58	0.296	0.172	1.146				
−40	0.300	0.179	1.148				
−22	0.304	0.187	1.152	256.09	6.53	61.71	5.28
−4	0.309	0.195	1.158	226.59	6.83	59.17	5.84
14	0.314	0.204	1.167	200.85	7.12	56.66	6.37
32	0.320	0.214	1.179	178.27	7.41	54.12	6.91
50	0.327	0.226	1.196	158.18	7.70	51.57	7.47
68	0.336	0.239	1.219	140.24	8.00	48.98	8.05
86	0.345	0.254	1.249	124.05	8.32	46.38	8.67
104	0.358	0.273	1.293	109.33	8.66	43.73	9.35
122	0.374	0.298	1.354	95.89	9.05	41.05	10.13
140	0.396	0.331	1.448	83.39	9.51	38.34	11.06
158	0.431	0.383	1.607	71.63	10.11	35.64	12.23
176	0.493	0.481	1.924	60.27	10.96	33.02	13.87
194	0.658	0.745	2.819	48.53	12.45	30.58	16.69

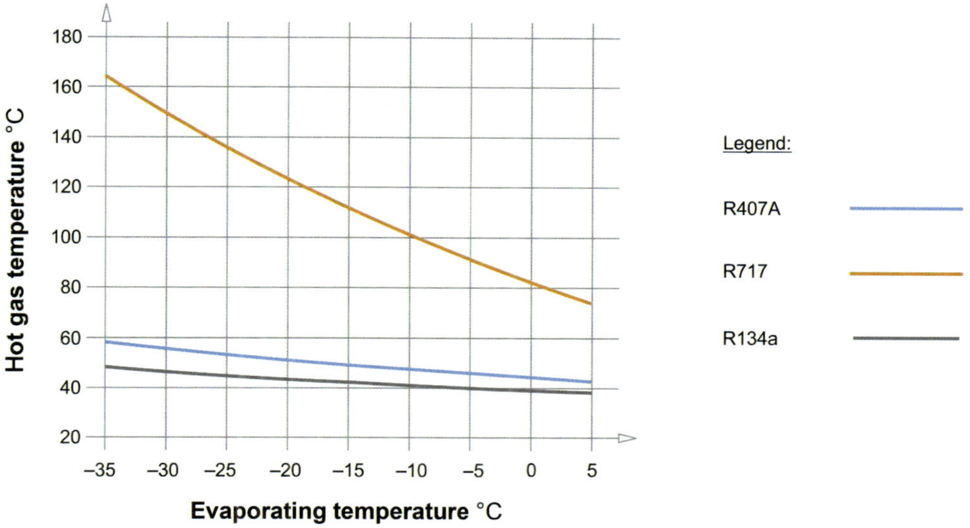

Figure 4.5 Hot gas temperature for isentropic compression with 0 K suction superheat and 35 °C (95 °F) fixed condensing temperature

Further values such as specific enthalpy, specific evaporation enthalpy and specific volume can be taken from the corresponding vapour tables or respective computer programs if required. SI values can be looked up in *Pohlmann – Taschenbuch der Kältetechnik* [4-5], in *VDI-Wärmeatlas (VDI Heat Atlas)* [4-6], as well as in *Newnes Refrigeration Pocket Book* [4-7] or in *ASHRAE Pocket Guide* [4-8].

4.2.1.1 Critical temperature and pressure

The critical temperature of the synthetic refrigerants is between approximately 70 and 100 °C (158 and 212 °F). The critical temperature of ammonia is 132,3 °C (270.1 °F). If the condensing temperature approaches the critical temperature, both the energy efficiency (COP) and the capacity of the system deteriorate considerably. For this reason, the operating points of the refrigeration systems used today – except for transcritical CO_2 systems (Chapter 5) – are usually well below the critical temperatures mentioned. For high-temperature heat pump applications, ammonia is therefore much more suitable than synthetic refrigerants.

The critical pressure of synthetic refrigerants is between 35 and 50 bar (507 and 725 psi). Ammonia has a critical pressure of 113 bar (1639 psi). For this reason, the volumetric cooling capacity of ammonia is higher in the case of high-temperature heat pumps than for synthetic refrigerants. Due to the high critical pressure, ammonia can also cover refrigeration applications with large temperature differences such as those in hot climates.

4.2.1.2 Compressor discharge temperature

Due to the high ratio of c_p/c_v (see Figure 4.3) the compression-end temperature for ammonia is very high. Figure 4.5 shows the compression-end temperatures of various refrigerants depending on the evaporation temperatures. The condensing temperature was set at 35 °C (95 °F) for this comparison. For ammonia, for example, this results in a theoretical compression-end temperature of 100 °C (212 °F) at an evaporation temperature of –10 °C (14 °F). The isentropic exponents of synthetic refrigerants are much lower, so the discharge temperature for a given condition is also lower. For R407A and R134a, the temperature is clearly below 60 °C (140 °F) under the same conditions. The much higher compression-end temperature of ammonia can be used particularly well in heat pump applications or when heat recovery is incorporated in a refrigeration system where hot water at 70 °C (158 °F) is being generated. However, the lubrication properties of the refrigeration oil used must be taken into account. The oil must be stable at high temperatures and have sufficient lubricating properties at those temperatures.

4.2.1.3 Normal Boiling Point (NBP)

Table 4.3 shows the normal boiling point and the saturation temperatures of various refrigerants at atmospheric pressure. The refrigerants used today are in a similar temperature range. At the same time, ammonia with a normal boiling temperature of –33,3 °C (–27.9 °F) at 1 bar (15.5 psi) enables its use in almost all common areas of refrigeration technology. For HFCs and HCs it is important to ensure that air and moisture are not drawn into the low-pressure side of the system, so they tend to operate with a suction pressure higher than atmospheric pressure. This constraint is less relevant for ammonia, which is not unreliable if air or water is entrained in small amounts.

4.2.1.4 Heat transfer

A comparative and balanced view of the heat transfer capability is difficult because the heat transfer coefficient is strongly influenced by the design and components of the refrigeration system as well as by the amount of oil present in the heat exchangers. The surface structure and the material of the heat exchangers in particular are decisive here.

As shown in Table 4.6, the thermal properties of ammonia are significantly more favourable than those of R134a.

Table 4.6 Thermal properties of ammonia and R134a at –10 °C (14 °F)

	R717	R134a	Ratio R717/R134a
Specific heat capacity at constant pressure			
for liquid in kJ/(kg·K)	4,556	1,316	3,5
for vapour in kJ/(kg·K)	2,510	0,854	3,0
Specific enthalpy of evaporation in kJ/kg	1297	206	6,3
Thermal conductivity (liquid) in 10^{-3} W/(m·K)	540,5	98	5,5
Dynamic viscosity in 10^{-6} Pa·s	196,8	298,9	0,6
Density (liquid) in kg/dm³	0,66	1,33	0,5
	R717	**R134a**	**Ratio R717/R134a**
Specific heat capacity at constant pressure			
for liquid in Btu/(lb °F)	1.09	0.31	3.5
for vapour in Btu/(lb °F)	0.60	0.20	3.0
Specific enthalpy of evaporation in Btu/(lb·°F)	310	49.20	6.3
Thermal conductivity (liquid) in 10^{-3} Btu/(h·ft·°F)	312.30	56.62	5.5
Dynamic viscosity in 10^{-6} lb/(ft·s)	132.24	200.85	0.6
Density (liquid) in lb/ft³	41.20	83.02	0.5

The following values for the heat transfer coefficient of ammonia in relation to R134a were determined from laboratory measurements:

- Condensation on the outer pipe surface: 4,1 times that of R134a
- Condensation on the inner pipe surface: 3,5 times that of R134a
- Boiling on the outer tube surface: 1,6 times of R134a
- Boiling on the inner tube surface: 2,0 times of R134a

However, these values are significantly influenced by the design and structure of the refrigeration system and the heat exchangers used and can therefore only be considered as a rough guide. Nevertheless, they clearly show that the use of ammonia has a positive influence on the heat transfer coefficient.

A high heat transfer coefficient means that less surface area is required to transfer a given amount of heat, so ammonia heat exchangers can be made smaller than those required for HFCs due to the significantly higher heat transfer coefficient (see Table 4.7). However, the incompatibility of ammonia and copper means that more expensive materials may be required, partially offsetting the cost advantage of the smaller units.

Table 4.7 Comparative heat transfer coefficient

	Ammonia	R134a
Condensation on the outer tube surface in $W/(m^2 \cdot K)$	7500–11000	1700–2800
Condensation on the inner tube surface in $W/(m^2 \cdot K)$	4200–8500	1400–2000
Boiling on the outer tube surface in $W/(m^2 \cdot K)$	2300–4500	1400–2000
Boiling on the inner tube surface in $W/(m^2 \cdot K)$ (pumping liquid)	3100–5000	1500–2800
Condensation on the outer tube surface in $Btu/(h \cdot ft^2 \cdot °F)$	1320–1940	300–500
Condensation on the inner tube surface in $Btu/(h \cdot ft^2 \cdot °F)$	740–1500	250–350
Boiling on the outer tube surface in $Btu/(h \cdot ft^2 \cdot °F)$	400–790	250–350
Boiling on the inner tube surface in $Btu/(h \cdot ft^2 \cdot °F)$ (pumping liquid)	550–880	360–500

A machinable alternative to steel with good thermal conductivity for ammonia refrigeration systems would be zinc and copper-free aluminium alloys, which can be safely used up to water contents of 10 %.

4.2.1.5 EER or COP

The EER (energy efficiency ratio) is based on the performance of a refrigeration system. The performance figure expresses the ratio of usable cooling capacity to the required operating power for the compressor. The term COP (coefficient of performance) or refrigeration coefficient is also used to express the same ratio.

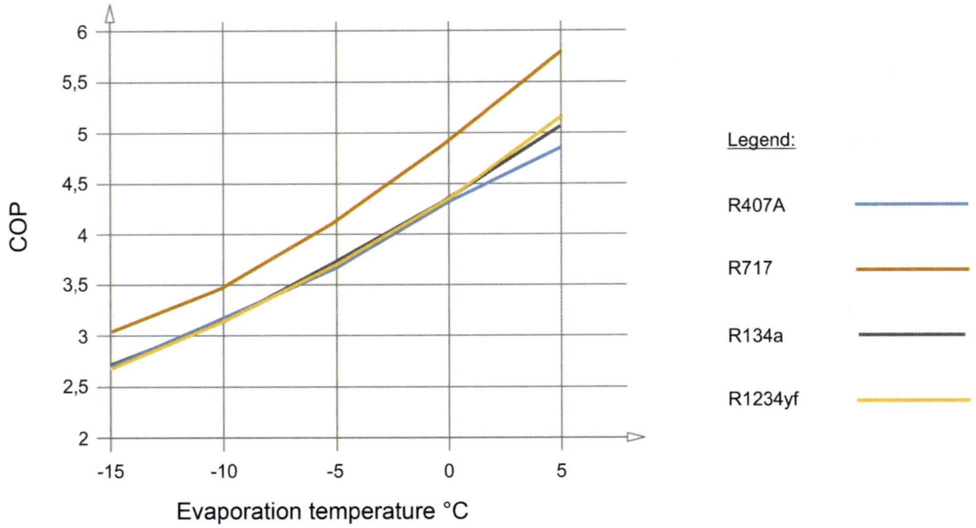

Figure 4.6 Comparisons of EER/COP at 35 °C (95 °F) liquefaction (values from Grasso Software)

Due to the better thermodynamic properties, the coefficient of performance of ammonia refrigeration systems is typically higher than refrigeration systems with synthetic refrigerants if everything else is equal. As shown in Figure 4.6, the performance figures of R134a and R407A are around 10 to 15 % lower than ammonia, depending on the application.

The comparatively poorer performance of R407A at higher evaporation temperatures is essentially due to the low critical temperature. In systems with screw compressors, this difference can be improved by using an economiser, but system complexity increases and additional heat exchanger are required.

4.2.1.6 Toxicity, threshold values

The information on the toxicity of ammonia is summarised in Table 4.8. These values are intended to serve as a guide for practical use. An international specialist group has set the specified guide values.

Ammonia is a rather atypical toxic chemical. The smell is pungent and very unpleasant. The nose detects ammonia at a concentration of approximately 3 ppm, which is 100 times lower than the IDLH value for ammonia in the U.S. which is 300 ppm. This gives a very good early warning effect. The tolerable breathing limit is 500–1000 ppm. This is almost 200 times the perception limit. In contrast, with most toxic chemicals, and also with synthetic refrigerants, the concentration of the chemical is so high when it is perceived that there is at least the risk of permanent health damage.

Chapter 4.4 deals with the toxicity and the resulting safety measures needed.

Table 4.8 Limiting values for toxicity at (exposure time) – the readers have to check for allowed values in their countries

Perception in nose from approximately	3–5	ppm
Long-term exposure limit (LTEL) (8 h)	25	ppm
Short-term exposure limit (STEL) (15 min)	35	ppm
Immediately dangerous to life and health (IDLH) (30 min)	300	ppm
Tolerability limit	500–1000	ppm
Specified level of toxicity (SLOT) (30 min)	3500	ppm
Significant likelihood of death (SLOD) (30 min)	6000	ppm

4.2.2 Impact on the components of the refrigeration system

From the thermodynamic and main properties stated in Chapter 4.2.2, there are some peculiarities regarding the selection, dimensioning and function of the individual components for the design of an ammonia system. These are shown below.

4.2.2.1 Reference circuit

With the help of a reference refrigerant circuit specified in Figure 4.7, the properties and performance data of various refrigerants can be compared more easily. The following key data are defined for this:

- cooling capacity: 200 kW (57 TR)
- condensing temperature: 35 °C (95 °F)
- evaporating temperature: –10 °C (14 °F)
- liquid temperature at the condenser outlet: 2 K subcooling
- refrigerant vapour at the evaporator outlet: 5 K superheat
- refrigerant vapour at the compressor suction nozzle: 7 K superheat

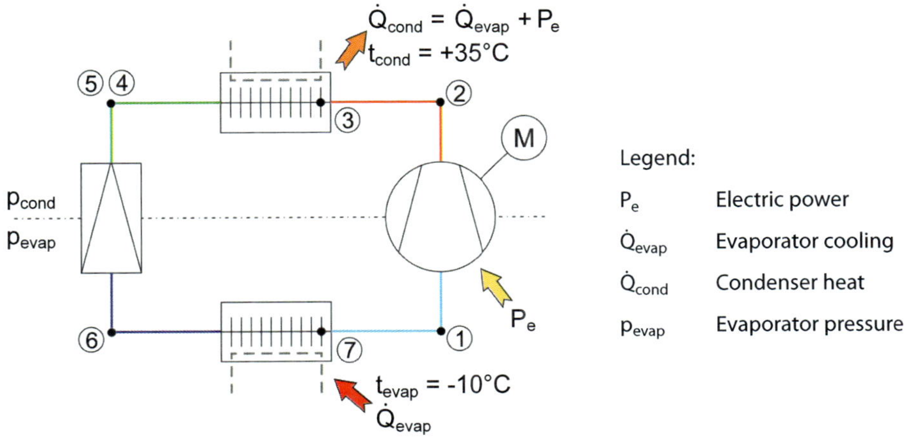

Figure 4.7 Reference circuit, see section 4.2.3.1

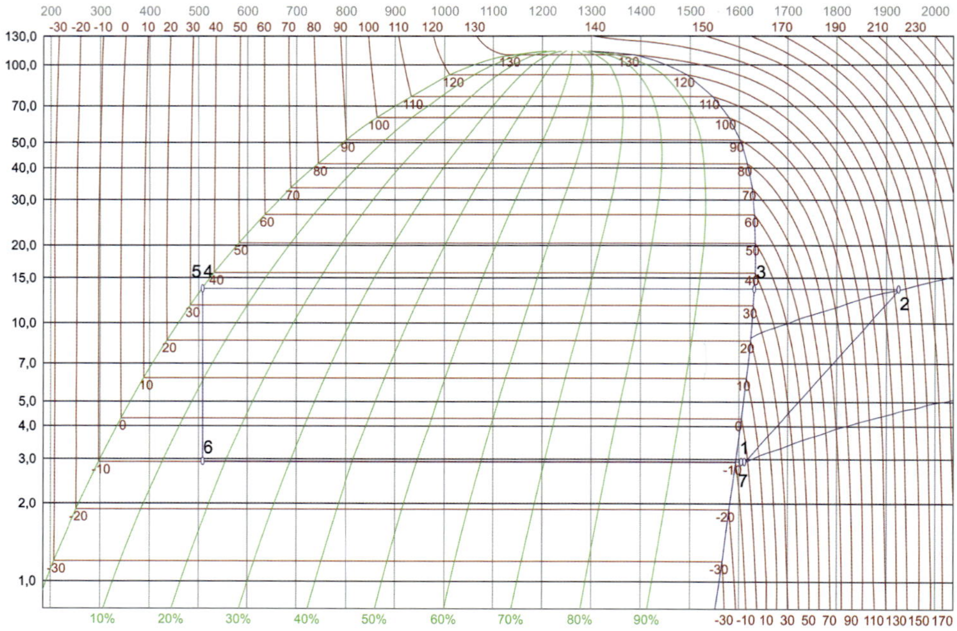

Figure 4.8 R717 Pressure-enthalpy diagram

Figure 4.7 shows the simplified refrigeration circuit. Figure 4.8 shows the associated p-h diagram for the ammonia circuit. The data apply to the isentropic compression process.

Note: Logarithmic pressure scale p in bar; vapour quality percentage and temperature in °C on lower scale; specific enthalpy in kJ/kg shown in grey at top of diagram. The calculations and p-h diagrams as well as the specific refrigerant values were carried out using Coolstar software.

The theoretical energy requirement, when taking into account the parameters specified in the reference refrigeration cycle for operation with the refrigerants listed in Table 4.9, are quite similar (see Figure 4.9). However, this only applies if the condensation temperature is not too close to the critical temperature. The end temperature of compression and thus the energy consumption of the compressed gas is considerably higher with ammonia than with the other refrigerants, but the energy loss during expansion is relatively low with ammonia.

4.2.2.2 Compressor

Due to the high evaporation enthalpy (see Table 4.3), the mass flow required for an ammonia system is small compared to other refrigerants. On the other hand, the low molar mass of ammonia causes a low-density gas, which would lead to a high volume flow for the compressor. However, the high enthalpy of vaporisation more than outweighs the low vapour density. Consequently, the volumetric cooling capacity (q_v) of ammonia is higher than that of R134a.

This means that comparatively small compressors can be used in ammonia plants. Table 4.10 shows the volume and mass flow as well as the cooling capacity based on the parameters of the reference cooling circuit. These performance data were obtained using actual compressor data (from Bitzer software v6.9.1).

Table 4.9 Thermodynamic performance: reference circuit (see Figure 4.7)

	State point	Unit	R717	R1234yf	R134a	R407A
Condensing pressure	2/4	bar	13,51	8,95	8,87	14,42
Evaporating pressure	6/1	bar	2,90	2,21	2,00	3,47
Vapour density at −10 °C	7	kg/m³	2,39	12,5	10,04	15,3
Vapour density at +35 °C	3	kg/m³	10,62	51,38	44,8	68,87
Suction temperature	1	°C	−3	−3	−3	−3
Discharge temperature	2	°C	110,5	38,8	47,6	54,1
Refrigeration capacity		kW	200	200	200	200
Drive power (isentropic)		kW	42,5	45,5	43,5	45,5
Efficiency ratio EER/COP (isentropic compression)		–	4,7	4,4	4,6	4,4

	State point	Unit	R717	R1234yf	R134a	R407A
Condensing pressure	2/4	psi	195.94	129.80	128.64	209.14
Evaporating pressure	6/1	psi	42.06	32.05	29.00	50.32
Vapour density at −10 °F	7	lb/ft³	0.14	0.78	0.62	0.95
Vapour density at +35 °F	3	lb/ft³	0.66	3.20	2.79	4.29
Suction temperature	1	°F	26.6	26.6	26.6	26.6
Discharge temperature	2	°F	230.9	101.84	117.68	129.38
Refrigeration capacity		TR	56.86	56.86	56.86	56.86
Drive power (isentropic)		TR	12.08	12.93	12.36	12.93
EER/COP (isentropic)		–	4.7	4.4	4.6	4.4

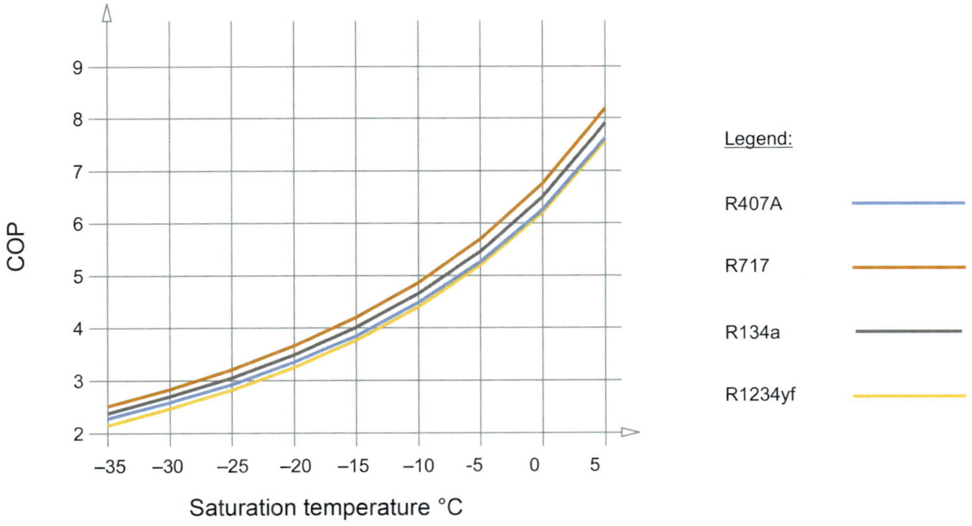

Figure 4.9 Comparison of EER/COP for isentropic compression to 35 °C (95 °F), no subcooling

Table 4.10 Reference circuit EER/COP for fixed cooling capacity and different refrigerants

		R717	R1234yf	R134a	R407A
Mass flow	kg/h	650,44	6186,14	4772,85	4663,85
Suction volume flow	m³/h	281,12	509,93	491,75	306,76
Cooling capacity	kW	200	200	200	200
Electrical power	kW	60,4	n.a.[1)]	61,3	62,9
EER/COP	–	3,31	n.a.[1)]	3,26	3,18
Mass flow	lb/h	1433.97	13638.09	10522.32	10282.02
Suction volume flow	ft³/h	7.96	14.44	13.93	8.69
Cooling capacity	TR	56.87	56.87	56.87	56.87
Electrical power	TR	17.17	n.a.[1)]	17.43	17.89
EER/COP	–	3.31	n.a.[1)]	3.26	3.18

[1)] not available

It can be seen that considerably larger compressors (by a factor of 2,3) must be used for refrigerants R1234yf and R134a when compared to R717.

Conversely, the cooling capacity of a specific compressor is largely dependent on the refrigerant used and on its volumetric cooling capacity.

4.2.2.3 Cooling during or after compression

Due to ammonia's high isentropic exponent, a very high temperature is reached during compression. When selecting the type of compressor, the required pressure ratio must therefore first be taken into account. High pressure differences between evaporation and condensing pressure lead to high discharge temperatures. The use of reciprocating compressors is only possible within limits. If higher pressure differences are required, the compression process can be cooled using a cylinder head cooling fan or water-cooled cylinder heads, and discharge temperature can thus be reduced.

If cylinder head cooling is not sufficient to maintain the permissible final temperature, an oil- or refrigerant-cooled screw compressor is more suitable. By regulating the injection of oil or refrigerant into the compression chamber, the discharge temperature can be lowered or regulated very precisely. This leads to very stable operating values, particularly in air-cooled applications or in heat pump applications.

The high isentropic exponent must also be taken into account with regard to the drive energy for the compressor motor: The superheating of the refrigerant at the compressor inlet has significantly more influence on the required drive energy for ammonia compared to other refrigerants: The lower the superheating, the better the EER of the compressor.

In this context, it must be taken into account that the available reciprocating compressors are available from a theoretical delivery volume of 20 m³/h (706 ft³/h), but screw compressors are only available from a delivery volume of approximately 80 m³/h (2825 ft³/h). In very small capacity ranges relying on reciprocating compressors, additional cooling measures may have to be implemented in the system to limit the discharge temperature.

4.2.2.4 Operating temperatures and pressures

Ammonia refrigerant is normally used in a range between –55 and 10 °C (–67 and 50 °F) evaporation temperature. The usual condensing temperatures are between 20 and 45 °C (68 and 113 °F). Piston compressor plants are usually operated in the range between 25 and 35 °C (77 and 95 °F). For special or heat pump applications, both the evaporation and the condensing temperature can be set to significantly higher values. If the system is to be operated with an evaporation temperature below –20 °C (68 °F), two-stage compression must be provided when using piston compressors. The background to this measure is that the hot gas temperature should not exceed 130 °C (266 °F) to protect the oil and the system. Plants with screw compressors can be operated in single-stage operation with a much higher compression ratio. This is possible with oil-cooled screw compressors due to the additional injection of cooling oil into the compressor.

If these operating envelopes for the compressors are exceeded for a certain refrigeration application, a multi-stage refrigeration system is recommended. Multi-stage systems with intermediate cooling between the compressor stages have the advantage over single-stage systems in that they are more economical to operate. They need less driving power and the operating costs are lower. However, the investment costs for two-stage systems with interstage cooling are higher. That is why they are a priority choice when high cooling demand, long service life and good economy are required. In today's heat pump applications, condensing temperatures of almost 100 °C (212 °F) are reached. However, it must be ensured that the machinery components on the high-pressure side are approved for high-stage pressures of 63 bar (913 psi).

The compressors currently available are usually designed for the following conditions:

- Shaft bearings: for pressure differences of max. 20 bar (290 psi) between suction and discharge side (screw compressor only).
- Compressor housing: for the saturation pressure at prevailing 60 °C (140 °F).

Constructions for special applications such as heat pumps or for CO_2 applications are exceptions. The housing for heat pump compressors should enable a maximum pressure of 63 bar / 913 psi (equates to 100 °C / 212 °F).

4.2.2.5 Operating limits for hot gas temperature

As already explained in Chapter 4.2.2.2, the hot gas temperature for ammonia as refrigerant is significantly higher than with other refrigerants due to the high isentropic exponent.

For ammonia compression, the hot gas must be cooled even at relatively low-compression ratios. If the temperature difference between the condensation and evaporation temperature is more than 60 K (saturation temperature in each case), a hot gas temperature of more than 150 °C (302 °F) will occur in a real-world compressor (non-adiabatic and with friction/isentropic), as assumed for the calculation in Figure 4.5). Synthetic oils can still be used at these temperatures, but it is advisable to reduce the hot gas temperature using cylinder head cooling when using piston compressors. Depending on the type of oil used, different limit values must be observed here: Systems that are lubricated with mineral oil should never be operated above 130 °C (266 °F). When using synthetic oil, the hot gas temperature should not exceed 150 °C (302 °F). The hot gas temperature should be limited for several reasons:

- The oil and its properties (in particular the lubricity) are strongly influenced (up to decomposition of the oil) at high discharge temperatures.
- The oil ages faster at high temperatures, and carbonising can occur.
- The oil separator's effectiveness reduces with increasing oil temperature because hot oil has a lower density and lower viscosity and more of the oil passes through the separator as gas.
- A lowering of the suction gas temperature at the compressor leads to a higher performance of the compressor.

Taking the aforementioned points into account, the high hot gas temperature in heat pump applications or upon heat recovery of a refrigeration system is of course an additional benefit for heating a medium (normally water).

When using oil-cooled screw compressors, the hot gas temperature is reduced with the help of an additional oil injection. The cooled oil is injected directly into the compression chamber of the compressor. Depending on the oil temperature and volumetric flow, the hot gas temperature can be cooled down very efficiently and in a controlled manner, which means that the screw compressor can have a much higher pressure ratio than piston compressors.

4.2.2.6 Pipes

The selection of pipe sizes is determined by the permissible pressure losses. Depending on the density ρ of the respective refrigerant and the mass flow, the corresponding volume flow can be calculated, which together with the pipe diameter d yields the flow velocity w in the pipes. Using the Darcy–Weisbach equation the pressure loss for a given pipe length l can be calculated to:

$\Delta p = \lambda \cdot l/d \cdot \rho/2 \cdot w^2$

The pressure loss in the refrigerant circuit depends on the density and the flow rate of the refrigerant. The higher the density of the refrigerant, the higher the pressure loss in the circuit. Higher pressure losses have to be compensated by additional drive energy for the compressor. In order to achieve the highest possible energy efficiency, low pressure losses must therefore be considered when designing the refrigeration circuit.

Due to the low density of the ammonia vapour, higher flow velocities in the suction line are possible without resulting in excessive pressure losses. This means that, compared to other refrigerants, smaller suction line pipe diameters can be used (see Table 4.13).

At the same time, the gas flow rate through the piston compressor or at the discharge port of screw compressors can be increased significantly when using ammonia. The maximum possible flow velocity depends on the speed of sound (see Table 4.11) of the refrigerant. Considering ideal gas behaviour, the speed of sound can be calculated using the formula:

$$c_{ideal} = \sqrt{\frac{\gamma RT}{M}}$$

Where γ is the isentropic exponent or compression coefficient, R is the universal gas constant [8,31446 J/(mol K)], T is the absolute temperature of the gas in Kelvin and M is the molar mass of the respective refrigerant in kg/mol. It follows that the speed of sound c_{ideal} and thereby the maximum flow velocity increases with a lower molar mass. According to Table 4.11, the speed of sound of ammonia vapour (at –10 °C / 14 °F) is 397,5 m/s (1303 ft/s). This is significantly higher than for other refrigerants.

Table 4.11 Speed of sound for reference circuit at −10 °C (14 °F) saturated suction gas temperature

Refrigerant	Molar mass	Volume flow	Density	Speed of sound
	g/mol	m^3/h	kg/m^3	m/s
R717	17,01	283	2,39	397,5
R134a	102,03	496	10,04	146,9
R1234yf	114,00	522	12,5	135,8
R407A	86,20	309	15,3	155,1
	lb/lbmol	ft^3/h	lb/ft^3	ft/s
R717	17.01	8.01	0.15	1303.80
R134a	102.03	14.05	0.63	481.83
R1234yf	114	14.78	0.78	445.4
R407A	86.2	8.75	0.96	509.06
Calculated with Refprop 10, database 23				

To avoid high pressure losses, make sure when designing the system that the flow velocity in the pipelines does not exceed 10 % of the speed of sound. The high speed of sound of ammonia gas allows for significantly higher flow rates compared to other refrigerants.

The velocities listed in Table 4.12 can be used as a guide for the pipe dimensioning of ammonia plants. These values have to be adjusted when the gas velocity needs to meet some additional criteria, for example for separating oil or liquid droplets, lifting liquid in a wet suction riser or oil return piping.

The guidelines mentioned here are based on practical experience. Values are based on the assumption that a pressure loss of 0,5 K per 10 m (32.8 ft) pipe length on the suction side and a pressure loss of 0,5 K per 30 m (98.4 ft) pipe length on the discharge side are permitted.

Compared to R134a, significantly smaller pipe diameters can be used when using ammonia. The cooling capacity of the suction line can be increased by two to three times with the same

Table 4.12 Flow velocities in pipelines

Pipeline	R717	R134a
Suction in m/s	20	8
Liquid in m/s	1,3	0,9
Condensate in m/s	3	2,5
Hot gas in m/s	15	6
Suction in ft/s	65.6	26.24
Liquid in ft/s	4.26	2.95
Condensate in ft/s	9.84	8.2
Hot gas in ft/s	49.2	19.68

Table 4.13 Comparative pipe diameters (in reference circuit. Temperature drops are shown per 10 m (32.8 ft) pipe length in reference circuit (DN – nominal diameter)

Pipeline	R717	R1234yf	R134a	R407A
Suction	DN 65 (0,21 K)	DN 100 (0,40 K)	DN 100 (0,43 K)	DN 80 (0,49 K)
High-pressure liquid	DN 20 (0,11 K)	DN 40 (0,18 K)	DN 40 (0,14 K)	DN 32 (0,14 K)
Hot gas	DN 40 (0,04 K)	DN 65 (0,37 K)	DN 65 (0,17 K)	DN 50 (0,43 K)
Low-pressure liquid	DN 50 (0,01 K)	DN 80 (0,06 K)	DN 65 (0,07 K)	DN 65 (0,04 K)

diameter and comparable pressure loss (see Table 4.13). This means that the pipe installation costs will be lower. If the pipe diameters were maintained as necessary for R134a refrigerants, then significantly lower pressure losses and thus higher evaporation temperatures or lower condensation temperatures would result. This would then lead to a higher coefficient of performance (COP) and lower operating costs for the plant. A rule of thumb is that the efficiency ratio (EER) of the overall system improves or deteriorates by 2–4 % per Kelvin temperature change. Increasing the evaporation temperature by 2 K and lowering the condensation temperature by 2 K would result in an efficiency improvement of approximately 8 to 16 %.

The effects of this on plant engineering can be seen in Table 4.13: The pressure losses stated in Table 4.13 relate to a pipe length of 10 m (32.8 ft). With longer pipelines, the required pipe diameters will differ even more. A typical value or rule of thumb for ammonia suction lines is to allow for a pressure loss equal to the eqivalent of 0,5 K lower evaporation temperature for every 30 m (98.4 ft) suction line length as the sizing criterion. The suction line velocity will then increase with increasing pipe diameter.

In the field of cold water generation, but also in most other refrigeration applications, planners and builders today demand that the systems be engineered for high condensation temperatures. The saturation pressure for ammonia will then vary between 15 bar / 217 psi (= approximately 40 °C / 104 °F) and 25 bar / 362 psi (= 58 °C / 136.4 °F). The materials for pipes, fittings and pressure vessels available at standard specifications can therefore be used for the refrigeration plant system. However, it is important to ensure that the materials meet the requirements of the regulations specified by the client.

4.2.2.7 Heat exchanger

The heat transfer coefficient (see Table 4.7) of a refrigerant decreases with increasing density. This is partly because a thicker liquid film forms in the evaporator or condenser due to the lower latent heat of the higher molecular weight refrigerant. The use of ammonia therefore achieves significantly more favourable values than with synthetic refrigerants. In addition, synthetic refrigerants have a lower thermal conductivity in the liquid phase.

First of all, this means that the heat exchangers of an ammonia system can be built more efficiently and therefore smaller. However, heat exchangers for synthetic refrigerants as well as for hydrocarbons and CO_2 can be built with more heat conductive materials (copper), and the better thermodynamic properties of ammonia are therefore partially offset (see Chapter 4.2.2.4). One option is to use aluminium in ammonia heat exchangers. Thousands of industrial aluminium ammonia evaporators have been installed and operated successfully since the 1970s [4-9].

4.3 Plant engineering aspects

4.3.1 Plant systems and types

4.3.1.1 Flooded and dry expansion systems / liquid separators

Ammonia refrigeration systems are mostly designed and built as flooded systems. Dry-expansion systems are mainly used when the refrigerant charge has to be kept low. At the same time, the system structure for dry-expansion systems is simpler, and it corresponds more closely to the known technology, which is also used for synthetic refrigerants. Special care has to be taken to keep the superheat after the evaporator(s) of DX ammonia systems as low as possible due to the high isentropic coefficient (index of compression) of ammonia vapour, see chapter 4.2.1. With the right plant design and control strategy, DX ammonia systems can be built very energy efficient. Another argument for using dry expansion is the much easier oil return to the compressor. Most ammonia DX systems use lubricants which are miscible with ammonia, but with the correct plant design also non-miscible oils can be used and are already being used in ammonia DX systems. Oils miscible with ammonia that are suitable for this purpose are significantly more expensive and must be used with special care due to their distinct hygroscopic properties.

Typically, a higher efficiency (COP/ EER) of the refrigeration systems can be achieved by the use of the flooded systems. Oil recirculation is more complex and the refrigerant charge is higher, but both oil recirculation and the reduction of the refrigerant charge can easily be made in a simple, manageable way. For example, the refrigerant charge of the flooded systems can be greatly reduced with the help of optimised heat exchanger designs and liquid separation systems.

For comparison, Figure 4.10 shows the simplified flow diagrams of the two systems.

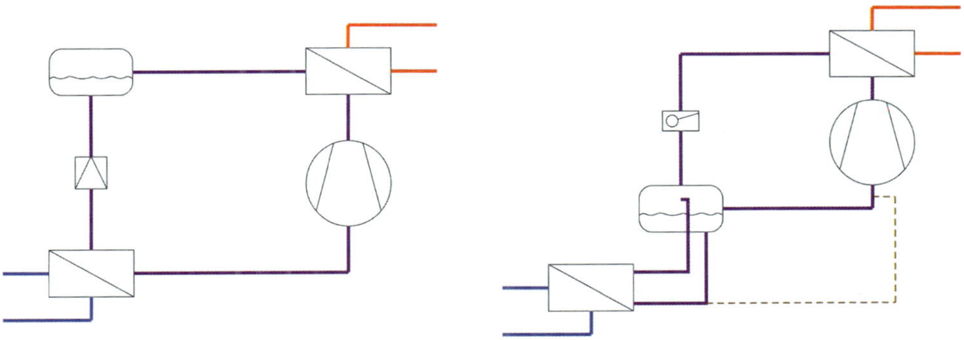

Figure 4.10 Dry expansion (left) and flooded evaporation (right)

4.3.1.2 Liquid separator

In flooded systems, the refrigerant liquefied in the condenser is expanded in the liquid separator using a high- or low-pressure float valve. The liquid separator protects the compressor by separating the liquid and gaseous phases coming back from the evaporator. Due to its density, the liquid falls to the bottom of the separator, and the gas can be extracted by the compressor.

When designing the liquid separator, it should be sufficiently large to accommodate the total volume of the system (receiver function) and also account for potential operating conditions which may lie outside the design conditions. This includes, for example, the start-up situation of

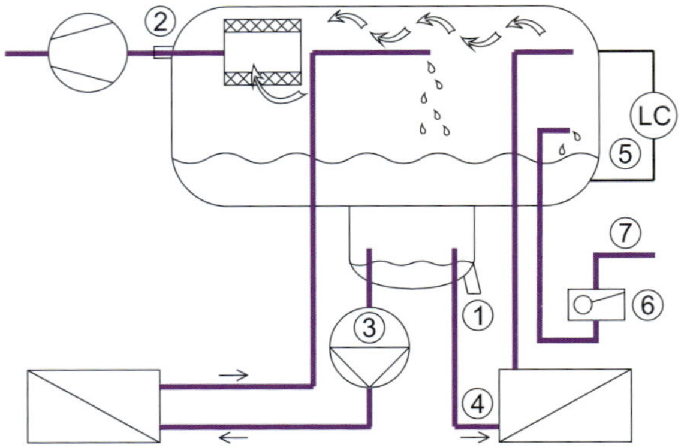

Figure 4.11
Schematic for liquid separator (also known as surge drum)

a refrigeration system with a warm secondary refrigerant (brine). Due to the higher temperature difference between the ammonia evaporation and the secondary refrigerant temperature at start-up, the performance of the evaporator is vastly increased. This leads to higher refrigerant volume flows on the return side of the evaporator. As a result, there may be liquid carry-over at the compressor inlet if the liquid separator is not large enough for these higher volume flows.

Figure 4.11 shows a schematic of a liquid separator. The left-side evaporator works with a refrigerant pump (3). The right-side evaporator (4) operates by gravity circulation. For illustration purposes, Figure 4.12 shows a compact liquid separator assembly with a built-in plate heat exchanger (evaporator).

Figure 4.12
Liquid separator (also known as surge drum) (© EM Polar)

4.3.1.3 Refrigerant pump circulation

When several evaporators are connected to an ammonia refrigeration system and are distributed over a large area (e.g., in a cold store), the system is usually designed for refrigerant pump operation, but some manufacturers also apply direct expansion for multi-evaporator systems. The evaporators are connected to a pipe network. This network consists of a supply and a return line and is connected to the low pressure (LP) liquid separator. With the help of one or more refrigerant pumps, the liquid refrigerant is continuously pumped through the pipes and is thus available to the evaporators. A simplified flow diagram is shown in Figure 4.13. The evaporators are fitted with an electrically controlled solenoid valve on the incoming liquid side.

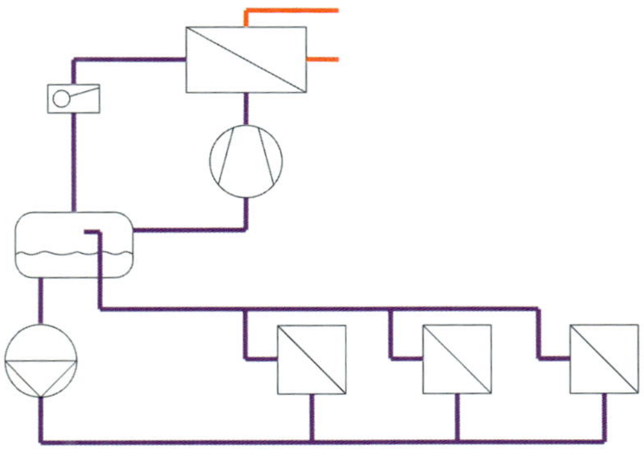

Figure 4.13 Pump recirculation

The solenoid valve opens when refrigeration is required, and the liquid refrigerant flows through the evaporator. The flow rate is often higher than the amount of refrigerant actually required. In practice, a multiplication factor of the mass flow between three and five times has proven itself and is called circulation rate. In this case, one speaks of a flooded evaporation system. The circulation rate should, on the other hand, not be too high, because it will unnecessarily increase the energy needed for the pump(s) decreasing the overall ammonia plant energy efficiency. Due to the very high evaporation enthalpy of ammonia, and despite the increased circulation rates, only relatively small refrigerant components are needed. This reduces investment costs for pumps, pipes, fittings etc., while at the same time, the operating costs of the circulation pumps are cheaper than for systems with synthetic refrigerants.

4.3.1.4 Two-stage systems

The use of two-stage systems is particularly useful when large temperature lifts (and therefore large pressure differences) have to be overcome.

To limit the compression ratio and the hot gas temperature, a two-stage compression process should be selected at low evaporation temperatures or at high external pressure conditions. This results in better system conditions and, above all, better performance figures. For the design and construction of the two-stage system, a distinction must be made between systems with reciprocating piston compressors and systems with screw compressors. In the case of reciprocating compressors, it is important that the hot gas (shown in Figure 4.14) is cooled after the first compression stage, and only then is it led into the second compression stage. Cooling takes place by injecting liquid refrigerant into a space between the first and second compression stages. It is important to ensure that no condensation can form in this space. One way to reduce the risk of condensation is to use an additional heat exchanger. This variation (Figure 4.15) is often used in systems with synthetic refrigerants, but is not common in ammonia systems. In systems that are designed with screw compressors, the intermediate cooling of the gas is less critical because the gas temperature is cooled by the oil injection into the compressor. However, it is recommended that any flash gas due to the first expansion stage be drawn off through the high-pressure (HP) compressor.

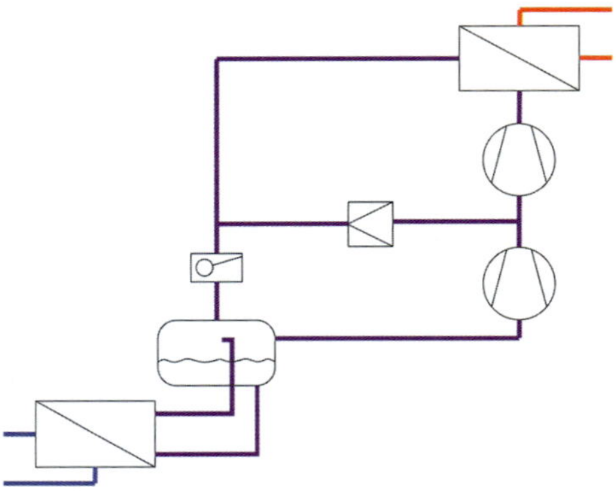

Figure 4.14 Two-stage compression with interstage liquid injection

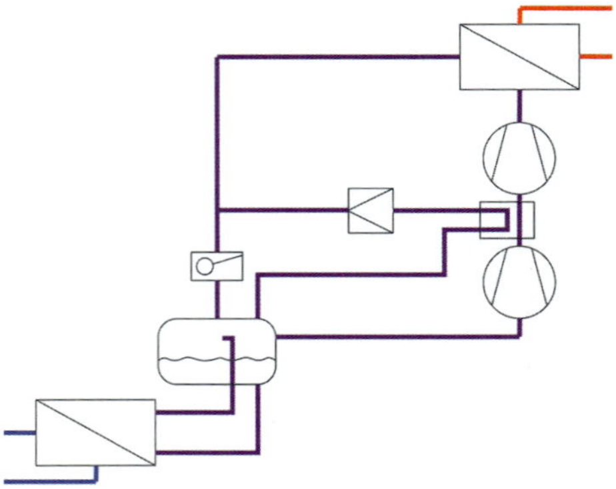

Figure 4.15 Two-stage compression with liquid injection in an interstage cooler

4.3.1.5 Two-stage expansion

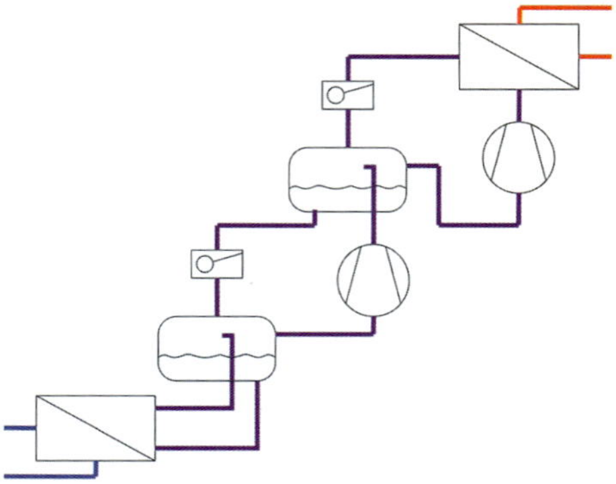

Figure 4.16 Two-stage expansion and compression with interstage liquid receiver (also called intercooler)

In the case of a two-stage expansion, the flash gas due to the first expansion stage is usually drawn off by the HP compressor and thus reduces the need for cooling of the suction gas in the HP stage. This circuit (shown in Figure 4.16) is often implemented using a medium-pressure drum between the LP and HP stages.

The pressure between the two compressor stages (intermediate pressure p_i) must be determined on a plant-specific basis and depends on the evaporation and condensing pressure as well as on the ratio of the two pressure stages, by which the overall plant compression ratio (LP to HP) is achieved. In case of additional evaporators being connected to the interstage liquid receiver, the intermediate pressure also depends on the temperature requirements of the services connected to this intercooler.

Regardless of the degree of thermal optimisation of any compression process, the energy requirement can also be reduced by multi-stage operation. Figure 4.17 shows a corresponding p-h diagram for a two-stage system (P6–P9, P9–P10) with an intermediate pressure drum (P8,9). One can clearly see the reduced flash gas content during the expansion from intermediate- to low-pressure level (P9–P10). The flash gas portion of the first expansion stage is drawn directly to the high-pressure compressor and therefore does not have to be passed through both compressor stages, thus reducing compressor energy and thereby increasing the energy efficiency of the refrigeration installation.

4.3.1.6 Screw compressor with ECO port

A similar, technically somewhat simpler process can be achieved with screw compressors. Some screw compressors are equipped with an additional "ECO" connection port. At full pressure

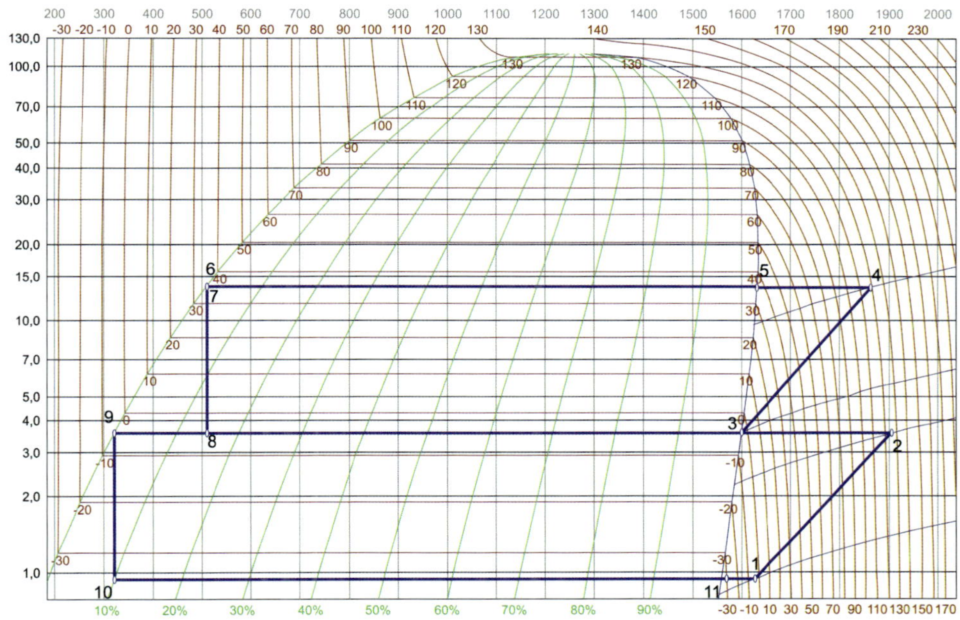

Figure 4.17 p-h diagram for two-stage compression and intermediate pressure liquid receiver

operation (LP to HP) of the compressor there exists an intermediate pressure also known as economiser pressure.

The gas drawn in at this port is mixed with the main gas stream and compressed in the compression chamber, up to HP. The additional drive energy required for this is low in relation to the additionally generated cooling capacity. Similar to the two-step process described in Chapter 4.3.1.4, an intermediate pressure is maintained here. This is called an economiser circuit. An "open flash" circuit is usually selected for ammonia systems (Figure 4.18). Alternatively, the liquid refrigerant can be subcooled using an additional heat exchanger (Figure 4.19). However, this variation is very rarely implemented in ammonia technology.

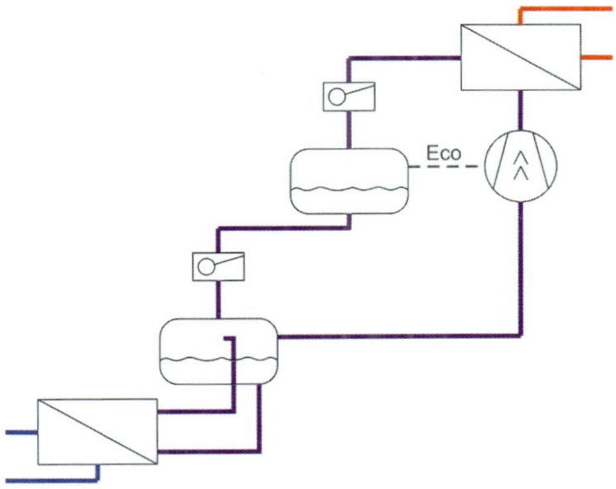

Figure 4.18 Two-stage expansion with "open flash" intermediate pressure receiver connected to economiser

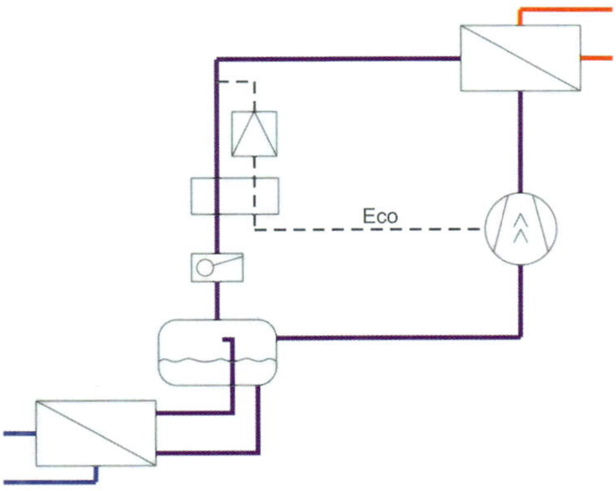

Figure 4.19 Liquid refrigerant subcooling using heat exchanger on economiser

Both variants have in common that part of the liquid refrigerant vaporises at the ECO port economiser pressure in order to cool down the rest of the refrigerant liquid. If large pressure differences between suction and high pressure have to be overcome, the efficiency can be increased by approximately 20 %. The drive energy increases by about 10 %, but the cooling capacity is increased by approximately 30 %. The same effect tends to occur at smaller pressure HP-LP differentials, but the effects are not so pronounced. The overall efficiency of the system is almost unchanged. It must therefore be weighed in individual cases whether the higher investment costs for an economiser system are worthwhile compared to the lower energy costs to run the system. In addition, the poor part-load energy performance of economised screw compressors has to be considered. Economised screw compressor plants servicing cold stores can use up to 2,5 times more energy annualised than systems with two-stage reciprocating compressors.

It should also be borne in mind that energy usage of many refrigeration installations is much higher than designed, due to poor maintenance. Therefore, highly efficient systems will only achieve the calculated and expected performance figures in the long term if they are properly adjusted and maintained.

4.3.2 Components

4.3.2.1 Compressor

Piston and screw compressors with an open shaft (often called "open compressors") (see Figure 4.20) are predominantly used in ammonia installations. The models available have a theoretical gas delivery rate of between approximately 10 and 15000 m^3/h (353 and 530000 ft^3/h). In the field of air conditioning, this corresponds approximately to a cooling capacity of 10 to 15000 kW (2.8 to 4265.2 TR). Large compressors – with a capacity exceeding 500 m^3/h (17657 ft^3/h) – are usually designed by manufacturers predominantly for ammonia use. Compressors in the performance range between 10 and 200 m^3/h (353 and 7062 ft^3/h) are often used also for other refrigerants. They can be adapted by the manufacturers for use with ammonia refrigerant.

Figure 4.20 Ammonia screw compressor set with horizontal oil separator (© EM Polar)

The selection of the compressor depends on many factors. First of all the required cooling capacity and the associated operating conditions, as well as the efficiency of the compressor, must be considered. However, many other factors, such as part-load behaviour or noise emissions, must also be taken into account.

Last but not least, the economic viability over the intended life must be considered when making the selection. The investment costs in comparison with the running costs for operation, service, maintenance and repair must be taken into account. The investment costs for a piston compressor are usually significantly lower than those for a screw compressor, but the maintenance costs for a piston compressor can be higher. In this context, the costs for necessary additional screw compressor components such as high-quality oil separators or oil cooling systems must also be considered.

The use of fully or semi-hermetic compressors designed for synthetic refrigerants is not possible for ammonia systems. This is due to the copper materials used (especially in the motor winding). Some manufacturers are now offering semi-hermetic screw compressors with aluminium motor winding for ammonia. Initial experience are promising, and it remains to be seen how these compressors will prove themselves in the long term.

The compressor manufacturers offer many special designs for ammonia applications. Compressors for heat pump applications (high condensation temperatures and thus high pressures) with reinforced casings are available. Another special design is the two-stage compressor, in which the refrigerant is compressed in two successive stages. If high pressure differences have to be overcome, these two-stage compressors in combination with some kind of interstage cooling are ideal.

4.3.2.2 Air-cooling heat exchangers (also known as air-cooling coils)

The specific heat transfer capacity of the materials used in the construction of the heat exchangers has a significant influence on the size and shape of heat exchangers.

For ammonia, there are three design principles available for air cooling coils: devices that are made entirely of galvanised steel tubes and fins, devices that are made of stainless-steel tubes and aluminium fins and those made of aluminium tubes with aluminium fins. In European practice, air coolers for ammonia applications are mainly built from stainless steel tubes with aluminium fins because the manufacturing process is similar to that of copper tube and aluminium fin heat exchangers used for other refrigerants, Figure 4.21. Elsewhere, galvanised steel tubes and fins are used. But internal surface enhancement for both, stainless steel and aluminium tubes are now being offered by several manufacturers for low overfeed and DX applications increasing the heat exchanger efficiency.

Figure 4.21 Piping manifold with hot gas defrost on ammonia air cooling coil (© Thermofin)

Table 4.14 Thermal conductivity for heat exchanger materials

Material	W/(m·K)	Btu·ft/(h·ft²·°F)
Steel	46	26.58
Aluminium	250	144.45
Zinc	16	67.02
Copper (not suitable for R717)	394	9.24

Internal "wetting" of tube surfaces is a function of thermal conductivity of the tube material [4-12]. As shown in Table 4.14, the thermal conductivity of stainless steel is lower than that of galvanised steel. However, the use of aluminium fins on stainless steel tubes often overcompensates the reduced thermal conductivity of the stainless-steel tubes. The heat exchanger performance of a stainless steel-aluminium air cooler is therefore typically higher than that of a galvanised steel unit. It should be taken into account that when using synthetic refrigerants, CO_2 or hydrocarbons, mostly copper materials are used for the heat exchangers. Due to the high heat conduction coefficient of copper, these cooling coils have a higher heat transfer capacity than the steel materials and can therefore be made smaller and lighter with the same output size. Aluminium tube and aluminium fin heat exchangers for ammonia achieve similar heat transfer performance than copper-tube-aluminium-fin heat exchangers. Compensation for poor internal wetting is possible by means of internal tube surface enhancement which is commonly available for copper and aluminium tubes.

Copper piping cannot be used in ammonia heat exchangers. It would theoretically be possible under appropriate conditions (100 % dry ammonia), but one never has completely dry ammonia inside an ammonia refrigeration system. A small water content is recommended in order to prevent stress corrosion cracking on steel.

When designing the heat exchangers, the challenge is that the pipe diameters should be small in order to achieve good heat transfer, such that the laminar flow can transition into a turbulent flow as quickly as possible. This transition will lead to more heat transfer around the entire circumference of the heat exchanger tube becoming available.

On the other hand, the pipe diameter should be large in order to keep the pressure drop low. Problems can arise with evaporators with stainless steel tube packages due to high-pressure losses. Low-temperature applications, especially those using dry evaporation or systems with a low liquid circulation, are affected by this.

Defrost systems for air coolers

In general, when defrosting air-cooling evaporators, care should be taken to ensure that the duration and frequency of the defrosting phases are adjusted to actual requirements. Time-controlled systems are energy and cost intensive and usually do not lead to good defrosting results. Ambient air defrost is an energy efficient defrost option in hot climates. Natural defrost is another energy efficient option for room temperatures well above 0 °C (32 °F). It is achieved by stopping the refrigerant flow to the evaporator and keeping the fan running. The resulting defrosting time is long.

Hot gas defrosting is a preferred and energy efficient way of defrosting. Electric heating element defrosting, common in many supermarket refrigeration systems, should not be used in combination with ammonia due to the risk of fire. It is important not to exceed a maximum of one third of all evaporators in a system being defrosted at the same time. This enables adequate hot gas output from the compressor(s).

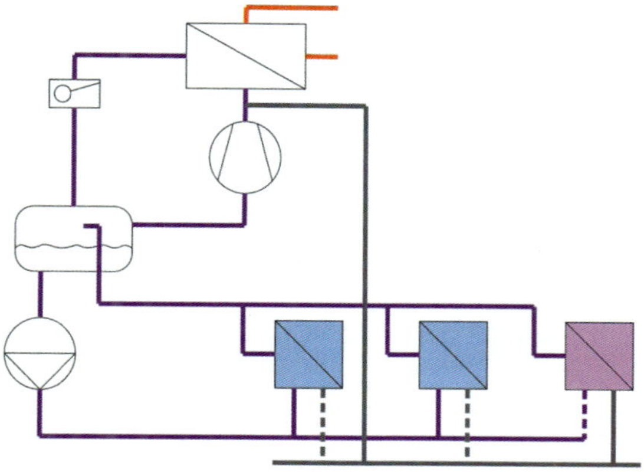

Figure 4.22 Hot gas defrosting

As shown in Figure 4.22, the hot gas is drawn off on the HP side of the compressor and led to the evaporator via a designated hot gas line. It is fed directly into the evaporator. The liquid line is kept closed during the defrost period. The hot gas is cooled in the evaporator during the defrosting, partially condensed and returned to the liquid separator via the suction line (see Chapter 4.3.5 “Risks due to hydraulic pressure surges”). Partial condensation of the hot gas is not necessarily desired and may be avoided by a high-pressure float valve for hot gas (condensate) return.

To ensure safe and energy-optimised defrost operation, the solenoid coils of the corresponding valves must be checked regularly and replaced in good time. As a result of the mounting position, ice often forms around the solenoids. Malfunctions such as the valves jamming permanently open or closed can occur. Along with this, the energy costs naturally rise considerably, and at the same time, there can be considerable damage to the system.

Around 50 % of the heat used to defrost is eventually ejected into the cold room. This implies that the cooling load for the room being defrosted is up to three times higher than during normal cooling. In order to reduce this extra cooling load to a minimum, possible optimisation options should be considered when planning the system. These include, for example, evaporators with defrost flaps or other designs that reduce defrost heat flowing into the cold room.

For large deep-freeze stores, an insulated penthouse cooler can be used (see Figure 4.23). This special design provides for the installation of the evaporator and the associated cold room fan in a separate, insulated box.

The complete housing is positioned against the cold store wall or on the cold store ceiling. The air flow is drawn from the cold store to the evaporator via a motorised flap with the help of the fan. The cooled air is then pushed back into the cold room. The flap is closed during defrosting. This significantly shortens the defrosting process and at the same time prevents defrosting heat and moisture from entering the cold store. Air pressure peaks may follow defrost cycles in cold rooms and refrigerated warehouses. Depending on the room size and existing ventialtion openings, additional air pressure relieve openings may be required.

4.3.2.3 Heat exchangers for liquid media

Stainless steel plate heat exchangers are predominantly used in ammonia plant construction. Various stainless-steel alloys are available. The selection is usually dependent on the medium to be cooled. Similar to synthetic refrigerants, the use of shell and tube heat exchangers is steadily decreasing in favour of the more compact and often more efficient plate heat exchangers.

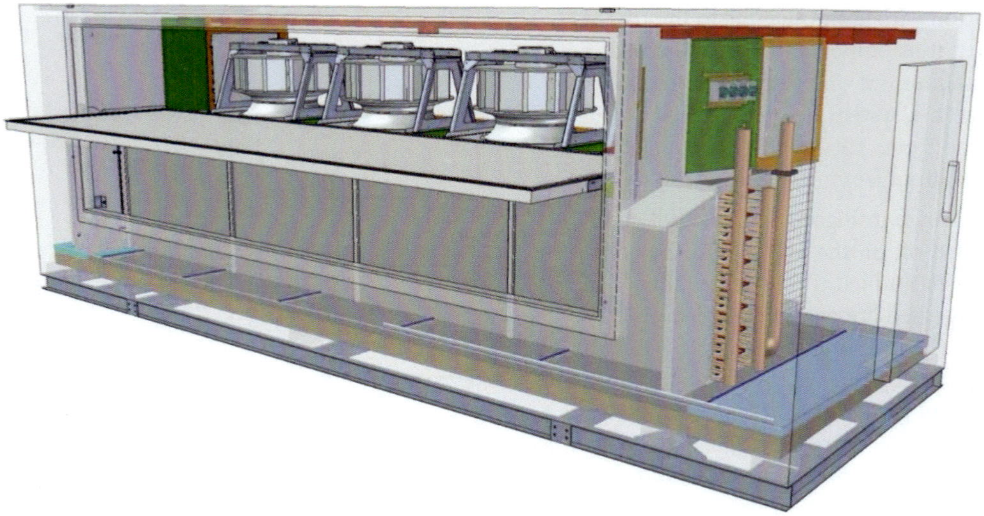

Figure 4.23 Evaporator and fan Penthouse cooler box (© Thermofin)

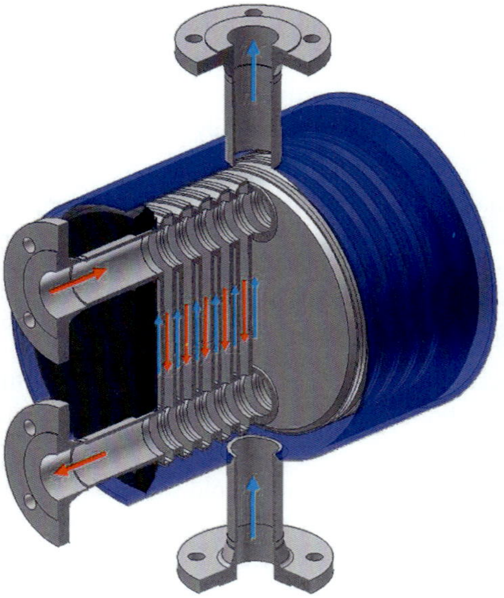

Figure 4.24 Circular welded plate-and-shell heat exchanger (© Gesmex)

Depending on the application, various different embossed plate designs for the plate assembly can be selected. Welded plate cassettes are used for standard conditions. These individual cassettes are available in various dimensions and flow patterns. The number and design of the individual cassettes depend on the cooling capacity required, the thermodynamic conditions and the medium used. The cassettes are bolted together in an appropriate frame. Rubber seals for liquid flow are inserted between the cassettes. The complete cassette assembly is then pressed together to a complete liquid cooling heat exchanger.

Fully welded plate-and-shell heat exchangers were introduced to the refrigeration market in the early 1990s and their market share in ammonia applications has grown steadily since then (see Figure 4.24). The first thing that distinguishes these cylindrical devices is that they are hermetically sealed. However, due to their compact design, they are space saving and easier to integrate into the chiller sets. It has to be taken into account that the cylindrical design is less favourable for the thermodynamic behaviour during refrigerant evaporation. As with the aforementioned cassette heat exchangers, a wide variety of plate sizes with different embossments is also available with this design. The plate thickness of both the sealed plate cassettes and the fully welded plates is determined by the pressure differences existing between the two fluids required in each case. Manufacturers have various plate thicknesses available for various embossed designs.

4.3.2.4 Oil circuit

Oil behaviour in the refrigerant circuit

The oil in the refrigerant circuit is used to lubricate and seal the compressor and its components. Sealing is especially important in screw compressors. The oil is also needed to cool the compression process in a screw compressor. Due to the cooling and sealing task of the lubrication oil in a screw compressor, such compressors have a high oil discharge and require effective oil separators. The oil content in the discharge from a piston compressor is relatively low provided the compressor is operating correctly. Nonetheless, the valves of piston compressors need oil for lubrication, so a small amount of oil has to leave the compression chamber.

Besides the type of the compressor, the amount of oil contained in the hot discharge gas depends on the operating conditions. In general, oil discharge increases with increasing hot gas temperature and increasing compressor speed. Oil outside the compressor is only needed in small amounts for lubricating some types of valves inside the refrigeration circuit. Oil has negative effects in heat exchangers where it degrades heat transfer and in pipes where it increases pressure loss, especially in the cold suction line. In order to keep the amount of oil that can get into the refrigeration circuit (termed oil throw, discharge or circulation rate) as low as possible, the installation of an oil separator is essential. In particular when using screw compressors, multi-stage oil separating systems (mostly integrated within one oil separator) are required. The efficiency of the separator depends on both the type and design, as well as on the operating conditions of the refrigeration system.

The separation effectiveness of the oil separator is – independent of its design and separation principal – determined by three influencing factors:

- Hot gas temperature: the oil separation effectiveness decreases with increasing temperature. The reason is the vapour pressure curve of the oil – at higher temperatures a higher percentage of the oil is in the form of gas mixed with the refrigerant gas and is unaffected by the oil separator, which work by separating liquid droplets.
- Oil type: compared to untreated mineral oil, the oil separation effectiveness with hydrocracked oils is significantly higher.

The gas velocity within the oil separator also plays an important role. Too high velocities have a negative impact on the oil separation efficiency of most oil separators, where too low velocities degrade the effectiveness of oil separators using centrifugal forces while coalescing oil cartridges are not affected by too low gas velocities; in fact, they might even become more effective.

Synthetic refrigerants usually show good miscibility with refrigeration oils and are for use in dry evaporation systems. Ammonia systems are usually designed as flooded systems and are operated with immiscible oils. This leads to a major difference in the design of the refrigeration systems: ammonia systems are equipped almost with an oil separator in order to separate the oil ejected with the hot gas.

Mineral oils (including hydrocracked oils), polyalphaolefins or oils based on alkylbenzenes are commonly used in ammonia plants. These three types of oil are immiscible. Despite an oil separation system, small amounts of oil may get into the refrigeration pipe network, especially in the form of a gas or a very fine mist. This oil fog condenses on cold surfaces in the refrigeration system and typically forms oil films on pipes and heat exchanger surfaces, especially in places with little flow velocity. The oil accumulates and must be drained regularly. When designing the pipe network, care must be taken to ensure that points of low flow are identified, or deliberately created, in order to enable simple and safe automatic oil return or manual oil removal.

The oil usually accumulates in the evaporator. The refrigerant flow rate in the evaporator manifold is generally low, and the viscosity of the oil is higher due to the lower temperatures. Of course, this has a significant negative impact on the evaporator's thermal performance.

In order to ensure the refrigeration system can be operated as economically and as independently as possible, the oil carry-over into the refrigeration system and thus the necessary cycles for manual oil removal should be kept as low as possible. This requires highly efficient, multi-stage oil separators. In compact systems, such as liquid chillers, manual oil removal can be completely dispensed with by using such oil separators and by using fully automatic oil return valves and piping systems.

Flooded evaporation systems and oil-flooded screw compressors are not always price competitive due to the costs of additional oil separator and the automatic oil return piping system. In order to reduce the efforts related to proper oil return, ammonia-soluble oils have been developed for ammonia cooling systems. These oils are based on polyalkylene glycol (PAG oils). Comparable to oil return in systems with synthetic refrigerants, PAG oils can be transported through the system. Additional partial flow compressor oiling may be necessary at very low evaporation temperatures. A very small refrigerant partial flow can be taken from the refrigerant pump pressure side and returned to the suction side of the compressor to provide some lubrication. To avoid liquid hammer on the compressor, this refrigerant partial flow is routed through a heat exchanger installed in the liquid line after the condenser.

It must be taken into account for flooded evaporation systems that oil return via the suction line (including PAG oils) from the LT surge drum will not take place. The LP surge drum acts like an oil trap. The oil will remain in the surge LP drum where it will accumulate at the lowest position due to the higher density of the oil compared to liquid ammonia as long as liquid ammonia and oil are immiscible. Miscible PAG oils are therefore used almost exclusively in systems with dry evaporation and in systems with very low evaporation temperatures. Automatic oil return from a LP surge drum is possible as long as the immiscible oil remains fluent enough, which means, it must not become too viscous due to the low temperature in the LP surge drum. The oil viscosity should be chosen properly.

Lubricating oil selection factors

The most important selection factors for lubricating oil are the oil's viscosity, the oil's viscosity index, the flash point, the aniline point, the pour point, the flame point and the vapour pressure. See also Chapter 9 – "Lubricants for natural refrigerants".

Choosing the correct oil has a significant impact on the function and longevity of the refrigeration system. It depends on a variety of factors. Primarily, the pressure and temperature operating conditions of the system as well as the required oil viscosity are essential knowledge, particularly during start-up and operation of the refrigeration compressor. Viscosity should be at least 7 cSt for each operating point, measured for oil temperature at compressor entry. The evaporation temperature (pour point) and the design requirements for the oil separator (flash point, vapour pressure) must also be taken into account.

A variety of oils are available for ammonia refrigeration systems. Depending on the application, oils were developed from different base chemicals.

The main types of oil, with their abbreviations, are listed below:

M	mineral oils
MHC	mineral oils (special treatment "hydrocracked oil")
AB	alkylbenzene
PAO	polyalphaolefin
POE	polyol ester
PAG	polyalkylene glycol

In addition, some manufacturers also offer blended oils from the above-mentioned list of oils. In principle, the compressor manufacturer should be consulted when selecting the refrigerating machine oil. Both the data sheets of the oil manufacturer and the technical documentation of the compressor manufacturer should be taken into account when selecting the oil.

Conventional mineral oils are used less frequently today. Due to the relatively short life of the oils, oil changes are required more frequently. By using the much more stable hydrocracked mineral (MHC) oils longer, more cost-effective maintenance cycles can be carried out. For this reason, many systems are charged with MHC oil when they are started up for the first time. Existing systems were also frequently changed over from mineral oil to MHC oil. Polyalphaolefins (PAO) are also increasingly used in ammonia plants. Some PAO oils are also mixed with alkylbenzene oils (AB). This can improve the sealing ability of materials, such as o-rings made of synthetic rubber. However, pure PAO oils are also offered with appropriate additives, which achieve the same results in terms of sealing ability.

Because of their good resistance to foreign gases (mostly air), PAO oils are very suitable for use in ammonia plants. Due to the low pour point, PAO oils are also suitable for low evaporation temperatures. In addition, they are thermally and chemically stable and have a low vapour pressure. This leads to a good degree of oil separation in the oil separator. Due to the low vapour pressure, the amount of oil in the hot gas is significantly lower than with other types of oil. However, it should be taken into account that the PAO oils are also more expensive than mineral oils. PAO oils have similar prices as MHC oils, but have a longer life, so in the end they are cheaper.

Some direct expansion ammonia plants use ammonia-soluble oils, other DX systems use non-miscible double hydrocracked mineral oils. Of the aforementioned oils, only polyalkylene glycol oils (PAG) are currently miscible and suitable. Oil return in systems with a dry direct expansion system is comparable to systems with synthetic or hydrocarbon refrigerants. However, it should be borne in mind that PAG oils are very hygroscopic and therefore absorb water or air humidity very quickly. This causes a significant change in the properties of the oil and can lead to saponification or greasing of the oil. In this case, the lubricating properties of the oil are completely lost, which can lead to a total failure of the compressor within a very short time. Special care is therefore required when performing maintenance or service work with these oils. Nonetheless, practical experience shows that ammonia has a stronger affinity with water than the PAG oil. Therefore, an oil sample taken from an ammonia system will not show contamination with water even though the PAG was initially (slightly) contaminated with water. The water content in the ammonia system will rise though. Contamination of PAG with water should therefore be avoided also in ammonia systems.

Further information on the selection of lubricants for ammonia refrigeration systems can be found in Chapter 9 "Lubricants for natural refrigerants".

Oil separation systems

Apart from polyalkylene glycol oils (PAO), the available oils suitable for ammonia refrigeration systems are very limited and not miscible with ammonia. First of all, this means that oil that enters the system circuit cannot be returned with the refrigerant. Additional measures for oil return are required in ammonia systems. Low oil carry-over of the compressor is the key element in order to keep these additional measures as simple as possible. It is therefore advisable to always equip ammonia compressors with very good oil separators since the oil throw of compressors cannot be completely eliminated.

Multi-stage oil separator systems (see Figure 4.25) in which the hot oil / refrigerant mixture passes after the compressor at first through a coarse separator and then via a fine separator system (e.g., fine separator cartridge) have proven successful. Such multistage oil separators can also be designed horizontally, Figure 4.26. A low oil discharge significantly reduces oil accumulation in the heat exchangers and in the piping system. This also leads to lower maintenance costs because the frequency and effort for manual oil drainage can be reduced. At the same time, the system can be operated in optimal stable operating conditions because the performance of the heat exchanger is not reduced by oil build-up.

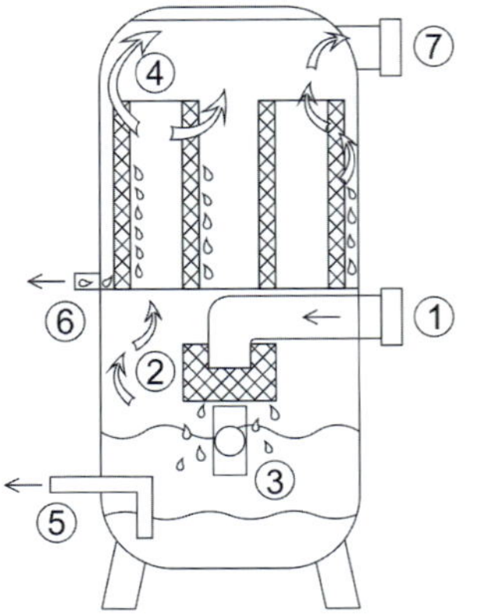

1 Hot gas inlet flange
2 Coarse separation
3 Oil level sight glass
4 Fine separation
5 Main oil return line
6 Oil return fine filtering
7 Hot gas exit

Figure 4.25 Components for a two-stage oil separator

It has been shown that very good oil separators make a significant contribution to reliable and trouble-free plant operation. In practice, oil carry-over rates of less than 10 ppm can be easily obtained with high-quality oil separators. Less than 5 ppm is recommended for DX NH_3 plant employing non-miscible oils.

Nonetheless, even these small quantities are entrained into the refrigerant piping and accumulate at the bottom of reduced-flow velocity sections. During the assembly of the piping system structure, such low points must be planned as oil traps. They must be equipped with a suitable oil removal system and oil drain valves. Due to the higher density of oil compared to liquid ammonia, the oil always collects below the ammonia liquid, so that it is easy to drain. However, the viscosity and hence the flow of oil is greatly reduced due to low operating temperatures. Therefore, the oil trap and drain valve may need to be externally heated.

Manually drained oil should not be reused. Depending on the mode of operation and the maintenance status of a system, this oil may be used for other applications.

Oil accumulation in the heat exchangers is limited when high-quality oil separators are used in the refrigeration system. This should be taken into account when designing the system's heat exchangers. In practice, it has been shown with older systems that air coolers in some cases have been significantly over-dimensioned to compensate for oil build-up. With better oil separation, hence less oil quantities fed into the evaporator, the temperature difference between the evaporation temperature in the heat exchanger and the air temperature could be reduced by 1 to 5 K. This means that the evaporation temperature could be raised by 1 to 5 K, hence saving on compressor drive power of up to 15 %.

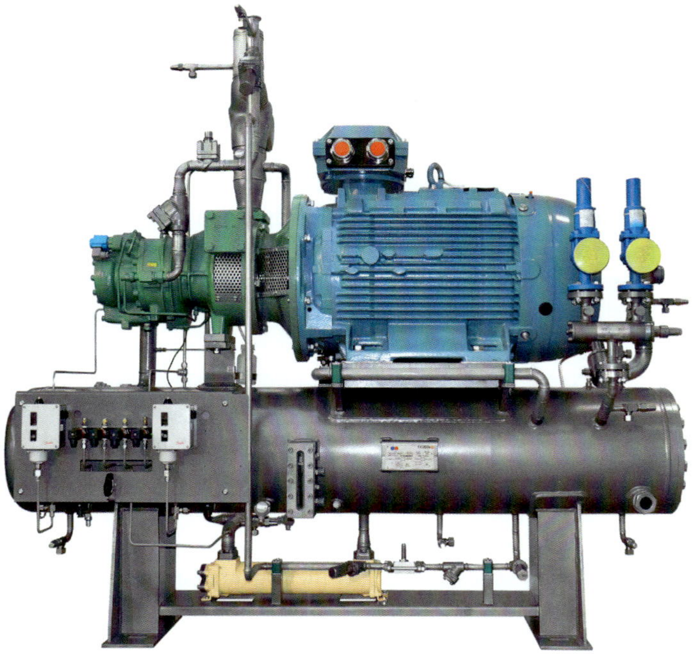

Figure 4.26
Screw compressor with horizontal oil separator (©EM Polar)

Oil return systems

Although mineral oils and their synthetic derivatives are not miscible with ammonia, due to their durability and low pour point, the oil return can be implemented using an automatic oil return system. The oil is then returned from the low-pressure side to the suction side of the compressor using hot gas pressure. It is helpful if there is an oil sight glass and an oil filter in this system for monitoring or cleaning the return oil, see Figure 4.27.

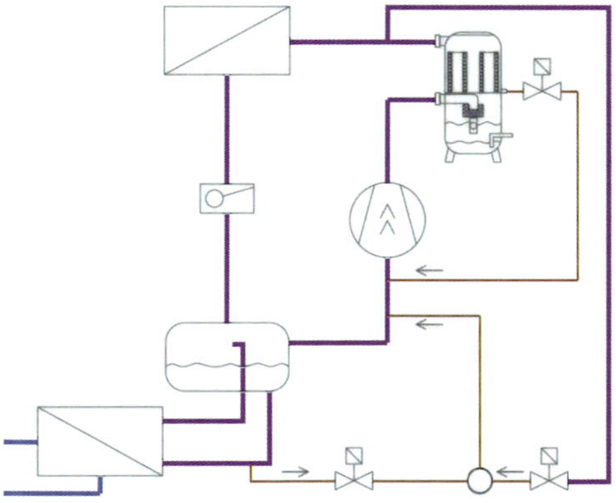

Figure 4.27 Schematic diagram of oil piping system using hot-gas powered automatic oil
return. Often, the oil feed from the oil separator is connected directly to the compressor, not into the suction line, especially if it has to feed the bearings and the shaft seal.

As shown in Figure 4.27, the oil that has collected in the evaporator is led to an outgassing chamber via a solenoid valve. The ammonia contained therein can evaporate due to the ambient heat. The vapour produced is extracted via the suction line. When the oil is almost free of ammonia, hot gas pushes the oil into the suction line or compressor housing and it re-enters the cycle.

It should be noted that the oil from the fine separation stage of the oil separator cannot be led directly into the oil sump of the oil separator, due to the pressure differences between the preliminary or main and fine separation stages. It must be led to the compressor via the suction line. If the refrigeration system contains large amounts of water, an automatic oil return should never be used.

4.3.3 Leakage monitoring

Even the smallest leaks are noticed immediately in ammonia systems, except when the leak occurs in an evaporative condenser, in a water cooled condenser connected to a cooling tower, or in an open water circuit where the leaking ammonia will be absorbed by the water. The pungent smell of ammonia is detected by the human nose at a concentration of 3 to 5 ppm. Operators and service personnel can therefore in most cases immediately notice leaks. Since the synthetic refrigerants are odourless, leakage monitoring for these refrigerants can only be implemented with special gas detectors. Leakages in such systems are often only detected when the cooling capacity of the system is no longer sufficient. In this case, however, a large part of the system refrigerant quantity has usually already leaked into the building and/or atmosphere. This often means expensive refilling of the system in addition to the environmental threats associated with leaking large quantities of HFC or HFO into the ambient.

Regardless of the cost of refilling, there is an often-underestimated risk when using synthetic refrigerants: The odourless synthetic refrigerant gases escaping at the leak are heavier than air and sink into lower areas. If these gases cannot flow away, the concentration above floor level increases, and serious accidents can occur. This danger usually does not exist with ammonia. Under certain circumstances, the ammonia forms an aerosol heavier than air and the danger of high ammonia concentration above floor level will exist in this case. Due to the intense and pungent smell of ammonia, an ammonia leak in any area that is normally accessible is recognised immediately by its smell. It triggers a natural early warning signal for those affected.

Regardless of this, EN 378 stipulates the leakage monitoring of ammonia refrigeration systems by using gas-warning devices due to the potential risk of fire. The refrigeration systems could be running unattended, in automatic mode, with no operators on site.

4.3.4 Contamination by water and inert gases

The ammonia as supplied by the manufacturer has a water content of approximately 0,3 % by mass, depending on the grade supplied. Industrial grade can contain as much as 5000 ppm water, whereas technical grade contains 280 ppm. The small amount of water is added to avoid stress corrosion cracking. Even though the heat treatment of welds after welding of components has a far more significant influence on stress corrosion cracking than the water content in ammonia.

4.3.4.1 Removal of water

In a refrigeration system that was installed according to good practice, it can be assumed that the system is free (or largely free) of water and inert gases. However, especially in refrigeration systems that already have been in operation for some time, the water content is often significantly higher. Water usually enters the circuit during maintenance work. However, the cause may also be that insufficient evacuation was carried out during commissioning. In systems with very low evaporation pressures (i.e. system pressure below ambient pressure), there is a risk that air moisture can penetrate into the system through leaks. As a result of the high affinity of ammonia

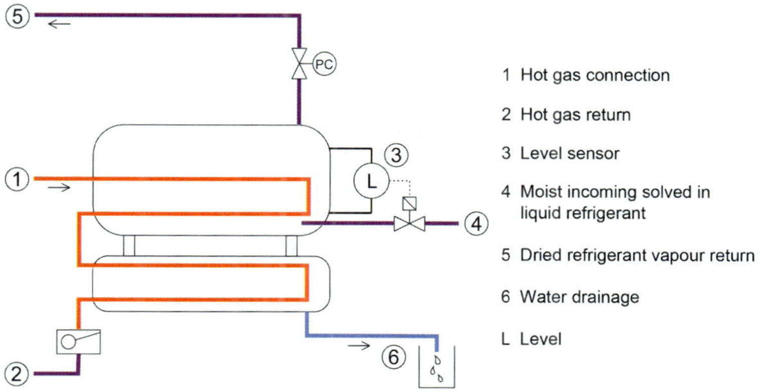

Figure 4.28 Ammonia dryer schematic

for water, the system does not need to be evaporated below atmospheric pressure for moisture to be drawn in during a leak, although the lower the evaporating temperature is, the worse the ingress will be. Experience shows that water can be drawn into ammonia systems, for example, during purging into a barrel of water, against a pressure of up to 8 bar.

The water ingress immediately dilutes within the ammonia in the refrigeration system. A high proportion of water leads to a significantly higher energy consumption due to the change in vapour pressures.

From a water content exceeding 500 ppm there is a risk of corrosion of steel components. Water in the refrigerant circuit also leads to chemical reactions and property changes in the compressor oil. Organic acids and sludge can be formed.

Water is significantly more soluble in ammonia liquid than in ammonia vapour. Water ingress therefore mostly accumulates on the low-pressure side in the liquid collector surge drum or in the separator. It can be removed from the system using special dryers. Separation and removal are, however, associated with great effort. As shown in Figure 4.28, the ammonia / water solution is boiled out. First, liquid refrigerant is fed from the system to the dryer (4). The refrigerant level in the dryer is kept stable via level sensors (3). The liquid refrigerant is then heated using hot gas (1) and begins to boil. The ammonia vapour thus generated is water-free and is returned to the suction side of the refrigeration system (5). The hot gas condensed in the dryer (shell and tube heat exchanger) is returned to the system's liquid receiver via a high-pressure float valve (2). The remaining liquid water collects in the dryer. It is manually discharged into an appropriate container and must then be disposed of properly as it contains traces of ammonia and oil.

4.3.4.2 Removal of inert gas

Air in the system has a major influence on the condensation temperature, which increases with the proportion of air in the system. Air and other gases increase the total pressure on the condenser side because according to Dalton's law, the total pressure is equal to the sum of the individual partial pressures. This means that the compressor not only has to generate the condensing pressure (saturation pressure) of the refrigerant, but also the higher total pressure in the condenser. The air in the condenser may also reduce the available heat transfer surface area for condensation of NH_3.

One of the following two measures usually removes air and other non-condensable gases:

Installation of a purger

One option for removing air and inert gases is to install a purger on the condenser or the receiver. Inert gas, which is not soluble in ammonia, accumulates at the condenser outlet and can be removed there. In order to separate the inert gas from the ammonia vapour, it is passed over a liquid evaporator coil inside a container. Ammonia is condensed out; the condensate is fed back into the refrigeration system from the purger tank via a high-pressure float valve. The gaseous inert gas remaining in the purger is then discharged into a vessel filled with water. This process is usually carried out continuously and automatically. Ready-to-connect purgers are available on the market for this application. These are already equipped with a control system for automatic operation. This installation is particularly useful in systems that are operated close to or below atmospheric pressure (low-temperature systems). Some people recommend purgers for systems with evaporation pressures below 3 bar.

Manual ventilation

Since it cannot be ruled out that air can enter the system during maintenance and service work, regular manual bleed-off of the system is recommended if no automatic purger is installed. Care must be taken to ensure that the bleed-off point is correct. The liquid manifold at the condenser outlet is ideal here. If the high-pressure float is located directly at the condenser outlet, the service valve attached to the float can be used for venting. Please note that this is not an effective procedure for manual ventilation – it is much more complicated than described here, and this procedure would get rid of a little air but also a great deal of ammonia.

4.3.5 Risks from hydraulic pressure surges in refrigerant piping system

Operating errors as well as errors in pipeline routing or in the valve control sequence of the system can lead to excessive pressures and pressure surges in the system. There are two mechanisms – "liquid hammer" (which can occur when solenoid valves open as well as when they close) and "condensate-induced shock", which is caused by a sudden rise in pressure resulting in the condensing of gas bubbles and voids in pipe systems. The problems caused in defrosting mode are often a result of the latter. Pressure surges increase the risk of hydraulic shock waves, which are particularly dangerous in the liquid lines. When a solenoid valve in the liquid line is opened or closed, violent shock waves can occur if the liquid is very subcooled, i.e. dense and the flow velocity in the line is high. Closing the solenoid valve causes the flow to stop abruptly. The resulting impulse of the liquid column (in the liquid line in front of the valve) can be so large that it leads to violent mechanical damage, such as cracks at fittings and valves.

Pressure surges and the associated shock waves often occur in freezers with hot gas defrosting. When configuring the piping system and the associated control system, special attention must be paid to the individual states during a defrost cycle. Two operational states have to be carefully considered here. Consider that at the start of the defrost process, by opening the hot gas solenoid valve, a shock wave can occur in the evaporator. The hot ammonia gas will condense on the cold surfaces of the evaporator, abruptly reducing its volume. This can lead to a pressure wave in the system. A second shock wave can occur when the suction valve is opened. The excess pressure prevailing in the evaporator, at the end of the defrost cycle can abruptly vent into the suction line, which is partly filled with liquid. This can also lead to a shock wave in the suction line.

Theoretically, hydraulic pressure surges can only occur and cause damage if the lines are completely filled with liquid. However, this only applies to a limited extent: A completely filled liquid line is a prerequisite for the propagation of shock waves, but the pipes do not necessarily have to be completely filled prior to pressure surge.

The sudden introduction of hot gas can lead to a pressure increase in the liquid. However, the gas bubbles will immediately condense due to the cold liquid; liquid fills the existing cavities almost immediately and is then stopped abruptly. This impulse leads to the development of the shock wave, which can then spread through the filled liquid line. The forces developed by condensate-induced shock are very high – they will even break, burst or crack welds and components that are sufficiently strong and have passed a strength pressure test. Weak welds are often blamed because that is where the failure occurs – this is only because it naturally fails at its weakest point. In order to avoid condensation-related shock waves and the associated dangers, the following points should be taken into account when planning and designing the system:

- Prevention of pressure surges
- Prevention of extreme pressure increases
- Prevention of cavities in which gas bubbles can form
- Prevention of cracks or nicks in the pipes

4.3.6 Direct and indirect systems

Indirect systems have the advantage that the refrigeration system can be built in a compact format and installed in a separate machine room. More detailed and specific information on secondary refrigerants and coolants can be found in Chapter 10. The refrigerant charge quantities of these systems are low due to the refrigerant pipe network required being only in the machine room. At the same time, service and maintenance work can be carried out within the machine room centrally and concisely. System operation can be better monitored thanks to the separate plant machine room installation. The effort to ensure safety with the central ammonia plant is reduced, and at the same time the group of people who have access to the system is smaller. The pressures in the coolant pipe network are also lower than in the ammonia system, which leads to a higher level of system safety. However, larger pipe diameters have to be laid for the coolant pipe network but can be greatly reduced when using ice slurry as secondary refrigerant. In contrast to direct expansion systems, energy costs for the circulation pumps have to be considered. If the energy and operating costs for the coolant pipe network can be kept within limits, the advantages of an indirect system outweigh the disadvantages, especially when the primary refrigerant used entails higher safety risks, as is the case with ammonia. The properties of the secondary refrigerants (viscosity, freezing point, thermodynamics, flammability etc.) must be checked depending on the refrigeration application. The selection of suitable coolants for use with indirect systems is particularly critical in low-temperature applications (amount and characteristics of freeze point additives); see Chapter 10.

Ammonia has always been used together with indirect systems in large distributed systems like supermarkets, some refrigerated warehouses, and ice rinks, or as chiller in air conditioning of large buildings. Ammonia as refrigerant, due to its favourable thermo-physical properties, offers energy efficiency advantages for such indirect systems, which is important considering the large refrigeration capacities required for such systems. However, indirect systems with ammo-

Figure 4.29 Ammonia liquid chiller for brine cooling for large kitchen (© Kälte Eckert)

nia liquid coolers have also been used for some years in supermarkets, canteen kitchens, and many other applications with rather low cooling capacities (Figure 4.29). The main coolants are ethylene or propylene glycol water solutions for medium temperature and potassium formate or acetate water solutions for low-temperature applications, see Chapter 10. The experience gained with these smaller systems to date has been consistently positive. It has been recorded that the use of ammonia is accepted positively by users and operators and is inconspicuous in everyday use. It is important that experienced personnel carry out the design planning, construction and subsequent maintenance of the refrigeration systems. This is necessary to ensure that the systems operate in a safe and optimal energy saving manner.

4.3.6.1 Advantages and disadvantages of indirect systems

Indirect systems are particularly suitable for food processing rooms and for air conditioning applications in office buildings. However, blast freezers and freezer stores (refrigerated warehouses) are usually designed with directly evaporating systems because, on the one hand, the energy costs are considerably lower and, on the other hand, the availability of suitable low-temperature coolants is limited.

In today's indirect ammonia systems, due to the increased use of plate heat exchangers, the required refrigerant charge quantity of the ammonia for the primary refrigeration systems can be significantly reduced. A charge of 0,1 kg/kW (0.23 lb/Btu/s) is state of the art today and charge optimized systems are available with ammonia charges as low as 40 or even 20 g/kW (0.046 or 0.093 $lb/Btu_{IT}/s$) refrigeration capacity. This means that a water-cooled ammonia refrigeration system with a cooling capacity of 500 kW (142.2 TR) can be operated with a refrigerant charge of only 50 kg (110 lbs). Regardless of this, the safety and tightness of the systems is further increased by using fully welded plate heat exchanger assemblies.

4.4 Safety precautions when handling ammonia

The distinctive smell of ammonia is normally no longer noticeable with the systems built today. It is precisely for this reason that it is particularly important for operators, maintenance and service technicians as well as for building managers to know the properties and characteristic behaviour of ammonia. In the event of a major leak, all these people must be able to act prudently and carefully.

The greatest danger is the inhalation of high concentrations of ammonia gas. However, this is often exaggerated in press reports and publications.

The actual danger of an ammonia leak depends largely on the type and size of the leak as well as on the pressure, temperature and phase of the ammonia escaping from the system. Everyone involved must have a clear understanding about this. Even the smallest ammonia concentrations are noticed immediately, which ensures greater vigilance. People in the vicinity feel uncomfortable. If the concentration increases further, panic and escape impulses can be triggered. There is a danger to health and life only at higher concentrations. With further increases, there is a risk of ignition or explosion.

Table 4.15 provides a quick overview of the dangers that can be expected at certain concentrations.

Most of the leaks can be repaired by experienced operating personnel before the concentration rises above the discomfort limit.

Table 4.15 Effects of ammonia exposure. Workplace limits may vary depending on the country. Consult, for example, the Database GESTIS for national limit values [4-11]. See also Chapter 4.4.

Volume ppm exposure	Physiological manifestations
3–5	hardly noticeable, smell threshold
20	indicative occupational exposure limit value (IOELV), eight hours' exposure, EU-value (corresponds to 14 mg/m³ / 1.4E-05 oz/ft³)
25	smell noticed by most people
50	indicative occupational exposure limit value (IOELV), short-term exposure, 15-minute average, EU-value (corresponds to 36 mg/m³ / 3.6E-05 oz/ft³)
100	inhalation for several hours may cause weak symptoms. For healthy persons this concentration has no effect yet.
250	beginning annoyance, still harmless at an exposure time of 1 h
250–500	accelerated breathing and increased heart rate
500–1000	tolerability limit; there are health risks for humans, animals and plants as well as property damage
2000	30 minutes of exposure is life-threatening
3000	even short-term exposure causes blindness
5000	death by respiratory arrest

4.4.1 Ammonia in the atmosphere

Ammonia liquid evaporates at atmospheric pressure (1,013 mbar) at a temperature of –33,3 °C (–27.9 °F), also called the normal boiling point (NBP). In order to store it in a liquid state, it must either be kept under pressure or at low temperature. Provided that there is sufficient heat for evaporation, 1 litre of liquid ammonia would evaporate to 750 l (165 gal_{IMP}, 198 gal_{US}) of ammonia gas at –33,3 °C (–27.9 °F).

If there is not enough heat available, the liquid ammonia will remain in the liquid state and will not evaporate further. For this reason, water should never be poured into liquid ammonia. As soon as the water dissolves in liquid ammonia, a considerable amount of heat is released. This heat causes the ammonia to boil, and large amounts of ammonia vapour are released.

In a flooded evaporator refrigeration system, most of the refrigerant charge is in liquid form on the low-pressure side. If there is a leak on the high-pressure side, only the compressed gas and approximately 10 to 20 % of the liquid charge quantity can evaporate if no additional heat is added.

Gaseous ammonia has a density of approximately 600 g/m^3 (0.599 oz/ft^3) and is therefore significantly lighter than air (approximately 1000 g/m^3 / 0.999 oz/ft^3). If ammonia escapes in gaseous form, it rises quickly into the atmosphere. The pure gas is colourless but forms a water vapour cloud in moist air. Gaseous ammonia can be dissolved very well in water: At room temperature, about 0,5 kg (1.1 lbs) of ammonia can be taken up in one litre of water. This corresponds to approximately 650 l (143 gal_{IMP}, 172 gal_{US}) of ammonia gas.

4.4.2 Flammability

At temperatures above 450 °C (842 °F), ammonia starts to decompose. At very high concentrations, ammonia vapour (at atmospheric air pressure) is flammable. The flammability limits in dry air at 20 °C (68 °F) and 1013 mbar (2116 psi) ambient pressure are:

lower flammability limit: 15,4 % by volume

upper flammability limit: 33,6 % by volume

Combustion can occur in closed rooms if ignitable mixtures have formed. The temperature required for ignition is approximately 630 °C (1166 °F). In comparison, the ignition temperature of propane or butane is around 490 °C (914 °F). The ignition energy required for igniting ammonia is very high. The heat generated by the combustion is not sufficient to maintain the combustion. If the ignition source is removed, the fire goes out immediately. Oil mist from the refrigeration oil prevalent in an ammonia/air mixture may alter the combustion behaviour.

Ignition or explosion is not possible outdoors because stable flammable mixtures cannot form. The heat of combustion of ammonia is around 19000 MJ/kg, which is low, compared to propane or butane (50000 MJ/kg). The flammability is classified as low (class 2) in the international standards EN 378, ISO 5149 and ASHRAE 15.

At a concentration of 23 %, the fastest burning rate is reached at 8 cm/s (3.1 in/s); the lowest burning rate (3 cm/s / 1.2 in/s) is at the lower flammability limit. These speeds are not sufficient to trigger a detonation because they are significantly below the speed of sound. The combustion is therefore classified as a deflagration.

Table 4.16 Explosion properties of ammonia and methane in atmospheric air

Explosive properties	Methane	Ammonia
Pressure peak in bar	7,24	4,14
Rate of pressure rise in bar/s	207	30,3
Pressure peak in psi	105	60
Rate of pressure rise in psi/s	3002	439
Type of combustion	detonation	deflagration

Nonetheless, in confined spaces, sufficient pressure can be generated in the event of a deflagration to cause structural damage. However, as can be seen in Table 4.16, the pressure rise and pressure peak are significantly lower than for hydrocarbons.

4.4.3 Effects on food

Due to the strong affinity of ammonia for water, products stored in cold rooms can be damaged by ammonia leaks. Both the level of the ammonia concentration in the area and the duration of this contamination are decisive for the degree of possible damage. Low concentrations or short contact times have no harmful influence on the stored food. The effects will vary, depending on the packaging of the products.

Raw meat absorbs ammonia, but due to its acidic pH, the effect of ammonia is buffered. After thawing and boiling, no traces of ammonia can be found. Fruits and vegetables that come into contact with ammonia turn brown. Depending on the degree of exposure, the colouring can be so pronounced that the items can no longer be sold. In principle, foods that have come into contact with ammonia should be subjected to a critical inspection by a specialist.

4.4.4 Effects on humans

4.4.4.1 Physiological effects

The typical strong smell of ammonia is perceived as unpleasant even at concentrations of approximately 5 ppm and is objectionable from a concentration of approximately 50 ppm. Experienced technicians may work on ammonia refrigeration systems up to a concentration of 100 ppm without experiencing any unpleasant side effects.

From a concentration upwards of approximately 150 to 200 ppm, irritation to the mucous membrane and the eyes is experienced. However, there are no long-term consequences. If the concentration rises to 500 to 700 ppm, the eyes are rapidly impaired. The eyes start to water within about 30 seconds, but air can still be inhaled.

At 1000 ppm, breathing is no longer possible, and vision is severely impaired. There is a risk of eye injuries resulting in permanent disability. A load of 1500 ppm or more damages or destroys the retina. Even experienced or trained personnel will succumb to the escape reflex and leave the room immediately. From a concentration of 2500 ppm upward, there is a risk to life. As the concentration increases, the risk of a fatal accident increases enormously.

Table 4.17 Acute ammonia exposure guide value (ppm) vs exposure duration

Personal impact	10 min	30 min	60 min	4 h	8 h
No discomfort	30	30	30	30	30
No impairment	220	220	160	110	110
Not fatal	2700	1600	1100	550	390

Being in an ammonia concentration of 5000 ppm will most likely lead to suffocation. In one study it was found that being in a 5000-ppm atmosphere for 30 minutes kills all life. The authors of a study compare recorded accidents and animal experiments and evaluated them statistically [4-10]. They conclude that the results from animal testing are transferable to humans. This resulted in an LC_{50}-value of 11500 ppm with a load duration of 30 minutes (LC_{50}-Value = value for a 50 % probability of death). The U.S. Environmental Protection Agency (EPA) has investigated the effects of acute ammonia exposure. The effects were divided into three areas. Table 4.17 shows the load classes, defined guideline values and associated load times.

The EPA values essentially reflect practical experience. In accordance with the STEL (short-term exposure limit), no effects are to be expected at an exposure of 30 ppm regardless of the duration. It is known from practice that at a concentration of 110 ppm it is possible to work permanently in ammonia installations, without impairments.

The tabulated guideline values for exposure without fatal consequences also correspond to the experience and requirements valid in Europe.

Another limit value from the USA is considered to be state of the art and therefore a requirement for handling ammonia. The IDLH value (immediately dangerous to life or health, determined by the National Institute for Occupational Safety and Health, NIOSH) specifies the atmosphere in which a person can remain for 30 minutes without long-term damage or death. The IDLH value for ammonia is 300 ppm. This limit value is also recommended in Europe as a useful guideline for wearing a respirator. At lower concentrations, even if the personal protective equipment (PPE) fails, operating and service personnel of the system remain unharmed.

4.4.4.2 Legal limits

The permissible concentrations in the workplace are regulated by national legislation. Different limit values and specifications are valid in various countries, but in general the international guidelines do not differ widely [4-11]. In Germany, the occupational exposure limit is 20 ppm for eight hours. For a short time (max. 4 x 15 min per shift) the value may increase to 40 ppm.

4.4.5 Accidents

4.4.5.1 Causes and reactions of accidents

The basic rule is that caution must be exercised when handling ammonia which is also true for all other refrigerants. All safety precautions and rules of conduct must be observed, without fail. Users, operators and service personnel must be continuously trained in the safe handling of ammonia and also be instructed in accident prevention for their respective refrigeration systems.

In order to take the right steps in the event of an accident, the type of accident or ammonia outbreak must first be known. A distinction is made between two main cases:

- Escaping ammonia liquid or gas, e.g., through a pipe break
- Formation of an ammonia gas cloud

If possible, spilled liquid ammonia should be blanketed with foam. The heat from the surroundings is required for evaporation. The foam blanket limits access of this heat and thus reduces evaporation.

Gaseous ammonia is lighter than air and will rise. In this case, good ventilation is recommended. Escaping ammonia gas "binds" to the humidity in the air, forming a weak nitric acid cloud.

The ammonia vapour then becomes heavier than air and looks like a fog that collects near the ground. The greater the air humidity in the area of the leakage, the more ammonia will remain near the ground. It may then also show typical heavy gas behaviour by sinking to the lowest floor level.

Usually, the cause of the accident can determine the type and number of possible injuries. A splash of ammonia into the eyes would have a devastating effect for the person concerned, but it would probably only affect a few people overall. On the other hand, in case of an ammonia cloud, the worst effects on the respiratory tract could affect a large number of people.

It must also be pointed out that when liquid ammonia escapes, it evaporates at ambient pressure and absorbs a great deal of heat. If the liquid ammonia comes into contact with human skin, the skin tissue is dried out and destroyed, resulting in frostbite.

4.4.5.2 Degree of accidents

For easier analysis, the physical effects on the accident victims are divided into three (arbitrarily defined) groups: A, B and C.

A) LIGHT, Exposure to concentrations of up to 5000 ppm for a few minutes.

Symptoms: effect is felt in the eyes, mouth and upper respiratory tract, burning sensation in the eyes and mouth, difficulty swallowing, noticeable hoarseness and tightness in the throat, slight cough.

Tell-tale signs: redness of the eyes, lips, mouth and tongue with swelling of the eyelids and oedema (swelling) of the throat, no medical signs of impaired lung function.

B) MODERATE, Exposure to concentrations of 5000 to 10000 ppm for a few minutes.
This group has a stronger effect within the airways (with bronchi and bronchioles)

Symptoms: tightening of the symptoms above, chest tightness, difficulty swallowing, occasionally complete loss of voice, cough with expectoration, occasionally with bloody mucous.

Tell-tale signs: pain, rapid pulse and rapid breathing, marked swelling of the eyelids with twitching and lacrimation, moderate oedema in the throat with burning mucous membranes and detachment of skin flakes, which leads to dark red, glossy spots on the skin, thoracic examination reveals reduced air entry with gurgling noises.

C) HEAVY, Exposure to concentrations of over 10000 ppm for a few minutes

Symptoms: throat and eye symptoms similar to those in group B, persistent cough with a lot of foamy expectoration.

Tell-tale signs: shock, restlessness and severe pain, oedema, faint but fast pulse, blueing of the skin and severe breathing problems, gurgling noises in the chest.

Complications: Group A sufferers who have no impairment of the lower airway region recover without further implications.

In group B and C patients, the airways are narrowed as a result of laryngeal swelling or pulmonary oedema. Survivors can develop pneumonia at a later stage, usually a few days after the accident. There are indications that an ammonia accident can lead to long-term respiratory illness. The degree of respiratory disease also depends on whether the person has smoked in the past.

4.4.5.3 Medical care

In this chapter, medical care is limited to the description of first aid measures and emergency medical care for affected accident victims. The eyes in particular are at risk from contact with ammonia. In order to avoid permanent injuries or disabilities, immediate action must be taken within the first ten seconds after the accident. The eyes and the affected skin areas should be continuously rinsed out with clean water for at least 15 minutes. The eyelids must be kept open.

Clothing that has been soaked in liquid ammonia should be removed immediately. The person concerned must then be showered at moderate water temperatures (approximately 25 °C / 77 °F). Cold water must not be used, as this could result in a fatal shock. First of all, the person concerned is showered to remove the ammonia residues. This is especially necessary for further treatment by the emergency services. If the patient smells too strongly of ammonia, it will not be possible to approach to treat him. Likewise, transportation in an ambulance will not be possible. Since various bodily reactions are delayed, the person concerned must be monitored for at least 12 hours.

If only individual parts of the body come into contact with ammonia, the contaminated clothing must be washed out and removed. It is important to ensure that the person concerned is protected from cold water. Affected skin areas should not be treated with cream or ointment to prevent ammonia from becoming trapped in the skin. In principle, a doctor should be called as soon as possible.

Serious ammonia accidents are very rare in the refrigeration industry. The early perception and prudent treatment of everyone involved allows risks to be identified and remedied in good time. Smaller quantities of ammonia, as well as improved safety technology and installation measures, reduce the risk of accidents. It is particularly important that knowledge about the handling and use of ammonia is passed on and expanded.

4.4.6 Technical safety equipment

Refrigeration systems must be designed and equipped in such a way that the danger they pose is reduced as much as possible. International standards specify which security organs and which calculation bases are necessary for this. The essential European requirements are laid down in EN 378-2. According to EN 378-2, ammonia refrigeration systems must be equipped with at least two independent safety switches on the high-pressure side of the compressor. The two safety switches trip at different pressures and must be reset manually. The switch with the lower switching threshold is equipped with a freely accessible reset button. The switch with the higher switching threshold can only be reset if the switch housing is opened using a tool. Depending on the refrigerant charge quantity and the individual vessel volumes of the system, additional equipment such as safety or overflow valves are necessary.

Additional safeguards, such as flow monitors, generally serve to ensure the functional reliability of the system rather than to avert danger.

Safety valves

Depending on the ammonia charge quantity, system volume and operating conditions, refrigeration systems must be equipped with high-pressure safety relief valves. These valves are used to protect the receivers and some pipe segments against unacceptable excess pressure and the associated risks of bursting or exploding. Excessive overpressures can be caused, for example, by fire, accidents or a system malfunction. The safety valves are necessary for receivers or sections of pipeline that can be isolated, whilst these sections are completely filled with liquid refrigerant.

When the safety valve responds, the refrigerant to be discharged is normally led away via a blow-off line. During planning and construction, special attention must be paid to the piping and to the point at which the refrigerant is to be vented into the atmosphere. Not only the people inside the plant area, but also the people outside and the people in neighbouring buildings and the surrounding area need to be protected.

Pressure relief valves should discharge to the atmosphere not less than 5 m above ground and 7 m from a window, ventilation opening or exit in any public building.

A particularly effective and safe way to discharge the refrigerant is to discharge the safety valve into a chimney. The gases can be blown off into the chimney at high speed. However, it is much cheaper and more common in practice to take the blow-off line up to a point where the gases are safely blown off into the atmosphere. The end of the blow-off line must be designed in such a way that rainwater cannot penetrate, and cross-sectional constriction by ice, dirt or animals must be prevented.

Instead of safety valves, overflow valves can also be used. The individual receivers are protected against overpressure by directing the refrigerant flow into another receiver as necessary. This can prevent ammonia from escaping into the atmosphere. The refrigerant remains within the system, and the receivers are protected. In practice, safety and overflow valves are often used in combination. The design of the safety valves is specified in EN 378-2. The safety valves are calculated and dimensioned according to EN 13136 and ASHRAE 15-2004 or the International Institute of Ammonia Refrigeration – IIAR.

4.4.7 Personal protective equipment

In principle, all persons who are involved in the operation, repair or maintenance of a refrigeration machine should have personal protective equipment (PPE) that corresponds to their task. It goes without saying that this equipment is kept in one location, is in a very good, ready-to-use condition, and is worn or kept ready to use in case of need.

It is also advisable to store emergency protective equipment in a freely accessible cabinet, in the vicinity of the refrigeration system, and outside the machine room. The design and scope of the protective equipment depends, among other things, on the size of the ammonia refrigeration system. It should be coordinated by the refrigeration system operator along with an experienced safety specialist.

The use of personal protective equipment is an additional safety measure, but not a hazard remedy! The better way is to record all measures to reduce the risk of accidents and undertake appropriate countermeasures as part of a hazard and safety-risk analysis. This includes, for example, that workplaces be well ventilated before and during the maintenance work. Appropriate information and warning signs should be placed in the vicinity of the work to indicate possible dangers. If possible, work should be carried out outside the machine room, preferably outdoors. The scope of the PPE to be worn then depends on the nature of the work to be carried out and the associated dangers.

For general maintenance work in which system components do not have to be opened (this includes, for example, checking the drive motor), the PPE generally includes suitable work trousers and jacket, protective goggles, protective gloves, steel-capped safety shoes and a bumpap. Hearing protection may also be required. If small amounts of ammonia are to be expected when performing service work (e.g., replacing a mechanical seal on the compressor), the PPE should be supplemented with a breathing mask. It is not necessary to wear the mask constantly during the work. It is important, however, that the mask and the associated filter are readily available.

Special care is required when operating shut-off valves. When the protective caps are removed, ammonia can be released because pressurised ammonia liquid or oil may have collected under the cap. Therefore, safety glasses and gloves should always be worn when working on valves. If work is to be carried out that is likely to result in a concentration of over 50 ppm, the mask should be worn permanently during the work. This could be the case, for example, when manually de-oiling an evaporator or when changing the oil filter.

If it can be assumed that the ammonia concentration will be higher than 300 ppm over a longer period of time, then a self-contained breathing apparatus must be worn. However, concentrations above 300 ppm should not occur during normal maintenance and service work.

Service companies and their employees are therefore generally not trained to use self-contained breathing apparatuses. Concentrations well over 300 ppm can occur during cleaning work or due to larger leaks. In this case, it is necessary to wear a self-contained breathing apparatus. These missions are usually carried out by the fire brigade or similarly experienced and well-trained personnel. In this case, a team must carry out the necessary work. The team consists of the repair technician and at least one assistant for the wearer of the breathing apparatus. Depending on the situation, additional respiratory protective device wearers may be necessary. At concentrations above 300 ppm, a tight chemical-resistant protective suit as well as chemical-resistant boots and gloves must be worn.

4.4.8 Work in plant room

The equipment specifications listed below are compiled in accordance with EN 378. These guidelines apply in principle to all refrigeration plant machine rooms, including machine rooms in which refrigeration systems with synthetic refrigerants are installed. Only the limit value and operating set point values of the individual components differ from the values given here, depending on the type of refrigerant.

4.4.8.1 Electric drive motors

The machine room ventilation has to ensure that the ammonia concentration cannot reach the lower flammability limit.

In addition, the European standard specifies that, at a concentration of 30000 ppm (this corresponds to approximately 25 % of the lower flammability limit), the power supply of all electrical devices in the machine room must be switched off. Sockets must also be de-energised. Devices that serve safety (this includes, for example, the drive motor of the fan, the warning lights and siren, the drive motors of louvre flaps) must be rated as an explosion-proof version. This regulation is used for systems with large refrigerant charge quantities. If measures are taken that reliably prevent a concentration of 30000 ppm from being achieved (for example, by appropriate ventilation of the machine room), the above voltage isolation can be dispensed with. In systems with small ammonia charge quantities, this requirement is generally not implemented, and an alternative measure is determined by doing a comprehensive risk assessment. It is advisable to coordinate the necessary implementation of safety measures with the competent authority (approved / delegated monitoring body), as early as possible during the planning stage of the project.

4.4.8.2 Ventilation

Under normal operating conditions, the machine room is ventilated using natural air exchange or using a mechanical ventilation system. The required air volumes are specified in EN 378-3 depending on the system refrigerant charge volume. The following applies to mechanical ventilation systems:

$$\dot{V}=14\cdot 10^{-3}\cdot \sqrt[3]{m^2}$$

With:

$\dot{V}$ air flow in m^3/s

m mass of the refrigerant charge in kg in the refrigeration system with the largest charge in the machine room.

Table 4.18 shows an example of the minimum air volume flows for different refrigerant charges. It should be taken into account here that according to EN 378-3, a minimum air exchange rate

Table 4.18 Minimum air flow rates (EN 378-3)

Refrigerant charge kg	Min air volume flow m^3/h
25	430
50	690
100	1085
200	1720
500	3175
1000	5040

Refrigerant charge lb	Min. air volume flow ft^3/h
55.12	15185
110.23	24367
220.46	38317
440.92	60741
1102.31	112124
2204.62	177986

of the room volume of 4 and a maximum air exchange rate of the room volume of 15 per hour is prescribed.

4.4.8.3 Gas detection system

According to EN 378-3, a gas warning device with appropriate ammonia sensors must always be installed in the machine room of ammonia refrigeration systems. Other national and international standards are in conformity with this, although differ according to the filling volume or the weight of ammonia in the system.

According to EN 378-3, sensors need not be installed outside the machine room. Nevertheless, under certain circumstances it can make sense to place a sensor at critical locations. In case of doubt, placement of sensors should be coordinated with the competent expert of the certified regulatory office during the planning phase. EN 378-3 does not require a sensor to detect toxicity in ammonia plants. Due to the very early perception by humans and the resultant early warning effect, this is not necessary with ammonia.

High-concentration warning

EN 378-3 requires two alarm values here:

Pre-alarm 500 ppm

If the pre-alarm threshold is reached, an acoustic warning signal and a warning light must be activated. At the same time, the ventilation system must be operated at full flow.

Main alarm (if technically attainable at all): 30000 ppm

If the threshold of the main alarm is reached, the ventilation system must be switched off, and the emergency extract ventilation system (fan certified for hazardous areas) must be switched on. At the same time, the power supply of all devices and connections in the machine room must be switched off.

Low-concentration warning for personal and product protection

A warning or notification of low threshold values is not prescribed in EN 378-3. The reason for this is that even very low concentrations of ammonia can be perceived with the nose and therefore people are warned in good time and can protect themselves.

Nevertheless, it is common today to install gas detectors to protect people. The level of the alarm threshold can be determined by the operators. Here it is advisable to coordinate with the responsible expert of the certified monitoring body as well.

The following practical values can be recommended in practice:

Pre-alarm: 100 ppm

If the pre-alarm threshold is reached, it is advisable to activate an acoustic warning signal and a warning light. At the same time, the ventilation system should be switched to full load extract operation.

Main alarm: 250 ppm

If the threshold of the main alarm is reached, the ventilation system should be switched such that no uncontrolled ammonia vapour distribution is possible. Access to the machine room is then only permitted for experienced people wearing protective equipment.

Regardless of the legal requirements, monitoring the ammonia concentration inside the refrigerated storage cold rooms can be useful to protect the products stored there from damage. Alarm values between 50 ppm and 100 ppm have proven their worth here. In the case of product protection, a suitable protocol for cold room ventilation or the relocation of goods must be considered.

Warning of increased ammonia concentration in the coolant media

According to EN 378-3, sensors must be installed in the cooling water or coolant circuit if the refrigerant charge exceeds 500 kg (1102 lbs). It must be ensured that the sensors do not react to other chemical additives in the secondary coolant circuits. In practice, pH sensors have proven their worth, although the maintenance effort should not be underestimated. If there is an ammonia leak, the pH value of the coolant will increase because ammonia is alkaline. In the case of a pH value alarm, a warning horn is not necessary, and there is no need to evacuate the building. However, the location and size of the leak must be found and repaired immediately.

4.4.8.4 Emergency exits and escape routes

Refrigeration machine plant rooms are to be equipped with at least two entrances to be in accordance with EN 378-3. One access door will be used for normal access for maintenance and service work, as well as an emergency exit, which acts as a replacement escape route in the event of danger. This specification applies to all refrigerant plant machine rooms – regardless of the refrigerant used. Alternative evacuation measures can be found in the course of the risk assessment if the local conditions are appropriate. These can be implemented after consultation with the local authorities. Figure 4.30 shows the required safety devices for the ammonia system machine room as well as the permissible limit values according to EN 378-3.

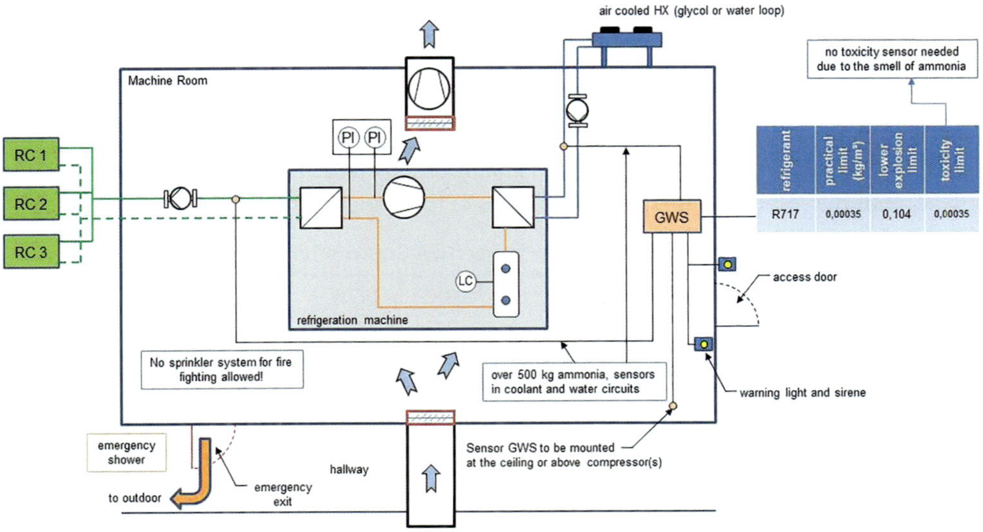

Figure 4.30 Machine room safety devices according to EN 378 (© Kälte Eckert)

4.5 Design criteria for ammonia plants

4.5.1 General basics

The main technical parameters for the design of ammonia compression refrigeration systems are described below. Depending on the use of the system, the purpose-specific conditions and requirements for this must also be taken into account and implemented.

4.5.2 Materials

Steel and stainless-steel materials are mostly used in ammonia plant construction. There are no restrictions on the compatibility of these materials. Copper and copper-containing alloys are not suitable. Aluminium can also be used for the construction of components, but care must be taken to ensure that the amount of water in the ammonia is reduced to a minimum (below 10 %). For compressors, the piston rod bearings are made of phosphor bronze. This is possible because the ammonia vapour in the area of the bearings is superheated and contains only a small amount of water. Corrosive behaviour therefore does not occur.

The construction of ammonia systems with brazed copper pipes is not possible. In theory, copper could be used as a construction material if one could ensure that the ammonia used was completely dry (not containing any water at all). In practice, however, it is not possible to keep the system completely dry during operation. Brass fittings cannot be used in ammonia systems either. They would corrode on contact with ammonia and cause leaks. In practice it has been shown that the corrosion points on the copper materials are not critical for the functioning of the plant. Due to the corrosion process, however, copper plating occurs on steel. This plating is particularly problematic for the bearings and valve plates of the compressors and can lead to the total failure of the compressor.

Most of the sealing materials used in refrigeration are compatible with ammonia. These include chloroprene rubber (CR), hydrogenated acrylonitrile-butadiene rubber (HNBR), ethylene propylene diene rubber (EPD) and nitrile seals. Aluminium seals can also be used. When choosing the sealants, attention should also be paid to the compatibility with the oil being used, especially when changing the type of oil.

Table 4.19 shows oils and their influence on the individual seal types. However, it is always advisable to consult both, oil supplier and seal manufacturers regarding their compatibility, taking into account the respective operating conditions of the refrigeration system.

Table 4.19 O-ring compatibility of R717 for oils (Source: GEA Grasso information C505511-12)

Oil type	M	MHC*	MHC* PAO	AB	POE	PAO	AB–PAO	PAG
O-ring type	CR/ HNBR	CR/ HNBR	CR/ HNBR	CR	-	HNBR/ CR*	CR	CR/ HNBR

*Only for Fuchs Reniso Synth 68 and Klueber Summit R100/R150/R200

M	mineral oils
MHC	mineral oils with special treatment "hydrocracked oil"
AB	alkylbenzene oils
PAO	polyalphaolefin oils
POE	polyol ester oils
PAG	polyalkylene glycol oils
CR	chloroprene rubber
HNBR	hydrogenated acrylonitrile-butadiene rubber

Components that come into contact with ammonia must not contain any mercury. Copper, zinc and alloys made from these materials must not be used. The use of steel and stainless-steel pipes is today's state of the art. Materials should have been tested for compatibility in accordance with EN 10204. A general test certificate should be obtained.

The corresponding test results and certificates of the materials used must be available for verification for all installed components in the refrigeration system.

For materials that are used for temperatures below −30 °C (−22 °F), the cold stress of steel must also be taken into account in order to prevent breakage or cracks.

4.5.3 Joining techniques

In order to largely avoid the risk of leaks and the associated accident risks, only non-detachable fittings should be used when constructing refrigeration systems. In practice, for ammonia systems these are primarily welded fittings. However, if maintenance and service activities require it, detachable fittings will be necessary.

Ammonia is a very powerful solvent. It is often also used in household cleaners. When the refrigeration system is initially filled with ammonia, dirt, deposits, sand and moisture residues from pipes, pressure vessels, valves, as well as fittings and all other components of the system, are cleaned out by the refrigerant. This cleaning effect is continued even during operation. Dissolved dirt is carried through the system and settles in areas with little or no refrigerant flow.

To prevent this, all components must be installed as cleanly as possible. Pipe systems must be cleaned mechanically or chemically after installation and then blown out with dry nitrogen. Welding processes must be selected such that the contamination of the system is reduced to a minimum.

Compressors and refrigerant pumps (and other components) can be protected against damage from debris using filters. The size and position of the filters must be defined in the planning phase. The filters will absorb larger amounts of dirt, especially in the commissioning phase.

4.5.3.1 Fittings for creating fixed connections

The common type of fitting in ammonia technology is the welded connection. Most of the welds are carried out using TIG welding.

It is advisable to coordinate the intended welding procedures and the weld integrity tests required with the responsible testing authority before the work starts on site.

To prevent accidents and refrigerant leakage, high-quality welded connections are essential. Severe accidents have occurred due to pressure surges in ammonia pipe systems, with cracks or breaks always leading to failure. Flaws in weld joints are usually due to human error or a lack of specialist knowledge.

In order to reduce incorrect work to a minimum, only certified and experienced welding personnel may be used for ammonia plant construction. For quality inspection and monitoring, X-ray and/or ultrasound tests are to be carried out. The system sections must be subjected to a pressure test after completion. The applicable standards and regulations are to be followed. The main standards are summarised below:

Standards and Codes of Practice: welding technology for ammonia refrigeration systems

- PED 2014/68 / EG or ASME or DNVGL
- AD2000 or EN 13445 for PED, ASME Sec. VIII Div. 1 for ASME or DNVGL (shipping)
- ISO 3834-2:2021-04 for QA (quality assurance)
- ISO 9001:2015 for QA (quality assurance)
- ISO 15614-1 for WPQR (workplace quality review)

Standard code / reference for NDE:

- RT - UNI EN ISO 17636-1:2013
- UT - UNI EN ISO 17640:2019
- MT - UNI EN ISO 17638:2016
- VT - UNI EN ISO 17637:2017

4.5.3.2 Detachable connections

In principle, one should avoid detachable connections as much as possible in ammonia refrigeration systems. If their use is necessary due to their function or other circumstances, cutting ring or compression fittings for small pipe diameters are available in the hydraulic technology field. These are available in different designs, such as a simple screw connection, double nipple and t-piece, or as an angle version. Alternatively, forged steel fittings with NPT threads are available to make the connection.

Flange connections are mostly used for larger pipe dimensions. It is important to ensure that only flanges with tongue and groove according to EN 1092 are permitted in ammonia refrigeration systems.

If using pipe connections with seals, the ammonia compatibility is of course important, especially the compatibility for rubber seals (see Table 4.19).

4.5.3.3 Stress corrosion cracking

Stress corrosion cracking is a trans- or inter-crystalline crack formation along the grain boundaries of the metal structure, see Figures 4.31 and 4.32. It occurs under the simultaneous influence of static stresses (tensile and residual stresses) as well as an attack fluid. As a rule, no corrosion products or traces can be seen. Stress corrosion cracking creates very fine and tiny cracks.

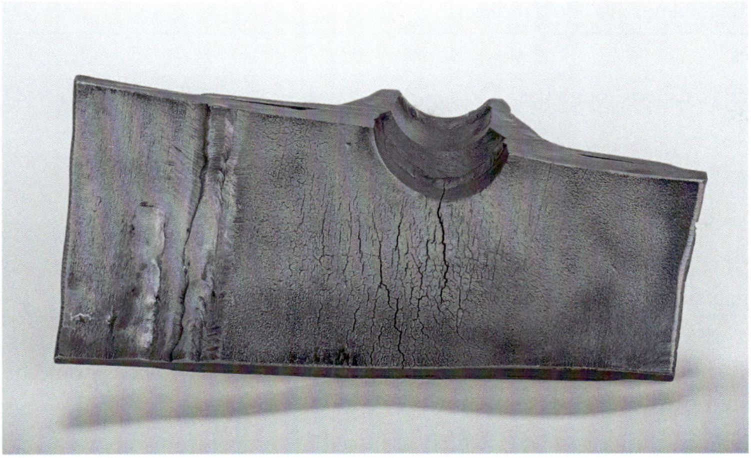

Figure 4.31
Stress corrosion cracking induced by welding stress on a branch reinforcement
(© CEphoto, Uwe Aranas)

Figure 4.32
Stress corrosion cracking on a pipe reducer
(© CEphoto, Uwe Aranas)

Stress corrosion cracking was found, for example, on ferritic steels in ammonia storage tanks. Stress corrosion cracking has also occurred in ammonia refrigeration systems. In particularly serious cases, the cracks were so severe that ammonia seeped out of them.

Experience from ammonia storage technology has shown an oxygen content of just a few ppm in liquid ammonia and of a few thousand ppm in gaseous ammonia promotes the formation of cracks in the steel. A minimum water content of approximately 0,2 to 0,3 % has proven to be an effective inhibitor for preventing this stress corrosion. The water content in an ammonia plant is usually below this concentration. Therefore, measures to prevent stress corrosion cracking must be taken. Here, the high-pressure side of an ammonia plant is much more at risk than the low-pressure side. A good way to reduce the oxygen concentration is to install a purger on the high-pressure side (see Chapter 4.3.4).

Stress corrosion cracking is unlikely to occur below –5 °C (23 °F). Receivers and pipe systems made from cold hardened steels do not tend to be affected by stress corrosion cracking. The

Table 4.20 Steel classification for ammonia liquid receivers.

P235GH	tubes
P250GH	forgings, flanges, round bar
P265GH	plates
P355NH	plates
P355NL1	plates down to –40 °C (–40 °F)
P355NL2	plates down to –50 °C (–58 °F)
P355NH	forging, flanges, round bar down to –40 °C (–40 °F)

yield strength of the steels should be less than 360 N/mm^2. Steel parts and pipe ends should be hot formed and heat-treated. Table 4.20 lists common steel grades.

To prevent stress corrosion cracking, receivers which are used at temperatures above –5 °C (23 °F) (e.g., high-pressure collectors or economisers), as well as receivers that are maintained for long periods at ambient temperatures (e.g., vessels in systems that are only used seasonally), should be stress relieved after welding.

4.6 Applications of ammonia as refrigerant

This chapter summarises the use of ammonia refrigeration plants in typical areas of application.

4.6.1 Refrigerated warehouses

Two temperature levels are typically considered in the food industry. A distinction is made between the low-temperature storage of frozen products (LT) at approximately –20 °C to –2 °C (–4 °F to 28 °F) and the storage of chilled products at temperatures between 0 °C and +12 °C (32 °F and 54 °F), commonly called medium temperature – MT. Depending on the product group, different storage temperatures are required in the medium-temperature area:

meat and fish	approximately 2 °C (36 °F)
dairy products	approximately 6–8 °C (43–46 °F)
beverages	approximately 6–8 °C (43–46 °F)
fruit and vegetables	approximately 8 °C (46 °F)
bakery products	approximately 8–15 °C (46–59 °F)

Some fruits, such as kiwis or ripening bananas or apples, sometimes have to be stored in separately controlled atmospheres. Controlled atmosphere storage has its own temperature requirements and gas content percentage peculiarities.

In large food cold stores, two-stage refrigeration systems with flooded direct evaporators are often installed for low and medium temperatures, see Figure 4.33. The refrigerant is circulated in the evaporators by means of refrigerant pumps. The evaporation temperature of –30 °C

Figure 4.33 Image of a two-stage screw compressor chiller system (© GEA Grasso)

(–22 °F) is the setpoint for low-temperature cold rooms, whilst medium-temperature cold rooms are maintained at –5 °C (23 °F) evaporation temperature. Evaporative condensers are often used to liquefy the refrigerant gas at around 35 °C (95 °F). Whenever possible, the heat from the condenser and possibly oil cooler of the refrigerated warehouse refrigeration system should be used for heating purposes rather than dumped into the ambient.

As described in Chapter 4.3.1, a screw compressor with economiser can also be used instead of a two-stage refrigeration system. For the selection of the compressor capacity, the usage of the cold rooms must be clarified with the cold room operator. It should also be clarified here at what times individual consumers are using the cold room. This results in simultaneity factors and therefore the operating concept of the refrigeration system.

A direct evaporating refrigeration system for a cold store with a storage area of 100 m (328 ft) in width and 250 m (820 ft) in length requires a refrigerant charge quantity of approximately 5 to15 metric tons (14.8 $tons_{IMP}$ / 16.5 $tons_{US}$) of ammonia in flooded operating mode with pump circulation. This amount of refrigerant is required to ensure trouble-free operation of the system even in the event of load fluctuations. A DX ammonia system servicing the same warehouse will have an NH_3 inventory of approximately 1 metric ton (0.98 $tons_{IMP}$ / 1.1 $tons_{US}$). There are close to thirty DX ammonia refrigerating plants servicing mixed warehouses in Australia. The largest facility is 250000 m^3 (8828675 ft^3).

In order to be able to operate the refrigerant pumps without cavitation, the amount of refrigerant in the liquid separators must be sufficiently large or the inlet connection must be sufficiently high and filled with liquid refrigerant. The evaporators in cold stores are usually defrosted using hot gas. For this purpose, part of the hot refrigerant vapour is removed from the compressor high-pressure discharge port and fed into the evaporator via a separate defrost hot gas pipe network. The gas partially condenses in the evaporator, which releases a significant amount of heat. Partial condensation is not necessarily desired and may be avoided with a high-pressure float valve for condensate return. This heat leads to quick, reliable and efficient defrosting of the ice layer frozen on the outside of the evaporator tubes and fins. Due to the sudden and considerable heat input into the evaporator, there is a risk of mechanical damage because of pressure surges or temperature strains. The defrosting process must therefore be properly controlled and its functionality checked regularly.

The refrigerant charge quantity for a direct evaporation system could be significantly reduced by using suitably sized expansion valves with dry expansion evaporators. However, practical experience has shown that dry-evaporating ammonia systems can be problematic for several reasons especially at low temperatures. But there are numerous DX systems installed which perform well.

- Most DX ammonia systems use non-miscible oils. This works fine as long as the right precautions are taken. Easier from the point of oil return could be the use of ammonia-soluble oil. However, these oils are relatively expensive and react with great sensitivity to water and dirt in the refrigerant. An excessively high proportion of water leads to saponification of the oil. This means that the oil no longer functions as a lubricant, and the compressor is severely damaged as a result. In order to reduce the risk of saponification, all work must be carried out cleanly and carefully even during the construction stage of the system. Later, during maintenance and service work, the same standards of care and cleanliness apply.
- Water in the refrigerant can freeze out on the expansion valve needle and impair the function of the valve or the expansion valve controlling the superheat after the evaporator is fooled by the change in thermodynamic properties of the refrigerat due to the water content in the ammonia. As a result, ammonia liquid can appear at the evaporator outlet. If this ammonia liquid reaches the compressor, it can result in liquid hammer and mechanical damage to the compressors, the couplings and the drive motors.

Liquid ammonia entering the compressor via the suction line can be remedied by suction-side liquid separator vessels (a suction line accumulator). The ammonia liquid that collects in the separator is evaporated via heat during operation and then drawn away by the compressor. The heat required is supplied via a high-pressure liquid line. For this purpose, a corresponding pipe loop is integrated into the separating vessel as a heat exchanger.

It must be taken into account that in the course of the liquid separation, the oil carried in the suction line is also separated, and the separator fills with oil over time. The separator must therefore be emptied regularly, and the oil returned to the compressor unit. An oil bleed hole in the pipe exiting the separator via a U-bend through the liquid reservoir might be used. Similar separation tanks can also be found in car air conditioning systems.

4.6.2 Supermarkets

In supermarkets that are equipped with ammonia refrigeration systems, indirect systems are today used almost exclusively, see also Chapter 10. However, it has been shown that the use of water-based refrigerants only makes sense in the medium-temperature (MT) range due to energy efficiency reasons. Various substances have been tested as a coolant in recent years. It has been shown that propylene glycol is a good MT-coolant for these applications, both technically and economically. Propylene glycol is also good to use from the point of view of toxicity. It is approved for use in the food sector.

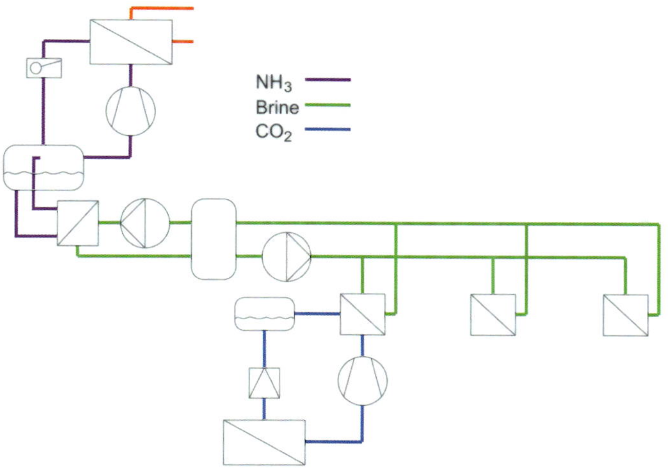

Figure 4.34
Ammonia–CO_2 cascade system with secondary loop MT system also cooling the CO_2 system(s)

With this application, it must be taken into account that regular defrosting of the heat exchangers is required. Defrosting in direct expansion supermarket systems (and also some indirect systems) is usually carried out using electrical heating elements. With an indirect system, the propylene glycol flow to the respective heat exchanger can simply be switched off while the air fans continue to run. In larger systems, however, defrosting can also be carried out with a warm brine heated by subcooling of the primary refrigerant. For this purpose, an additional warm brine network with corresponding feed pumps and brine storage is installed. This warm defrost loop contains the same additive as the cooling network, e.g., propylene glycol. Depending on the application, the additional installation costs for this warm brine network can be amortised within a few years thanks to significantly lower operating costs. This saving effect is achieved through the re-use of the already existing available condenser heat. Only excess energy is then released into the atmosphere via a cooling tower or air-cooled condenser. It is standard in today's supermarkets to use this existing available condenser heat for heating the supermarket and adjacent buildings or for hot water generation. Planners, builders, and project developers alike are required to think about the intelligent and practical use of this available condenser heat capacity. Clever utilisation of heat recovery often makes a conventional gas- or oil-fired heating system obsolete.

Particular attention should be paid to the location for the refrigeration system. It has been shown that installation in the roof area is easiest regarding potential ammonia leaks, because this has less impact on customers and employees. Obviously, there should not be any ventilation air intake grills near the ammonia system.

The use of indirect systems with liquid refrigerants in the low-temperature (LT) area has not proved to be successful. The available substances for the secondary refrigerants require essentially more complex technical solutions such that the efficiency of the system due to the high-energy demand of the circulation pumps is usually not desirable. In recent years, subcritical direct evaporation CO_2 systems have therefore been increasingly established for the LT area. As shown in Figure 4.34, these (distributed) CO_2 systems are supplied via the secondary refrigerant circuit of the medium-temperature (MT) area.

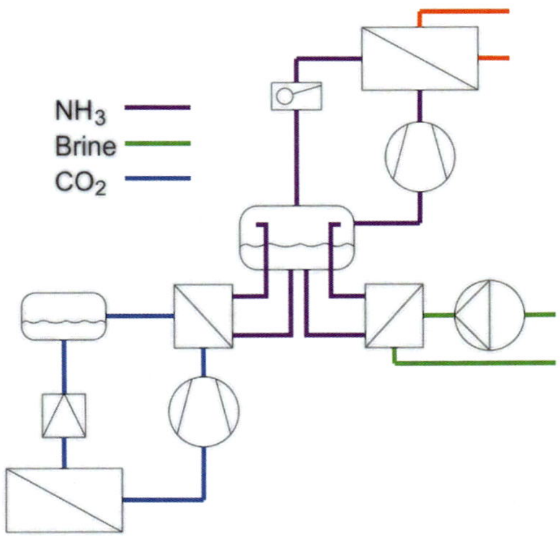

Figure 4.35
Ammonia-CO_2 system with gravity-fed brine (secondary refrigerant) circuit, in parallel

Depending on the application and refrigeration system, the condensation of the CO_2 cycle can also be done directly into the NH_3 circuit of the high-pressure stage (see Figure 4.35). The advantage of this application is that losses due to the necessary temperature differences at the heat exchangers (approximately 5 %) can be saved. The disadvantage of this design is that a leaking cascade heat exchanger (ammonia evaporator/CO_2 condenser) can lead to total failure of the ammonia plant: Due to the entry of CO_2 into the ammonia cycle, the pressure in the CO_2 system is higher than that in the ammonia system, and salt crystals (ammonium carbonate) are formed, which are distributed throughout the ammonia network (see Chapter 4.6.8).

4.6.3 Commercial kitchens

In the past few years, the use of ammonia systems to cool the kitchen cold rooms as well as the public sales and service counters has also become state of the art for commercial kitchens. The systems used are very reliable and durable. Figure 4.29 shows an example of a refrigeration system with a cooling capacity of 20 kW (5.7 TR). Ammonia is used in the medium-temperature circuit and CO_2 in the low-temperature circuit (Figure 4.34). Even when converting existing systems that were previously on synthetic refrigerants, it is often practical to use ammonia and CO_2. Several commercial kitchens in Australia use DX NH_3 in the freezers.

Similarly to supermarkets, indirect cooling systems are also used for kitchens. Depending on the local conditions, both water- and brine-cooled condensers are used. The plants are often built as flooded systems. Due to the relatively low cooling capacities (approximately 15–100 kW / 4.3–28.4 TR), open piston compressors with cylinder head cooling are mainly used. The capacity control is usually done with frequency-controlled drive motors. The ammonia charge quantities of small plants are approximately 0,1 to 0,5 kg/kW (0.23 to 1.16 lb/Btu/s).

For reasons of operational reliability, these small systems are usually built with two compressors in parallel. A brine (secondary refrigerant) network is used to distribute the cold. The use of ethylene glycol has proven itself. Both, pump manufacturers and the manufacturers of heat exchangers and fittings are now very experienced in the selection and use of the components. In the medium-temperature area, brine-flow temperatures of approximately -5 °C (23 °F) are usually maintained. This can significantly reduce dehumidification in the cold rooms and display cabinets. This leads to less product drying and thus to a better product quality. At the same time, the evaporation temperature of the refrigeration system can be raised – this leads to lower operating costs of the refrigeration system – approximately 2–3 % per degree higher evaporation temperature.

If natural refrigerants are used, the low-temperature rooms and cold stores are usually operating on a subcritical, directly evaporating CO_2-system. The condensation then takes place rejecting heat into the brine circuit, at the medium-temperature level.

Figure 4.36 Buffer storage and pump station for an indirect brine system for a commercial kitchen (© Kälte Eckert)

This system technology has proven itself. The required main components, such as ammonia chiller, brine pump station (Figure 4.36) and CO_2-compound systems, are offered by well-known manufacturers. These would include components that are ready for installation and start-up and also include the necessary control technology. Refrigeration contractors can therefore install ammonia plants without having to master the special ammonia manufacturing techniques themselves.

4.6.4 Ice rinks

Either direct or indirect cooling systems are used in ice rinks. In direct systems, the ammonia is pumped through a steel pipe network installed in the concrete slab of the floor. In indirect systems, a coolant is conveyed through the pipes in the floor. Considerable pump power is required for this, but the overall energy analysis shows that the indirect systems also work very efficiently: the energy consumption of both systems is almost identical. The high pressure drop in the refrigerant pipe network of the direct evaporation system probably causes the relatively high-energy requirement of such a refrigeration system. In direct systems, the steel pipes are embedded in a concrete floor and are therefore well secured against damage. At the same time, the floor is considered to be an adequate separation between the refrigerant-carrying pipes and the public. There are no additional costs for any safety measures for the public in the freely accessible ice area (in accordance with EN 378-3). Leaks in the concreted-in refrigerant pipe network are almost impossible as long as the concrete is not subject to any acid treatment. CO_2 is also increasingly used for cold distribution and used as evaporating coolant underneath the ice rink. The temperature distribution and thereby the quality of the ice tends to be more even in ice rinks with evaporation of ammonia or carbon dioxide in the concrete slab.

In order to keep the amount of refrigerant small, indirect systems are today mostly equipped with plate heat exchangers on the evaporator and condenser side. The condensing takes place via open or closed cooling towers with the help of ethylene glycol as heat transfer media. Whenever possible the heat from the condenser and possibly oil cooler of an ice rink refrigeration system should be used for heating purposes rather than dumped into the ambient.

4.6.5 Air conditioners

Office and administration buildings are usually air-conditioned via supply and return air distribution systems. The cooling coils of the air-handling units required for this are generally designed for chilled water systems. This means that indirect refrigeration systems are mostly used for such air conditioning systems. A chilled water inlet temperature of 6 °C (43 °F) and a return temperature of 12 °C (54 °F) are common if this low feed line temperature is required for a reliable air dehumidification. In modern air handling plants, however, the water temperatures are increasingly raised by 2 K, sometimes by up to 5 K. This can reduce energy costs by approximately 10–25 %. If one takes into account that the refrigeration system in the building has by far the highest electrical energy requirement, it becomes clear that the building operating costs are significantly reduced by increasing the water inlet temperature.

The air conditioning water chillers are usually built in a power range between 250 and 2000 kW (71 and 569 TR), although significantly larger or smaller types are also available. Both, air-cooled and water-cooled chiller versions are standard for the manufacturers. Whenever possible the heat from the condenser and possibly the oil cooler of the ammonia chiller should be used for heating purposes rather than dumped into the ambient.

Ammonia is increasingly used for water chillers today. Previously, however, synthetic refrigerants were predominantly used for water chilling. The reasons for this were the lower procurement prices and, above all, the lack of experience of the contractors and building operators in handling the ammonia refrigerant. But due to the better efficiency of the ammonia compressors and the strict requirements in Europe of the F-Gas Regulation, modern buildings are now increasingly being equipped with ammonia chillers. In this context, it should also be pointed out that the ammonia chillers have a very long service life due to their solid construction. The most common arguments of flammability and toxicity can be completely invalidated by the operational experience of the past 25 years. Likewise, when viewed from a neutral point of view, it becomes clear that the installation conditions for ammonia systems, which are often described as complex and expensive, are essentially the same when using EN 378-3. They are not more complex or expensive when compared to using synthetic refrigerants.

4.6.6 Heat pumps / heat recovery

Ammonia is very suitable for heat pump applications. The critical point of ammonia is slightly over 130 °C (266 °F). This means that condensation temperatures of around 100 °C (212 °F) are theoretically possible. Only a few refrigerants can be used at such a comparatively high temperature level – isobutane, for example. With ammonia, however, it must be taken into account that the condensing pressure rises sharply at high temperatures. Due to the high compression temperature, ammonia heat pumps can achieve a higher end point temperature at the hot gas inlet

Table 4.21 Comparison of thermal parameters between ammonia and R134a (isentropic compression) for two t_{evap}/t_{cond} temperature scenarios

	R717	R134a	R717	R134a
Evaporation temp. / condensation temp. in °C	24/64		52/92	
Condensing pressure in bar	28,79	18,46	53,28	33,76
Evaporating pressure in bar	9,73	6,46	21,41	13,85
Compression ratio	2,96	2,86	2,49	2,44
Pressure difference in bar	19,1	12,0	31,9	19,9
COP	7,22	6,5	7,44	5,52
Compressor discharge temperature in °C	107	67	127	98
Evaporation temp. / condensation temp. in °F	75.2 / 147.2		125.6 / 197.6	
Condensing pressure in psi	417.56	267.74	772.76	489.65
Evaporating pressure in psi	141.12	93.69	310.5	200.88
Compression ratio	2.96	2.86	2.49	2.44
Pressure difference in psi	277.02	174.05	462.67	288.63
COP	7.22	6.50	7.44	5.52
Compressor discharge temperature in °F	224.60	152.60	260.60	208.40

of the condenser. The prerequisite is that this higher temperature can actually be transferred to the medium being heated (usually water). The use of ammonia is therefore particularly suitable in the heat pump sector due to this very useful thermodynamic characteristic.

Typically, the temperature difference between the heat source and heat sink of a heat pump is around 30 K. In one study, the refrigerant R134a and ammonia were compared with regard to the possible uses in heat pump applications. The behaviour in an isentropic compression refrigeration cycle without superheat and without subcooling was examined. The results were as follows (see Table 4.21).

The compression ratio for ammonia is somewhat higher than for R134a (approximately 3 %). However, the COP is significantly higher for ammonia: approximately 11 % for the 24 / 64 °C (75 / 147 °F) heat pump operating temperatures and approximately 35 % with the 52 / 92 °C (126 / 198 °F) heat pump. The energetic advantage of ammonia over R134a is clear.

However, it must be taken into account that the operating pressures in the ammonia circuit are quite high. The ammonia components must be designed for these high pressures. Particularly with regard to the compressor, it must be able to produce a sufficiently high pressure level. Compressor manufacturers do offer corresponding high-pressure models for operating pressures up to 50 bar (725 psi). With these compressors, condensing temperatures of up to 80 °C (176 °F) are possible. Due to the increasing use of CO_2 compressors for significantly higher operating pressures (up to 63 bar / 914 psi), such units are now also available. With these compressors, condensing temperatures of almost 100 °C (212 °F) are possible using ammonia as refrigerant.

Establishing heat recovery equipment on existing refrigeration systems is comparatively easy to implement. It is always advisable to carry out a careful check of the local financial constraints. If the available heat can be used, the investment usually quickly pays for itself. The critical question here is whether the available temperature level is sufficiently high for the customer. Floor heating, concrete core heating or preheating service water are very suitable for utilising heat recovery from refrigeration systems. The use of heat for defrosting heat exchangers in cold rooms (hot brine defrost) has also proven itself in practice. If required, an additional heat pump can transform the condenser heat of the refrigeration system to a more useful level.

Modern building technology systems increasingly rely on hybrid chillers. The chillers are used for cooling in summer and for heating in winter. In this manner, cooling or heating as well as simultaneous heating and cooling is provided. Hybrid refrigeration systems are particularly suitable for processes in which heating and cooling are required at the same time and are often used in conjunction with geothermal refrigeration systems for office buildings and data centres.

4.6.7 Chillers for process cooling

Fluids can be cooled by ammonia chillers, see Figure 4.37. Generally chilled water networks are cooled by using chillers. Increasingly these chillers also cool other liquid coolants (also called secondary refrigerants or brines) such as mixtures of water with alcohol, glycol or salt, various oils and other chemical solutions or even pure substances (see Chapter 10 for more details). Various designs and operating concepts are offered. The liquid chiller must be constructed along well-established guidelines with the experience of the component manufacturer and equipped with the required control technology.

Factory assembled liquid chillers also offer a very good opportunity for refrigeration system contractors who have less experience in handling and building ammonia systems to use ammonia as a refrigerant in a refrigeration system.

Liquid chillers are available in water- and air-cooled versions – if required, also with integrated heat recovery. They are offered as a single and composite system with one or more compressors. The chillers are mainly designed as a flooded system, but there are also dry evaporation systems on the market. Some manufacturers offer customised models. The cooling sets can be adapted to structural building conditions and space requirements at the installation site. This can expedite the physical installation.

The chillers can be designed for various installation requirements or, for example, special designs in explosion-proof design.

4.6.8 CO_2 and NH_3 cascade systems

The combination of ammonia and CO_2 is not new and has proven itself in various combinations. CO_2 is increasingly used here as a very efficient refrigerant for lower temperature applications.

A combination of a subcritical CO_2 / ammonia cascade system is a common application already described above. The advantages of both refrigerants can be combined here. This type of plant construction is perhaps the most suitable approach for the construction of environmentally

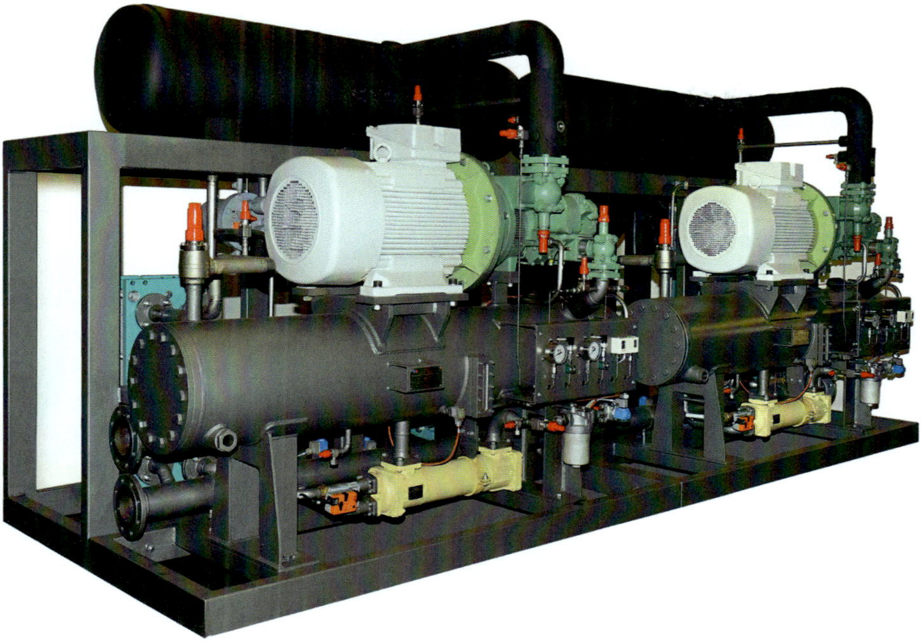

Figure 4.37 Liquid chiller set (© EM Polar)

friendly and energy-efficient refrigeration plants. The two common systems for condensing the CO_2 are:

- re-liquefaction via an ammonia circuit (evaporation temperature < 10 °C / 50 °F)
- re-liquefaction via a brine (secondary refrigerant) circuit (brine temperature < 10 °C / 50 °F)

Direct re-liquefaction in the ammonia evaporator has been implemented frequently in recent years. However, it must be taken into account that a leak in the combined heat exchanger of the ammonia evaporation / CO_2 condensing process can lead to a total failure of the ammonia system: The CO_2 – being at higher pressures – penetrates the ammonia cycle which leads to salt formation (ammonium carbonate) in the ammonia cycle. (Depending on the water content in the ammonia circuit it will form methyl carbamate or ammonium carbonate or both as the reaction of CO_2 with ammonia forming methyl carbamate also yields water. This in return will aid the formation of ammonium carbonate.) If the salt crystals are spread out in the refrigeration circuit, they can cause mechanical damage to the components of the system. The CO_2 / ammonia heat exchanger must therefore be selected with particular care.

The chemical reactions that lead to salt formation can be represented by:

$$2\,NH_3 + CO_2 \rightarrow CO(NH_2)_2 + H_2O$$

for methyl carbamate and

$$2\,NH_3 + H_2O + CO_2 \rightarrow (NH_4)_2CO_3$$

for ammonium carbonate.

As an alternative, there is the option of removing the heat of condensation via a brine-cooled condenser. The installation of such an additional heat exchanger increases the energy requirement by approximately 3 to 5 %, but a leak in the heat exchanger will at least not lead to a total system failure. At the same time, the brine circuit can act as a buffer (thermal storage) during brief shutdown of the ammonia system. In this way, continuous operation of the system is better enabled.

4.6.9 Small capacity installations

Due to the stricter applications of the current EU F-Gas Regulation, ammonia plants with very low capacity (cooling capacity < 20 kW / 5.7 TR) are in demand. The system size and cooling capacity is limited only by the availability of suitable compressors. All other required components, such as oil separators or high-pressure float valves, have been previously designed by the refrigeration component manufacturers for these small cooling capacities and are also available. Chillers currently available on the market have a cooling capacity from 10 kW (2.8 TR). Small liquid chillers have proven their worth in recent years, especially in food refrigeration (see Chapters 4.6.2 and 4.6.3).

4.6.10 Container solutions

Especially if the local conditions at the installation site of the chiller are restricted and there is no good way to create or upgrade a suitable plant room, it is advisable to use a container system, see Figure 4.38. Especially if the refrigeration system is to be used at a different location at a later point in time, a container system can be used to create a very cheap and flexible professional refrigeration solution. The container is designed as a plant room in accordance with EN 378-3 and is set up in accordance with local and safety-related requirements. Refrigeration machine(s),

Figure 4.38 Air-cooled ammonia refrigeration system 250 kW (71 TR) for indirect cooling (© Kälte Eckert)

control and regulation technology, as well as the required hydraulic elements such as pumps and storage vessels, are installed in the factory or by the refrigeration system contractor within the container, completely piped and electrically connected.

A major advantage is that the entire system can already be conformity marked (e.g. CE conformity marking in Europe or given respective national compliance approval) in the factory. After connecting the required media (electrical, cooling and heat transfer media), the commissioning can be carried out immediately.

4.7 Standards and regulations

Existing European and international standards and regulations that apply when dealing with ammonia refrigeration systems are summarised below. Due to the very different circumstances and installation conditions, not all legal requirements can be fully listed and taken into account here. In each individual case, it must always be checked with all involved parties (in particular with the responsible government approval authorities), which further regulations have to be considered.

For the European area, the overarching standards and guidelines have been developed by CEN / TC 182 (European Committee for Standardisation, Area 182: Refrigerating systems, safety and environmental requirements). In the meantime, the CEN has drawn up many European standards for refrigeration technology and introduced a large amount of common technologies throughout the European Union. However, health and safety regulations are generally still set at the local and national, and not at the European, level and must be complied with.

The regulations listed below generally apply to all refrigeration systems and are independent of the type of refrigerant used. Nonetheless, different specifications are set out for individual refrigerants within the standards. The existing standards and regulations can be divided into two categories:

- Design specifications for the construction of the plants and for plant safety. These are standards that the designer, manufacturer and installer must observe when designing and constructing the refrigeration system.
- Operator specifications: These specifications are aimed primarily at the operator and the maintenance support service provider of the refrigeration system. Environmental, personal and occupational safety aspects predominate here.

4.7.1 European regulations

The most important European standards for the design, installation and operation of refrigeration systems are listed below.

EN 378 Refrigerating systems and heat pumps – Safety and environmental requirements

This standard has four parts:

- part 1: Basic requirements, definitions, classification and selection criteria
- part 2: Design, construction, testing, marking and documentation

- part 3: Installation site and personal protection
- part 4: Operation, maintenance, repair and recovery

Refrigerants are classified according to their flammability and toxicity. Ammonia is assigned to category L2 (lower flammability, increased toxicity) in EN 378-1 Table EI.

Pressure equipment directive 2014/68 / EU	Requirements for pressure equipment
Low voltage directive 2014/35 / EU	Requirements for electrical devices
Machinery directive 2006/42 / EG	Machine requirements
EMC directive 2014/30 / EU	Electromagnetic compatibility of equipment

The following regulations must also be consulted for the design and manufacture and the associated testing of pressure equipment and parts of equipment, in particular pipelines and pressure vessels:

EN 13480	Pipeline requirements
EN 13445	Container requirements

Both sets of rules are based on or meet and substantiate the requirements of the Pressure Equipment Directive. It should be noted that the set of rules once selected must be applied constantly and consistently for the respective product or project. A change between the regulations is not permitted. Also, any work done on a previously approved project implies that, after upgrade, the whole of the project must meet the current Equipment Directive.

4.7.2 Safety training of employees

As shown above, a great deal of information, technical bulletins, best practices and standards are available. However, users and operators must be familiar with these documents and understand and master procedures and processes in order to be able to implement them in emergencies.

It is important that employees who are responsible for the maintenance and operation of ammonia refrigeration systems are familiar with the properties of ammonia. In the event of a defect in the system and especially in the event of an ammonia leak, one must be familiar with the necessary protocols and procedures in order to be able to protect oneself and others from possible dangers.

Employees who work on or near an ammonia plant (e.g. in production lines, forklift drivers and office employees) must be informed of the consequences of an ammonia leak so that they react correctly in the event of a leak and not panic. In this regard, employees should also be able to distinguish between a fire alarm and an ammonia alarm. It is advisable to familiarise the local first responders with the basics of first aid for ammonia injuries. It lies within the responsibility of the facility owner to ensure the recurring and extensive training and education of the employees. Furthermore, fire brigades and other emergency services must be informed about the location, structure, type and function of the system. They must become involved in joint evacuation and medical stabilisation exercises. Again, this is the responsibility of the facility owner.

Table 4.22 Advantages and disadvantages of ammonia compared to R134a or other synthetic refrigerants

Advantages	Disadvantages
• Higher critical temperature • Higher heat transfer coefficient • Lower density • Higher efficiency in almost all temperature ranges • Smaller pipe diameters • Lower refrigerant purchase price • Leakages are recognised immediately • Greater tolerance for contamination with water • Sustainable, very low environmental impact • No requirements regarding leakage control within the framework of the EU F-Gas Regulation	• Incompatibility with copper and certain copper containing alloys • High hot gas temperature[1)] • Low toxicity limit • Personel training indispensable • Immiscible with oil (mineral oils) • Higher plant costs

[1)] However, the higher hot gas temperature can also be a significant advantage in heat pump or heat recovery applications.

4.8 Summary of the advantages and disadvantages of ammonia

Table 4.22 shows the advantages and disadvantages of ammonia as a refrigerant. Weighing the advantages and disadvantages shows that the use of ammonia as a refrigerant brings many advantages as long as the plant installation contractor has the necessary knowledge and procedures for erecting and maintaining an ammonia plant. It is essential that experience of steel piping systems and good welding knowledge are available. Both refrigeration plant operators and service and maintenance personnel must be trained and certified in the use of ammonia.

Economic consideration

The good thermodynamic properties in particular promote the selection of ammonia as a refrigerant. In most cases, the initial capital costs for an ammonia installation are higher compared to systems with synthetic refrigerants.

However, the generally higher COP or EER of ammonia refrigeration systems leads to lower operating costs. Furthermore, the leakage control intervals are significantly longer than with comparable HFC systems (EU F-Gas Regulation). Depending on the operating behaviour of a plant, the higher capital costs of an ammonia plant often pay for themselves within a very short period of time.

At the time of writing, the price per kg of R134a, R407C and R404A has been approximately five to twelve times higher than that of ammonia. It can be assumed that synthetic refrigerants will continue to become extremely expensive in the next ten years – especially in Europe. The reason for this is the looming shortage of HFC refrigerants, partly due to the EU F-Gas Regulation (517/2014). This will most likely be happening worldwide due to the Kigali Amendment to the Montreal Protocol calling for a worldwide phase-down of HFC refrigerants.

Regardless of the purchase price, the refrigerant charge quantity for a refrigeration system – due to the comparatively lower density of ammonia (especially in the liquid form) – is lower than with synthetic refrigerants. This leads to an additional savings.

4.9 Outlook

The further development of ammonia refrigeration technology will essentially depend on the general technical developments as well as on future political requirements. The decisive factors will be the requirements for environmental protection and the unavoidable increases in energy and CO_2 emission prices.

In recent years it has been shown that the use of ammonia is possible for almost all technical applications. If the systems are designed and built professionally and carefully and the operators and service personnel are well trained in handling ammonia, the dangers and risks in handling ammonia are very low. With the enactment of EN 378, a reasonable and easily implementable European standard for the construction and handling of refrigerants was created. If the designers and plant contractors adhere to its content and specifications, owners and operators will continue to have increasing confidence in ammonia refrigeration technology in the future.

With the promulgation of EU F-Gas Regulation 517/2014, European legislators have once again shown very clearly that a shift from synthetic to natural refrigerants can be achieved. Given the negative characteristics of F-gases, it can be expected that more far-reaching restrictions with regard to the use of F-gases will be prescribed in the very near future. Due to the HFO-TFA issue (see Chapter 1), the future for ammonia as refrigerant looks even brighter.

In parallel to the published ordinances and government programs to promote environmentally friendly and energy-efficient refrigeration systems, the legislators have been and are setting out further restrictions. It can be expected that the costs of gas and electricity will increase in the course of providing a sustainable and CO_2 emission-free energy supply in the coming years. The high energy efficiency of ammonia plants will therefore play an even more important role in the selection of refrigerants and plant technology and will lead to a further growing market share of ammonia. Heat recovery in particular is an important part of sustainable energy management, which is why heat pump applications and hybrid systems will increasingly be using ammonia in the future. Large refrigeration capacity cooling systems and industrial applications are the strong points for the future of ammonia.

With the "rediscovery" of ammonia as a refrigerant, the development efforts of the component manufacturers have intensified. This has led to new and improved ammonia-compatible components, particularly in the area of heat exchangers. The refrigerant charge quantity and efficiency of the heat exchangers have been continuously improved in recent years. It can be assumed that further optimisations will be achieved through the use of new materials and improved manufacturing options. At the same time, the compressor manufacturers are constantly working on improving their designs. For example, the compressor casings can be used for ever higher operating pressures, and the first semi-hermetic ammonia compressors are already on the market.

The trend to use ammonia for small capacity cooling will continue in the coming years. Small capacity ammonia compressors are already offered today. There are many applications for which both the components and the system technology are available.

Although the capital costs for an ammonia system can be significantly higher than for an equivalent system using synthetic refrigerants, ammonia systems will prevail due to the significantly lower operating costs, the high reliability and the longer service life.

Ammonia systems will increasingly be used in the future. Offices, hotels, hospitals, shopping centres, universities and conference centres will soon discover the high reliability and very economical generation of chilled water, by using ammonia in their air conditioning systems.

Indirect systems, in combination with ammonia refrigeration systems, will also be increasingly used in the food sector. In addition to the classic applications, such as in breweries, systems with smaller cooling capacities, such as those required in commercial kitchens or in supermarkets, will also be increasingly used.

Whether ammonia systems can be produced in the future in large numbers, and therefore cost-effectively, will depend on the extent to which service technicians and operators are familiar with the ammonia technology.

For this reason, vocational schools, universities and tertiary training institutions will have a particular impact on the acceptance of ammonia.

The safe and reliable operation and the simple, cost-effective and energy-efficient design and construction of ammonia systems remains the most important prerequisite for further growth of this industry. The confidence in ammonia technology, which has been built up over the past years, must not be lost by carelessness or negligence on the part of component manufacturers, contractors or operators of the refrigeration systems.

4.10 System examples

Three typical systems are described below. Operating constraints and potential refrigerants are stated. These systems clearly show what level of information is required to develop a workable solution.

4.10.1 Refrigeration supply for a fruit-and-vegetable wholesaler

For the storage and processing rooms of a wholesaler, certain corresponding specifications for room temperature and humidity must be maintained, depending on the type of products stored and their specific processing methods. Moisture in particular plays an essential role in maintaining the quality of the goods when storing fruits and vegetables. The basic heat load of the storage rooms is dissipated via a brine circuit. The brine is cooled by an air-cooled ammonia system. A chilled water circuit with ice storage was installed for the processing rooms. In this way, peak loads can be covered, and – if needed – very high dehumidification performance can be achieved as required.

For space and cost reasons as well as with regard to cost-effective maintenance and service work, the entire refrigeration system, consisting of brine circuit and ammonia refrigeration system, was housed in a container, see Figure 4.39.

A high level of economy and a particularly environmentally friendly and sustainable system concept were required.

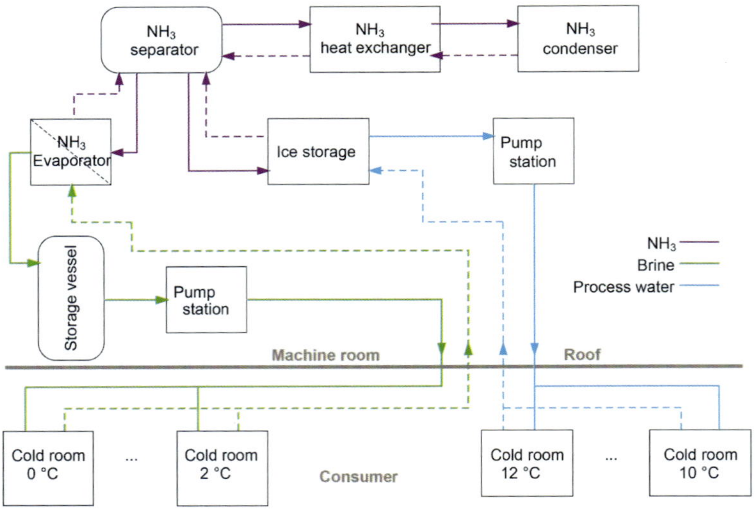

Figure 4.39
Schematic representation of refrigerant circuits and fruit and vegetable storage

System details:

Total cooling capacity	300 kW (85.3 TR)
Cooling capacity; storage rooms	169 kW (48.1 TR), 11 cold rooms
Cooling capacity; climate control	140 kW (39.8 TR), seven climate rooms
Refrigerant charge	46 kg (101.4 lbs) of ammonia
Operating conditions t_{evap} / t_{cond}	–11 °C / 35 °C (–2 °F / 95 °F)
COP of the refrigeration system	3,4

- flooded ammonia system with three parallel reciprocating compressors
- air-cooled NH_3 condenser
- chilled water circuit +7 °C / 0,5 °C (45 °F / 33 °F) with ice storage, 250 kWh (853200 Btu) for processing machines and the cooling or dehumidification of the processing rooms
- brine (secondary refrigerant) circuit, ethylene glycol –8 °C / –4 °C (18 °F / 25 °F) with buffer storage for cooling the storage room
- circulating air defrosting of the cold room heat exchangers without additional electrical energy
- control and visualisation via programmable logic controller (PLC)
- remote monitoring via the internet
- peak load management through ice storage
- container solution: machine room in accordance with EN 378
- complete system to comply with CE approval on delivery

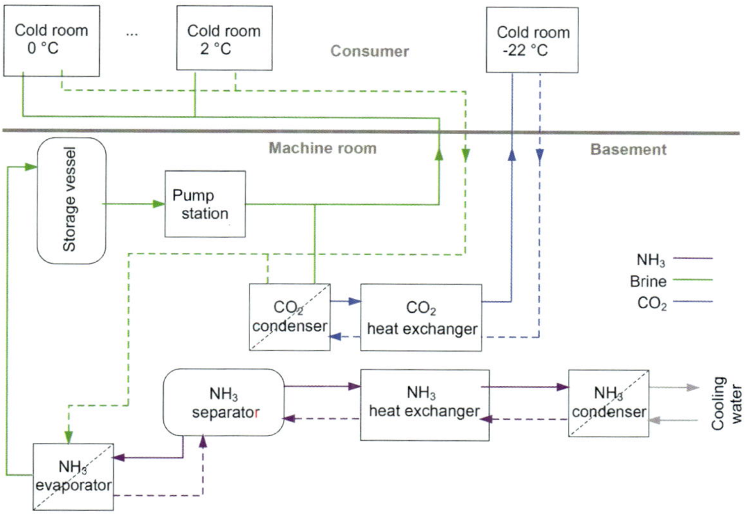

Figure 4.40
Schematic representation of both refrigerant circuits, commercial kitchen

4.10.2 Cooling provision for a commercial kitchen of an automobile manufacturer

Medium- (MT) and low-temperature (LT) cooling are required for the storage and processing rooms. Cooling points are required in the kitchen, in workplaces, and in the buffet and counter areas.

For medium-temperature cooling, a two-circuit system consisting of a water-cooled, flooded ammonia system in combination with an ethylene glycol circuit was installed, see Figures 4.40 and 4.41. The freezer rooms are cooled using a subcritical CO_2 refrigeration system with condensation against the ethylene brine circuit.

For operational safety reasons, the ammonia system is designed with two compressors in parallel. The storage temperatures must observe legal requirements for food products and have to be continuously recorded.

The operator's planning requirement was to use only natural refrigerants, to ensure that the system was highly economical and to create a user-friendly and operator-friendly system concept. Particular attention was paid to ensuring food quality even with long storage times. The plant is located in the basement in a separate machine and technology room with appropriate mechanical ventilation.

System details

- total cooling capacity: 70 kW (19.9 TR)
- medium-temperature capacity: 50 kW (4.2 TR) (13 cold rooms, 18 cold points, four quick coolers)
- operating temperatures, at t_{evap} / t_{cond}: –7 °C / 38 °C (19 °F / 100 °F)
- COP for chiller plant: 3,6
- low-temperature cooling capacity: 16,3 kW (4.6 TR) (two freezer rooms, three cold spots)

- operating temperatures at t_{evap} / t_{cond}: –28 °C / –4 °C (–18 °F / –25 °F)
- COP for freezer plant: 4,9
- refrigerant charge quantity: 22 kg (48.5 lbs) Ammonia, 13 kg (28.7 lbs) CO_2
- flooded ammonia circuit with two piston compressors in parallel
- water-cooled NH_3 condenser, cooling water temperature rise from 30 °C to 35 °C (86 °F to 95 °F)
- ethylene glycol / water at –5 °C / 0 °C (23 °F / 32 °F) with storage for cold stores
- non-electric defrost air circulation for all cold rooms
- subcritical CO_2 circuit with two piston compressors in parallel for frozen goods applications
- control and visualisation by means of programmable logic controller
- remote control via internet

Figure 4.41 Water-cooled NH_3 plant (© Kälte Eckert)

4.10.3 Liquid chiller for air conditioning office and server rooms

The requirement of the client and operator was for a water-cooled chiller to re-cool a chilled water network used for air conditioning in a bank building. It was required to use a chiller with natural refrigerants, adapted to the local conditions. Emphasis was placed on very high operational reliability. In particular, because of the need to cool a small data centre, the refrigeration technology must not fail. The new chiller was set up to replace a defective chiller. It was complicated, as the new chiller had to be brought into the existing building in separate parts to then be installed on-site. The chiller was installed on the second basement level in a plant room with two similar machines. The water connections of the new chiller had to be adapted to the existing piping and

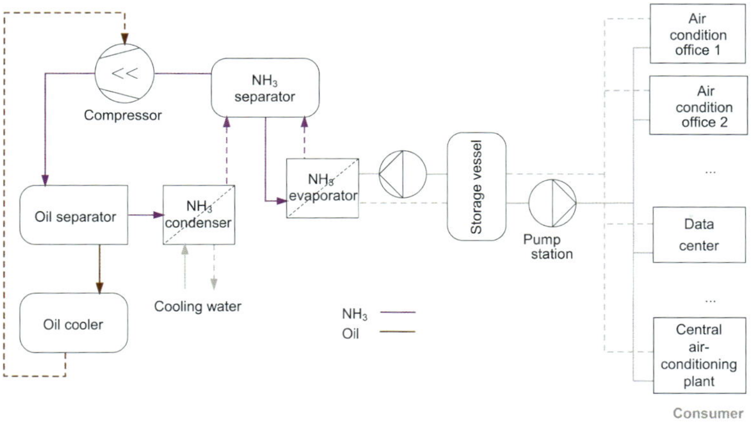

Figure 4.42 Schematic representation of the cooling supply, air conditioning and server room

Figure 4.43 NH_3 liquid chiller plant (© Kälte Eckert)

valves. A master controller was adapted to the plant room installation requirements. The three chillers can now be operated in an economically viable manner. See Figures 4.42 and 4.43.

System details

- total cooling capacity: 650 kW (184.8 TR)
- operating temperatures: 6 °C / 35 °C (43 °F / 95 °F)
- COP: 5,9
- refrigerant charge quantity: 36 kg (79.4 lbs) ammonia

- flooded ammonia using screw compressor
- frequency controlled drive motor
- water-cooled NH_3 condenser, cooling water at 30 °C / 35 °C (86 °F / 95 °F)
- air conditioning water circuit at 8 °C / 14 °C (46 °F / 57 °F) with storage vessel
- control and visualisation with programmable logic controller
- interconnection with occupier's building management system (BMS)
- remote control via internet

4.10.4 Cooling supply for a company restaurant

The planner specified that a new refrigeration system with natural refrigerants should replace the existing refrigeration system that was operated with synthetic refrigerants. Efficiency, high-operational reliability and the quietest possible operation were the operator's specifications.

A two-circuit system consisting of a flooded reciprocating piston ammonia compound system in combination with an ethylene glycol secondary refrigerant circuit was newly installed, see Figures 4.44 and 4.45.

It was a particular challenge to install the new brine pipes required for the brine circuit in the existing building and to place the new heat exchangers in the existing cold rooms while the kitchen was in use. To do this, the new chiller first had to be set up and put into operation. During the entire construction period, both the new and the old chiller had to be kept in operation. The new chiller was given a new location in the building on the first basement level of the publicly accessible underground car park.

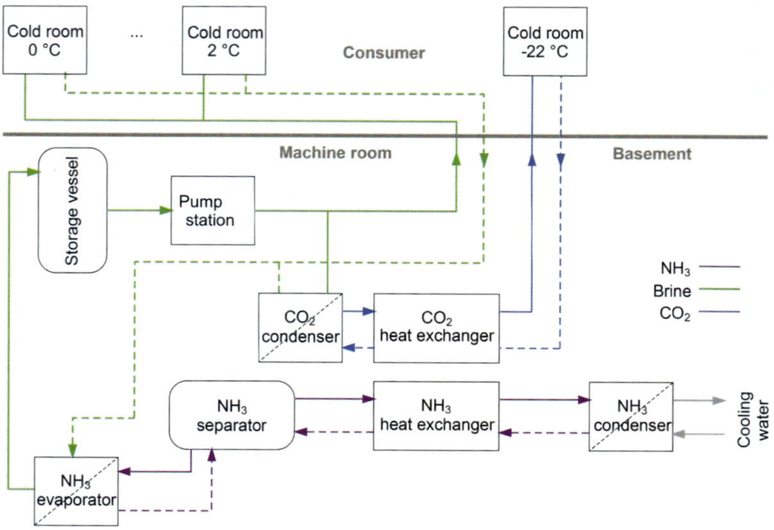

Figure 4.44
Schematic representation of the cooling supply, company restaurant

Figure 4.45 Engine room for CO_2/secondary refrigerant cooling supply of a company restaurant (© Kälte Eckert)

To increase the economic efficiency of the system, an additional heat exchanger was installed in front of the condenser for heat recovery. The complete waste heat from the refrigeration system is used in winter to heat the building using the concrete core activation concept.

System details

- system cooling capacity: 22 kW (6.3 TR)
- cooling capacity (MT): 22 kW (6.3 TR) (7 cold rooms, 13 cold spots, 1 quick-cooler)
- heat recovery: 15 kW (4.7 TR)
- operating conditions t_{evap}/t_{cond}: –6 °C / 37 °C (22 °F / 99 °F)
- COP of the refrigeration system: 3,5
- refrigerant charge quantity: 11 kg (24.3 lbs) of ammonia
- flooded ammonia system with two reciprocating compressors in a parallel operation
- water-cooled NH_3 condenser, cooling water 30 °C / 35 °C (86 °F / 95 °F)
- heat recovery for the use of waste heat
- brine circuit, ethylene glycol –4 °C / 0 °C (25 °F / 32 °F) with buffer storage for the cold rooms
- circulating air defrosting of the cold rooms without additional electrical energy

- control and visualisation via programmable logic controller (PLC)
- remote monitoring via the internet

4.11 Literature

[4-1] Pearson, A.B.: *Ammonia as Refrigerant*, International Institute of Refrigeration – IIR. Paris, France: 2008

[4-2] Harris, G.F.P., MacDermott, P.E.: *Flammability and Explosibility of Ammonia. I. Chem. E. Symposium Series No. 4*, 1964

[4-3] Wood, S., Cowie, A.: *A Review of Greenhouse Gas Emission Factors for Fertiliser Production*. IEA Bioenergy Task 38, June 2004

[4-4] Baral, A., Minjares, R., Urban, R.A.: *Upstream climate impacts from production of R-134a and R-1234yf refrigerants used in mobile air conditioning systems*. International Council on Clean Transportation, 2013

[4-5] IKET GmbH (Ed.): Pohlmann – T*aschenbuch der Kältetechnik*, 22. A., Berlin, Offenbach: VDE VERLAG, 2018

[4-6] Stephan, P., Kabelac, S., Kind, M., Mewes, D., Schaber, K., Wetzel, Th. (Ed.): *VDI-Wärmeatlas*, 12. A., Berlin, Heidelberg, Wiesbaden: Springer-Verlag, 2019

[4-7] Boast, M.: *Newnes Refrigeration Pocket Book*, Butterworth Heinemann – London UK, 1991

[4-8] Comstock, W.S.: *ASHRAE Pocket Guide for Air Conditioning, Heating, Ventilation, Refrigeration (SI Edition)* ASHRAE – Atlanta, Georgia, US, 2005

[4-9] Nelson, B. I.: *The Aluminum Advantage – Comparing Aluminum vs. Galvanized Steel Ammonia Evaporators*. Colmac Coil Manufacturing Inc. Technical Bulletin, 2016

[4-10] Committee on Toxicology, National Research Council: *Acute Exposure Guideline Levels for Selected Airborne Chemicals*, Volume 6, Committee on Acute Exposure Guideline Levels, 2007

[4-11] IFA Institut für Arbeitsschutz der Deutschen gesetzlichen Unfallversicherung: *GESTIS International Limit Values*. https://limitvalue.ifa.dguv.de

[4-12] Cotter, D.J.; Missenden, J.F.; Maidment, G.G.: *An experimental investigation of two phase ammonia in tubes and bends*. IIR Gustav Lorentzen Conference for natural refrigerants, Copenhagen, 2008

Additional

Caplin, M.: *Ammonia-gas poisoning – Forty-seven cases in a London shelter*. Lancet 241 (July 26, 1941), pp. 95-96

Alder, G.: *Heat, Energy and Refrigeration*. – G C Alder, South Africa, 2004

http://www.eurammon.com/natural-refrigerants/fact-sheets/ammonia-natural-refrigerant

http://www.iifiir.org

5 Carbon dioxide as refrigerant

Armin Hafner

This chapter is a revised edition of the "KOMPENDIUM CO_2 (R744) som kuldemedium", published in 2016 by the Norwegian Society of Refrigeration NKF. The Norwegian edition was prepared under direction of Hans. T. Haukås.

5.1 History of the refrigerant CO_2

In 1850, Alexander C. Twining first described carbon dioxide (CO_2, R744) as a potential refrigerant in a patent application. Until the 1940s, CO_2, together with ammonia (NH_3, R717), were almost the only refrigerants for industrial refrigeration systems and marine refrigeration systems. Ammonia was, and is, a very energy-efficient refrigerant, but it is also toxic. CO_2 on the other hand, is not toxic in low concentrations and was used previously in refrigeration units, when ammonia leaks were a potential danger for people, for example on ships. The reasons listed here ultimately led to the remaining first-generation CO_2 refrigeration systems being decommissioned in the 1970s:

- The former CO_2 systems usually had leakage problems, since initially only leather was used as the sealing material on valves, joints, etc.
- The developments regarding the crankcase pressure ratings and higher rotational speeds (50 Hz) were only applicable to compressors for low-pressure working fluids.
- It was challenging to use water-cooled systems in the United States, because water was scarce in many places.
- American safety standards and pressure vessel regulations further reduced the spread and maintenance of CO_2 refrigeration systems as the high working pressures were considered problematic.
- Likewise, the aggressive marketing by chemical companies in the 1930s–1950s for their newly developed, synthetically manufactured CFC and HCFC refrigerants (e.g., R12 and R22) sealed the fate of CO_2 as a refrigerant.

In the 1980s, CO_2 was rediscovered as a refrigerant when Professor Gustav Lorentzen, with other Norwegian Institute of Technology (NTNU) and SINTEF researchers, developed car air conditioning systems and CO_2 heat pumps for water heating [5-1]. At that time, particular attention was paid to the architecture of the new transcritical refrigeration systems and their control. The main motivation for reintroduction was the very high ozone depletion potential of CFC / HCFC refrigerants (Montreal Protocol, 1987) and the high greenhouse effect of HFC refrigerants (Kyoto Protocol, 1997; EU F-Gas Regulation, 2006/2014). At the same time, a safe and environmentally neutral refrigerant was rediscovered that is neither toxic, like ammonia, nor flammable, like the hydrocarbons.

Over the past two decades, CO_2 refrigeration technology was the subject of considerable research efforts. The result has been a defining of the fundamental knowledge about the use of CO_2 as a refrigerant for various refrigeration and heat pump systems, as well as for air conditioning applications. In parallel to this research work, which focused on the thermodynamic and fluid dynamic properties of the working fluid, well-known manufacturers developed the first components for various cooling and heat pump applications. This enabled the first CO_2 pilot installations to be set up and presented to the market around the year 2000. The first CO_2 refrigeration systems were on a par with conventional refrigeration systems in terms of energy efficiency, and in many cases system reliability was also equivalent. With this knowledge and the adapted components, new refrigeration installation concepts could be developed. These take into account the physical and thermodynamic advantages and disadvantages of CO_2. Today, CO_2 installations represent a global, energy- and cost-effective alternative for a variety of applications.

5.2 CO_2 effects on health, system reliability, and the environment

Health

In the event of a leak, CO_2 displaces air. Unlike with conventional HFCs and HCFCs, CO_2 leakage does not cause a direct lack of oxygen. However, the escape of CO_2 and concentration in the immediate vicinity of people is a challenge.

CO_2 is produced during metabolism in human cells, which is released into the environment via the lungs. If the air inhaled has a higher than normal concentration of CO_2 than in the environment (about 400 ppm), the CO_2 content in the blood will increase, and the pH of the blood will decrease. This affects, among other things, respiratory rate and brain function. Even at a low concentration, and even if there is sufficient oxygen in the air, this can be fatal.

There are many descriptions in the literature of the effects of inhaling air containing CO_2. The effect is dependent on the concentration, as shown in Table 5.1.

CO_2 is not toxic in low concentrations[1]; it is approved for usage in food processing and does not need to be declared. According to EN 16798-3, a good indoor climate is defined as when the CO_2 indoor air concentration stays below 800 ppm. The maximum occupational exposure limit (8 h) is 5000 ppm (0,5 % by volume). As a peak limit, a concentration of 10000 ppm (instantaneous value) is permitted three times per shift for the duration of 60 minutes each. From 50000 ppm, it becomes deadly if people cannot leave the area within 30 minutes. Therefore, the limit for Immediate Danger for Life and Health (IDLH) is at 5 % (50000 ppm).

In refrigeration standards, as published by ASHRAE[2], CO_2 is assigned to class A1 and classified as non-toxic and non-flammable. Therefore, it may be used everywhere, in appropriately designed systems, in accordance with regulations (e.g., EN 378). Due to the A1 classification, it is particularly suitable for direct evaporation systems that are located in public areas (supermarkets, transport etc.) or in households (hot water heat pumps) and for medium- or low-temperature systems that contain particularly large quantities of perishable foods (cheese, fish, meat etc.).

[1] Concentrations of CO_2 in the air we breathe above 10 % are lethal (see Table 5.1).
[2] ASHRAE classification of refrigerants: www.ashrae.org

Table 5.1 Health effects of various CO_2 concentrations

	Health conditions
417	Average content in atmosphere 2020
< 800	EN 16798-3: good indoor air quality
5000 (0,5 vol%)	Exposure limit in workspace 8-hour limit – weighted average
10000	Short-term exposure limit 60 min, max 3 times per 8h shift
20000	50 % increase in respiratory frequency! may affect breathing capacity, cause irritation and disturbance of central nervous system
30000	100 % increase in respiratory frequency after short-term exposure
50000	Immediate danger to life and health Leave area within 30 min to avoid irreparable damage of health
100000	Lowest lethal concentration, minutes before loss of consciousness
200000	Fatalities have been reported
300000	Rapid unconsciousness, cramps, death

A person exposed to high concentrations of CO_2 will usually recover quickly when they again are exposed to fresh air. Nevertheless, it is preferable, and in severe cases imperative, to consult a doctor. Poisoning by CO_2 does not normally have long-term effects.

CO_2 is practically odourless; however, it carries a pungent smell at very high concentrations. When designing CO_2 systems, the need for a gas alarm unit must be assessed using the same designs as for HCFC / HFC systems, paying attention to the following:

- It must be determined whether the gas concentration can exceed a defined limit (practical limit, see below) in case the entire refrigerant charge escapes into the room. If this is not the case, the CO_2 alarm is not required.
- If it is possible that the concentration limit can be exceeded, CO_2 detectors or an alarm system must be installed in rooms that are public rooms, including medium- and low-temperature storage rooms. If the CO_2 concentration is at life-threatening levels, the gas detector and alarm must also be installed in rooms that are not considered to be public rooms.
- In the event that rooms are well ventilated, this should be considered when assessing the need for an alarm.

In the European standards (EN 378), the practical limit for CO_2 of 10 g per m^3 (0.0099 oz/ft^3) room volume is defined. This corresponds to approximately 5,4 % by volume that equals 54000 ppm at 1 bar (14.5 psi) atmospheric pressure and 20 °C (68 °F).

The set point for the gas detector must be selected with a sufficiently large safety margin in relation to the standard. It is recommended to reduce the maximum set point to 50 % of the practical limit. Lower set points can be used to detect and reduce refrigerant loss if necessary. However, the set points should always be set high enough to avoid false alarms. The recommendation here is a low alarm level ("Warning") at 5000 ppm = 0,5 vol% and a high level ("Alarm") at 10000 ppm = 1 vol %. If the alarm system allows only one level, 5000 ppm is selected.

CO_2 is heavier than air, so the detectors should be installed at a low level in rooms and protected against possible damage.

Gas detectors used in conjunction with CO_2 installations measure concentration directly. An oxygen detector that will sound an alarm due to a lack of oxygen can have an alarm limit of, for example, 18 % oxygen. The CO_2 concentration at this oxygen level (about 15 %) can be fatal.

Because CO_2 as a refrigerant is environmentally neutral, there are no regulations prohibiting the venting of CO_2 directly from the system into the environment. However, the refrigerant should only be vented outdoors or to the outdoor area. When venting larger quantities, check where and how far the gas can get from the discharge point. It must be ensured that it cannot run off into the basement or accumulate in low-level areas in the surroundings. The venting must be carried out in such a way that no dry ice forms inside the system and no dry ice can block the blow-out pipe or the drain hose, see Chapter 5.9.2.

After a CO_2 discharge, the room must be well ventilated. In addition, the concentration must be measured before granting open access to the room. The CO_2 content must be below 5000 ppm (long-term exposure limit) and preferably below 1000 ppm.

Environmental impact

Carbon dioxide is an environmentally friendly natural refrigerant. It has no negative effects on the ozone layer.[3)] The proportion of CO_2 used as a refrigerant (R744) has an insignificant impact on global warming in relation to the CO_2 quantity emitted globally. CO_2 is used as a reference fluid to indicate the relative greenhouse potential of alternative substances ($GWP_{CO2} = 1$).[4)] R744 is largely obtained as a side product from industrial processes, which is why it can be classified as environmentally neutral. The CO_2 emissions to provide one kilogram of R744 for use in refrigeration are in the range of 1 kg (2.2 lb) of CO_2.

CO_2 is not flammable and is often used as a fire extinguishing agent. At a concentration of 10 % by volume, it extinguishes fire.

Installation safety

Working pressures of over 100 bar (1450 psi) can occur on the high-pressure side during operation of the refrigeration systems [5-2]. This requires that the applicable safety rules and operating standards be observed when designing the installation, selecting the components and operating the system. By overcharging a system, high-pressure levels inside the system can be reached even at a standstill if the unit is exposed to elevated ambient temperatures, which trigger the safety valves.

Dry ice forms when a safety valve is opened. Dry ice has a surface temperature of –78,9 °C (–110 °F) (sublimation point at 0,981 bar / 14.2 psi). It occurs when liquid CO_2 is exposed to a pressure of less than 5,18 bar / 75.1 psi (triple point at –56,6 °C / –69.9 °F). Dry ice can block valves and pipes and therefore requires corresponding experience and expertise in charging and maintaining the CO_2 installations. Dry ice is very cold under atmospheric pressure, and there is a risk of brittle fractures forming in the steel pipes and components. Frostbite can result from skin contact with dry ice [5-3].

3) ODP – ozone depletion potential, $ODP_{CO2} = 0$.

4) GWP – global warming potential.

5.3 R744 refrigeration system characteristics and their effect on design and operation

Note: All pressure data refer to the absolute pressure in bar, unless otherwise stated. All pressure-enthalpy diagrams refer to logarithmic pressure scale.

5.3.1 Introduction and definitions

CO_2 can be distinguished from other fluids in many areas. This is of great importance for design, operation and maintenance. The main differences are:

- low critical temperature (31,1 °C / 87.9 °F)
- triple point pressure is above atmospheric pressure (5,18 bar / 75.1 psi)
- generally higher pressures than other refrigerants at similar temperatures
- particularly favourable heat transfer properties
- high thermal expansion coefficient in the liquid phase

The pressure-enthalpy diagram is often used as the basis for process analysis in the fields of refrigeration technology. The pressure-enthalpy diagram in Figure 5.1 shows the physical states of CO_2. These are dependent on pressure and temperature.

CO_2 phase diagrams for various pressure ranges are available.

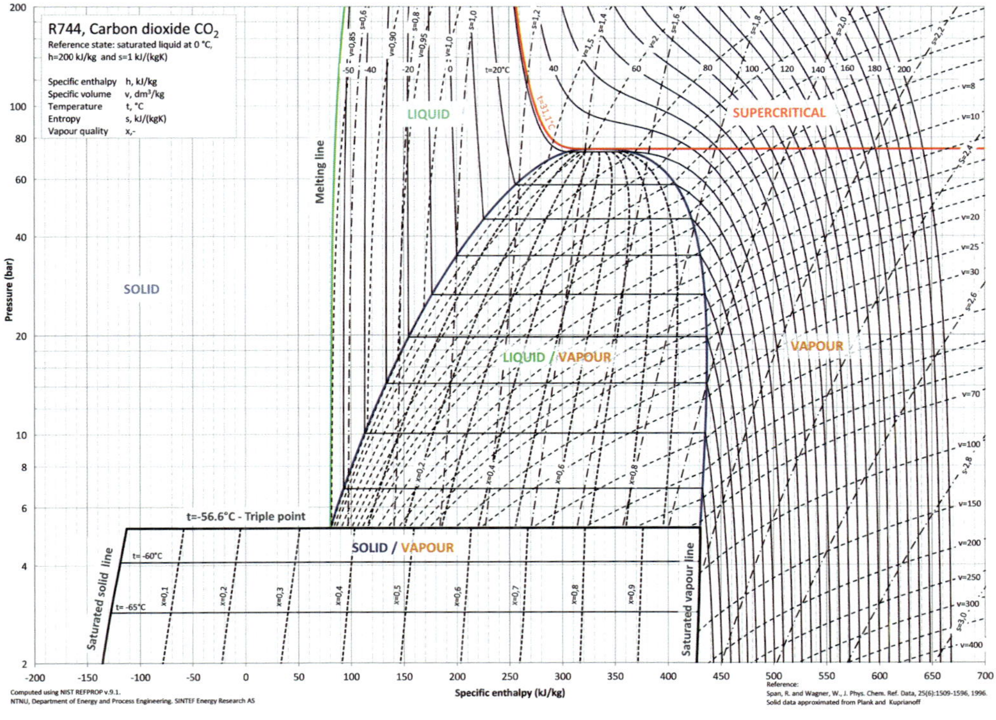

Figure 5.1 Logarithmic pressure-enthalpy diagram for CO_2 incl. phase regions

The boundary lines for saturated vapour and liquid meet at the critical point. The critical point is a combination of pressure (critical pressure p_{cond}) and temperature (critical temperature t_{cond}), where the liquid and vapour phases meet. CO_2 has a critical pressure of 73,8 bar (1070.4 psi) and a critical temperature of 31,1 °C (87.98 °F).

At subcritical pressure, the liquid and vapour phases, and possibly also the solid phase, occur either simultaneously or separately. There is only one phase in the supercritical area.

If the pressure is above the critical pressure and at the same time, the temperature is above the critical temperature, the CO_2 is in the *supercritical* area. In Figure 5.1 the supercritical area shows as the marked area to the right of the 31.1 °C isotherm and above the critical pressure of 73,8 bar (1070.4 psi). In the supercritical area, CO_2 is a gas with a very high density. It behaves more and more like a liquid when the fluid temperature drops.

The *triple point* of a substance is the only possible state in which all three states – solid, liquid and vapour – are in equilibrium. At lower pressure levels than the triple point, solid and vapour are again in equilibrium; above the triple point, there is liquid and vapour. The triple point of CO_2 is at a pressure of 5,18 bar (75.1 psi) and a temperature of –56,6 °C (–70 °F). Solid CO_2 is also called "dry ice" because it sublimates, it changes directly from solid to gas, at atmospheric pressure.

In the pressure-enthalpy diagram the triple point appears as a line that describes a different distribution of the three phases (= –56,6 °C/–69.9 °F) line at 5,18 bar (75.1 psi) in the diagram). The triple point with its potential solid content usually plays a subordinate role in refrigeration technology since this area below the triple point is often avoided.

The two-phase area of vapour and solid is below the triple point line. As in the two-phase area (liquid/vapour), the vapour fraction is determined or specified as an "x" value (x = vapour fraction). The saturation temperature of the solid CO_2 depends on the corresponding saturation pressure and the sublimation line (bottom left in the diagram). At atmospheric pressure, the saturation temperature for solid CO_2 is at –78,5 °C (–109 °F), which corresponds to the sublimation temperature of dry ice.

5.3.2 Importance of the relatively low critical temperature

5.3.2.1 Transcritical process

Unlike conventional refrigerants, the critical temperature is rather low at 31,1 °C (87.9 °F), which is a common temperature for condensing the refrigerant when the heat is rejected to air or water. This difference is essential for the design and the operation of R744 systems that transfer heat in this temperature range (or higher) on the high-pressure side.

In the supercritical area, there is no longer any condensation when the heat is rejected. Therefore, refrigeration systems and heat pumps with CO_2 in the conventional (subcritical) process cannot have higher condensing temperatures than ~30 °C (86 °F). In practice, an even lower condensing temperature of around 27 °C (81 °F) has actually to be considered as the upper limit for the conventional subcritical vapour compression cycle with CO_2.

However, the refrigeration or heat pump process can also be maintained if the heat sink temperature rises above 31 °C (88 °F). Then the heat is rejected on the high-pressure side of the system in the supercritical area. A transcritical cycle occurs, as shown in the pressure-enthalpy

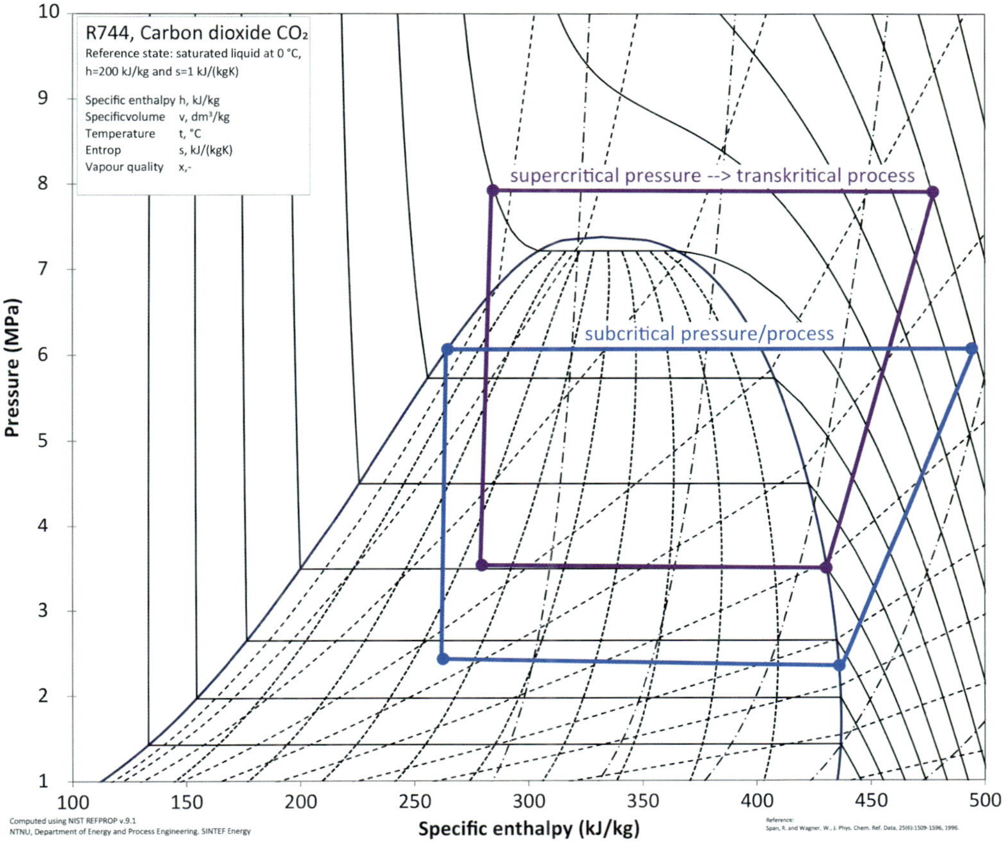

Figure 5.2 Subcritical and transcritical CO_2 processes within the pressure-enthalpy diagram

diagram in Figure 5.2. Instead of condensation, not only is the usual superheated part rejected in the so-called gas cooler, but the entire heat load is transferred at gliding temperatures on the refrigerant side. The heat absorption in the evaporator takes place as with conventional systems in the two-phase area. In efficient and well-designed systems, however, superheating in the evaporator is deliberately avoided since the ratio $\Delta t_{evap}/\Delta p_{evap}$ is close to 1 K/bar (1 K/14.5 psi); i.e., in order to achieve a 1 K difference in the saturation temperature in the evaporator, the saturation pressure must be lowered by 1 bar (14.5 psi). The transcritical process is described in more detail in Chapters 5.4 to 5.6.

If the external conditions allow a conventional subcritical process, this usually is the most energy-efficient variant. However, if a very high temperature lift is desired on the heat sink fluid side, the gliding temperatures in the gas cooler of the transcritical process are advantageous. These make it possible to use the heat more effectively than in a conventional condenser. It is thus possible to heat water (or other fluids) in a gas cooler from, for example, 10 °C (50 °F) to relatively high temperatures (of max. 90 °C / 194 °F). This makes heat recovery with R744 refrigeration systems both very rewarding and flexible with regard to the application of waste heat utilisation.

If the temperature level of the heat sink medium varies greatly, as can be the case, for example, with outside air, the lowest energy consumption of the refrigeration system is achieved by actively controlling the high-pressure level according to the heat sink temperature, see Chapter 5.4.5. The refrigeration system then alternates between transcritical and subcritical operation. In the case of active use of waste heat, i.e. most waste heat is used for heating purposes, the R744 system could operate permanently in transcritical mode.

5.3.2.2 Lower effective coefficient of performance

A simple R744 refrigerant circuit with an air-cooled condenser works – in moderate climates – on the high-pressure side close to the critical pressure. This causes larger losses during the throttling through the expansion valve. (A lot of liquid evaporates when throttling; see Figure 5.2.) The same challenge applies to the transcritical process at elevated heat sink temperatures.

These additional internal losses mean that the theoretical coefficient of performance (COP) of the reference process for R744 refrigeration systems is lower than for conventional refrigerants, as shown in Figure 5.3.

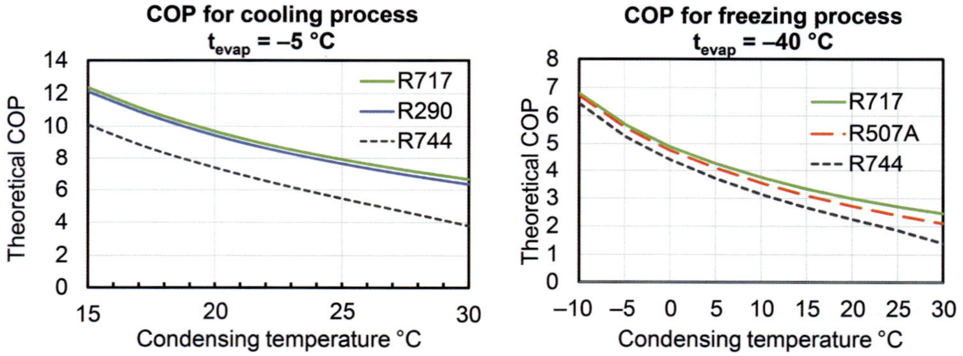

Figure 5.3 Comparative theoretical performance COP for medium temperature –5 °C/–23 °F (left) and low temperature –40 °C/–40 °F (right)

Figure 5.3 shows the theoretical performance figures for a medium- and low-temperature process with R744, R507 or R290, and R717 at an evaporation temperature of –40 °C (–40 °F) or –5 °C (23 °F) with increasing condenser temperatures. (In practice, these graphs and comparisons look different for several reasons, as will be explained in the following chapters.)

5.3.2.3 Other conditions

A low critical temperature can also be of practical importance in other contexts. Some of the following examples are used to determine the pressure in a system, part of a system or a component exposed to high ambient temperatures (above the critical temperature). They also show how to calculate the amount of refrigerant discharged if the safety relief valve opens.

The specific heat capacity of CO_2 changes in the supercritical range. Near the critical point, it varies very strongly with temperature changes. For illustrative purposes, consider the distance between the isotherms in this area in the pressure-enthalpy diagram in Figure 5.1. This relationship is important for the design and operation of heat exchangers for heat rejection in this temperature range; see also section 5.4.

5.3.3 Significance of high pressure at the triple point

As mentioned, the pressure at the triple point of CO_2 is at 5,17 bar (74.9 psi), i.e., clearly above the ambient pressure. This means that CO_2 – for example during maintenance work – can change into its solid form (dry ice).

Dry ice is mainly created by throttling liquid CO_2 under certain conditions. However, throttling can start from the vapour phase or from the supercritical area; see also Figure 5.4.

The change of state in the pressure is shown in the pressure-enthalpy diagram by a vertical line (= constant enthalpy) from the initial state. When the line reaches the two-phase area of solid and vapour, part of the medium is converted to dry ice. The x value represents the mass fraction of CO_2 vapour. For example, 32 % of the refrigerant becomes dry ice when the saturated liquid is expanded from 20 °C (68 °F) to a value below the triple point; see change of state "a" in Figure 5.4.

The pressure-drop for the formation of dry ice from the supercritical area and from the (saturated) vapour phase can also be found in the diagram. It is described by the state changes "b" and "c". An expansion of saturated vapour at 60 bar (870 psi) produces approximately 5 % dry ice; see line "b". The same amount of dry ice is created when expanding from the supercritical state of 80 bar (1160 psi) and 40 °C (104 °F) or 120 bar (1740 psi) and 63 °C (145 °F). The theoretical limit for the formation of dry ice is described by the line "c". It assumes saturated vapour at approximately 35 bar (507.6 psi) (approximately 0 °C / 32 °F). At 80 or 120 bar (1160.3 or 1740.5 psi), dry ice is no longer formed as soon as the corresponding CO_2 temperatures exceed 48 °C / 118 °F (80 bar / 1160 psi) or 73 °C / 163 °F (120 bar / 1740.5 psi).

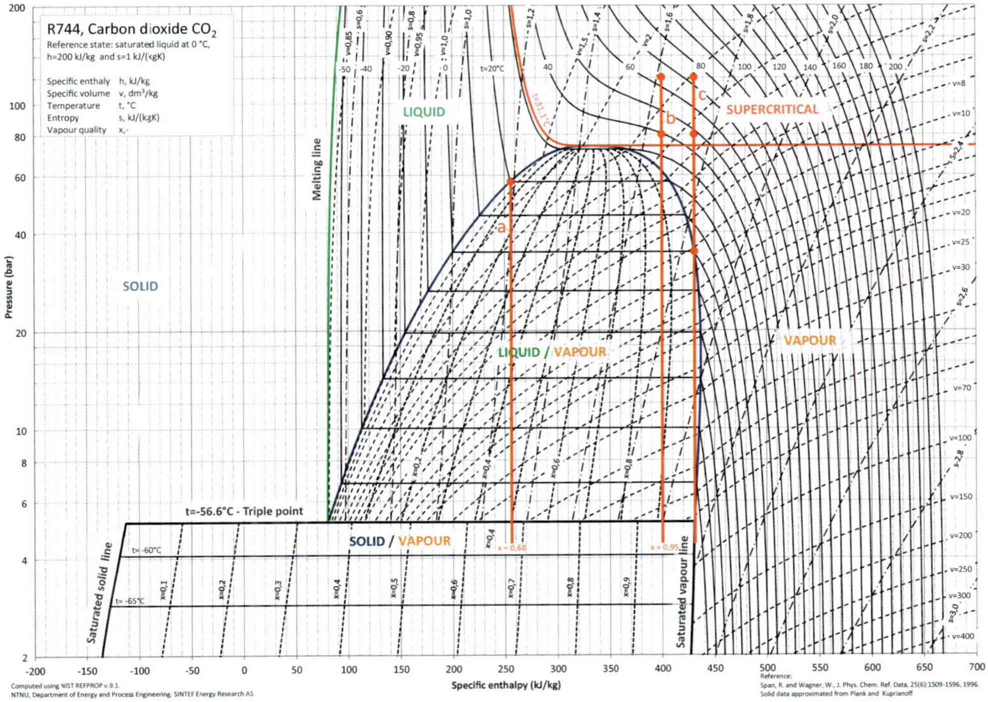

Figure 5.4 Examples for changes of state which produce dry ice

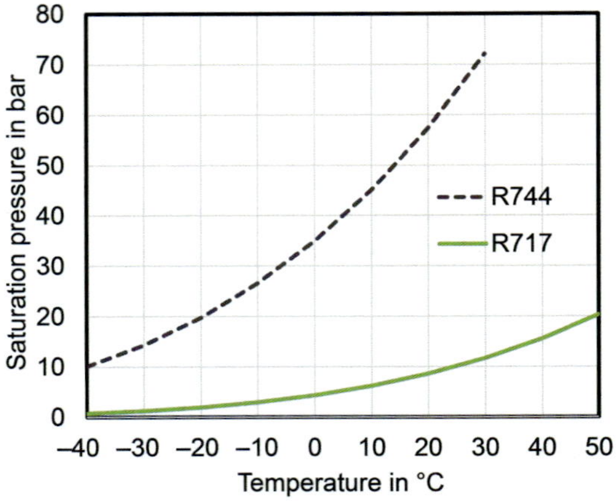

Figure 5.5 Saturation pressures for R717 and R744

Dry ice can form by throttling due to charging or emptying of installations and installation vessels as well as by the opening of safety relief valves etc. Icing creates a risk of clogging inside the pipelines. For this reason, special care must be taken when designing the drainage relief system of safety valves, and special procedures must be used during the maintenance of the systems; see Chapter 5.9.2.

The formation of dry ice limits the theoretically achievable evaporation temperature in a normal evaporator to –56.6 °C (–69.9 °F). In practice, the limit is somewhat higher, between –54 °C (–65 °F) and –55 °C (–67 °F) [5-4]. However, special evaporators (or sublimers) are presently being developed, in which heat is absorbed by subliming dry ice. In such "evaporators", surface temperatures of –78 °C (–108 °F) are possible.

5.3.4 Significance of a generally high-pressure level

5.3.4.1 Typical pressure levels during operation and at standstill

The pressure in a R744 refrigeration system is significantly higher than in systems with other refrigerants. Figure 5.5 shows the saturation pressure for CO_2 and ammonia as a function of temperature.

When operating a low-temperature system, the evaporating pressure t_{evap} = –35 °C (–31 °F) will be approximately 12 bar (174 psi) and, for a cooling system, at t_{evap} = –5 °C (23 °F) and 30 bar (435 psi). At t_{cond} = 27 °C (80 °F) (≈ highest condensing temperature in subcritical operation) the pressure in the condenser will rise to approximately 67 bar (972 psi). The high-side pressure of a system that has been designed for transcritical operation can be operated with pressures up to 100 bar (1450 psi). At extreme heat sink temperatures, the high-side pressure can be raised further if the safety chain[5)] is calculated and installed adequately for protection on the high-pressure side.

The stand-still pressure level inside an enclosed component, in a part of the system, or in the entire system depends on the one hand on the temperature and on the other hand on the internal volume of the system and the amount of refrigerant in it, i.e. the specific volume or its reciprocal, the density. When heat is added, the system pressure follows the saturation curve until the two-phase region ends, where the refrigerant changes to the supercritical state. These pressure changes can be followed along the constant specific volume line in the pressure-enthalpy diagram; see Figure 5.6. When the temperatures increase, the lines go through the two-phase area and also through the vapour area. The corresponding system pressure can be read from the

[5)] Safety chain: maximum high pressure = maximum permissible work pressure of the components – tolerance (bandwidth) in the pressure switch – tolerance of the safety valve.

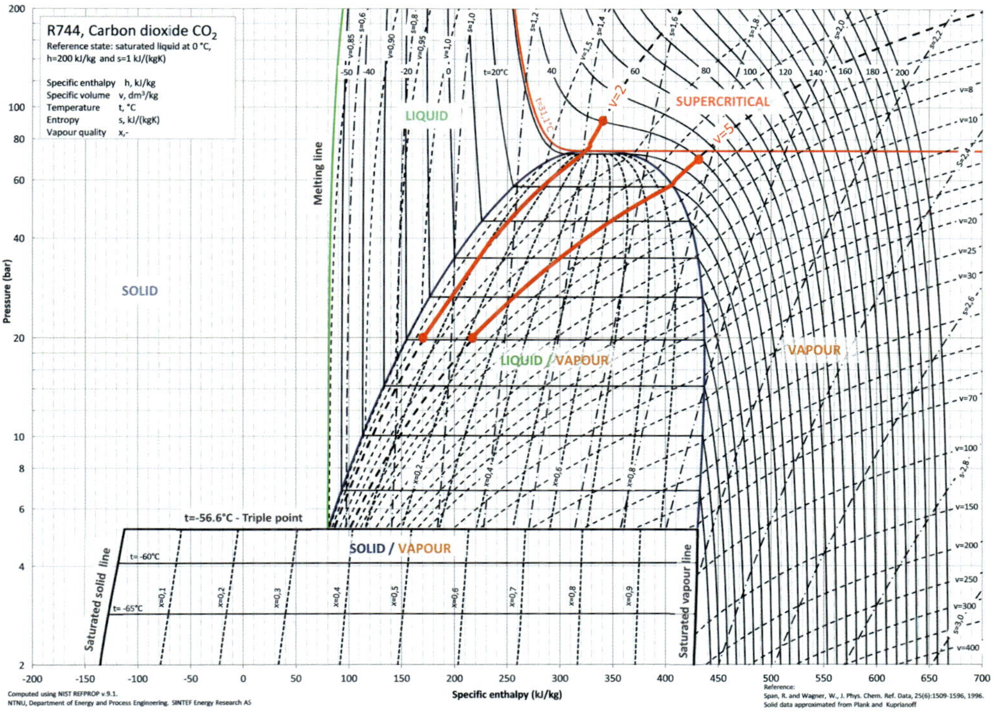

Figure 5.6 Typical pressure increases for systems with different internal volumes

point in the diagram at which the line of constant specific volume (v = constant) intersects the line of the ambient temperature around the system.

The diagram in Figure 5.6 shows the development of the stand still pressure of two refrigeration systems, each with a different refrigerant charge. Left: 0,5 kg (1.1 lb) of refrigerant per litre of system volume, v = 2 dm³/kg (v = 0.032 ft³/lb) and right: 0,2 kg/dm³, v = 5 dm³/kg (0.0032 lb/ft³, v = 0.08 ft³/lb). The initial temperature is –20 °C (–4 °F), at which both systems are at standstill. At an ambient temperature of 40 °C (104 °F), the final pressure in the two systems is approximately 90 bar / 1305 psi (left) or 70 bar / 1015 psi (right).

Due to the high density of CO_2, the system pressure rises relatively quickly when heat is added.[6] It is therefore necessary to calculate the pressure rise when the temperature rises in gas-filled components and system parts.

In the same refrigeration circuit, these high pressures have several advantages, which are explained below. One disadvantage, however, is that if there is a leak, the loss of refrigerant can be relatively high. Another challenge in such a circuit is the use of organic sealing materials (o-rings etc.). CO_2 diffuses into the o-rings at high pressure and can literally "blow" them apart if the pressure in the system is relieved too quickly. Special care must be taken when selecting such sealing materials.

[6] Pressure increase with rising temperature is proportional to the density of the gas (according to the gas laws).

5.3.4.2 Pressure classes

In the past, a pressure class of 25 bar (362 psi) gauge (26 bar / 377 psi absolute) was adequate for conventional refrigeration systems. With R744 systems, one has to design and check each individual CO_2 system very carefully.

Table 5.2 shows an overview of the common pressure classes for different applications within R744 systems. However, other pressure classes can also occur. For example, if CO_2 is used as a secondary fluid / coolant (see Chapter 5.5.1); the same pressure class can be used as on the corresponding low-pressure side (LP side).

Transcritical systems are usually designed in such a way that they can allow a stand-still pressure of 81 bar (1175 psi) even on the low-pressure side, without the safety valves triggering. Large systems with high thermal masses are often dimensioned to withstand stand-still pressure over a (short) period to prevent the refrigerant from being released. A pressure of 61 bar (885 psi) is often used here.

Table 5.2 Frequently used pressure classes for refrigeration systems and heat pumps with R744

	LT side,	MT side	HP side, subcritical	MP side, transcritical	HP side, transcritical	Hot gas defrost
bar	26/41/61/(81)	41/61/(81)	41	80–90	80–120	53
psi	377/595/885/(1175)	595/885/ (1175)	595	1160–1305	1160–1740	769

LT = low-temperature; MT = medium-temperature cooling (max. evaporating pressure), MP = middle pressure in transcritical process, HP = high pressure (pressure at heat injection)

For larger industrial refrigeration plants (cooling capacities >50 kW/14.2 TR), the dimensioning of the entire system for high pressure is often not economical. Specific system parts can be designed for pressures below the stand-still pressure of the main part. To limit the stand-still pressure, small reliquification systems (cold fingers) are used, which are continuously cooling the refrigerant in the storage devices and thus limiting the pressure level. This is also dealt with in detail in Chapter 5.10.1.

In the low-pressure stage of a refrigeration system, normally 26 bar / –10°C (377 psi / 14 °F) is sufficient as a dimensioning pressure class for the evaporator. However, this means that there is a small safety margin only. In the event of a system failure, the risk of losing the refrigerant charge is quite large. Increasing the maximum pressure level to 41 bar / 6.3 °C (595 psi / 43 °F) offers a significantly higher margin of safety, without any noteworthy extra costs. In larger industrial R744 refrigeration systems, the general dimensioning and protection of the low-temperature side at 41 bar (595 psi) is preferred, such as when CO_2 is applied in the lower stage of a cascade refrigeration system or as an evaporating coolant in an indirect system.

A limiting factor in some cases can be the lack of availability of industrial compressors with a sufficiently high standstill (idle) and suction pressure. If the compressor casing is not designed for such effective standstill pressures, the compressor must be separated from the system automatically. An external safety valve should also be provided for the compressor.

Hot gas defrosting is recommended when R744 refrigeration systems are used in low-temperature cooling rooms. The system parts should be designed for 61 bar / 22.7 °C (885 psi / 72.9 °F).

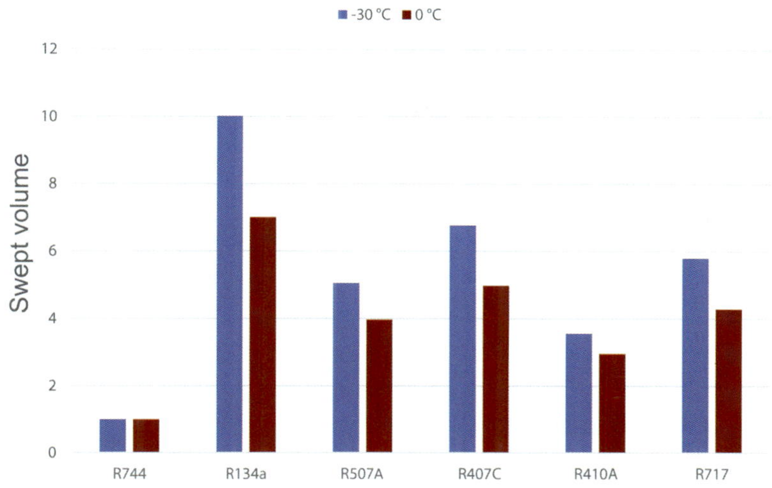

Figure 5.7 Relative required swept volume for various refrigerants compared to CO_2 (CO_2=1). Evaporation at −30 °C (−22 °F) and 0 °C (32 °F), respectively, and condensing at 20 °C (68 °F)

Experience has shown that there is enough time for maintenance work on the system before the safety relief valve will open. In addition, such systems should include a pressure vessel (receiver), which is designed for 80 bar (1160 psi) or more. This can serve as refrigerant storage during longer maintenance work.

5.3.4.3 Displacement volume and efficiency of the compressor

The high gas density causes the normally high operating pressure of the CO_2 in refrigeration systems. This results in a significantly reduced swept volume requirement for the compressor. The swept volume of an R744 refrigeration system is typically only 15–20 % of what a conventional refrigeration system, with a similar cooling capacity, would need. Figure 5.7 shows the comparison of theoretical required swept volumes (relative values; CO_2 =1) for an R744 refrigeration system and a number of other refrigerants. The evaporation temperatures are −30 °C (−22 °F) and 0 °C (32 °F); the condensation temperature was assumed to be 20 °C (68 °F) for each case.

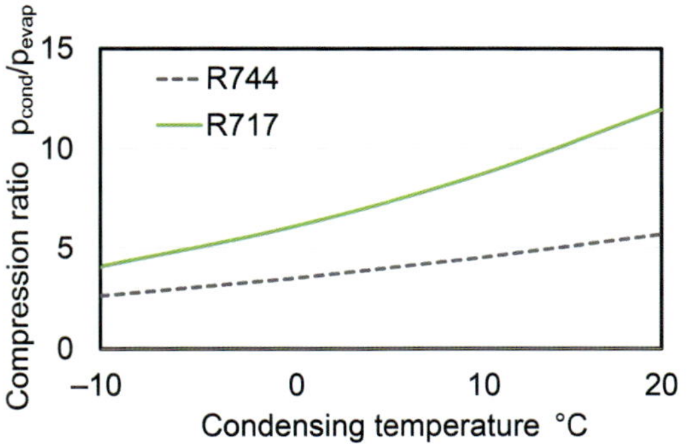

Figure 5.8 Compression ratio versus condensing temperature for low-temperature cooling installation with R744 and R717. Evaporation temperature is constant at −40 °C (−40 °F)

In addition, the generally high pressure level means that the pressure ratio in the R744 refrigerant circuit is much lower than in systems with other refrigerants. This contributes significantly to a higher efficiency of the compressor. Figure 5.8 shows the pressure ratio p_{cond}/p_{evap} for an R744 and an R717 low-temperature unit. Figure 5.9 shows the typical isentropic efficiency of industrial reciprocating compressors for both refrigerants as a function of the pressure ratio.

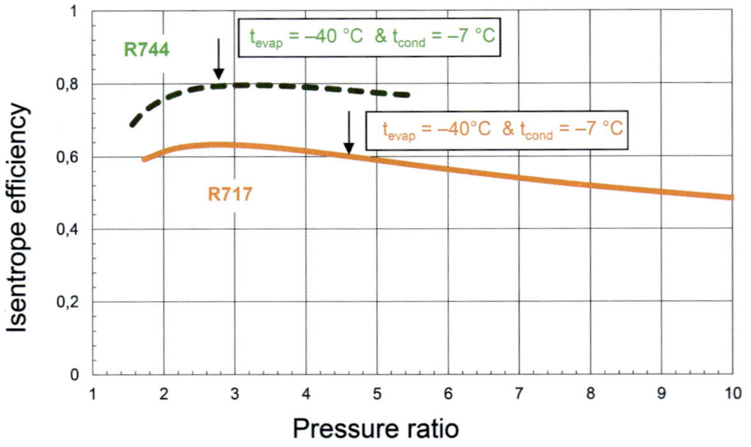

Figure 5.9
Typical isentropic efficiency versus pressure ratio for CO_2 (R744) and ammonia (R717)

The isentropic efficiency curves shown in Figure 5.9 were calculated using a commercial compressor selection program from the manufacturer. A condensing temperature of –7 °C (19.4 °F) at an evaporation temperature of –40 °C (–40 °F) was chosen as an example; the two arrows indicate the corresponding operating points. At these pressure conditions, the isentropic efficiency of the CO_2 compressor is approximately 30 % higher than that of an ammonia compressor. This reduces the disadvantage of the theoretical COP from Figure 5.3. It can therefore be shown that R744 low-temperature systems have a distinct energy advantage over R717 systems.

5.3.4.4 Pressure and temperature losses in the vapour and two-phase area

Pressure loss due to friction in single-phase and two-phase flows in refrigeration and heat pump systems reduces the suction pressure and increases the discharge pressure of the system's compressor. At the same time, it reduces the available temperature differences in the condenser and evaporator. This decreases the cooling capacity and the COP of the process.

The process parameters change (roughly) in proportion to the relative change in the suction pressure. Such pressure drops affect performance less at higher operating pressures. The performance loss due to the suction-side pressure losses with CO_2 therefore turn out to be lower than with other refrigerants.

In terms of energy efficiency, the process is directly affected by the change in saturation temperature due to the pressure drop (i.e. "temperature loss"). (The specific power requirement increases in proportion to the temperature difference between the high- and low-pressure side; see also Carnot process.)

The high pressure level causes the saturation pressure curve of CO_2 to be steeper than with conventional refrigerants; see Figure 5.5. A pressure drop therefore leads to a smaller temperature loss than with other media and at the same time this has less influence on the efficiency of the process. Figure 5.10 shows the ratio of temperature drop per pressure loss ($\Delta t/\Delta p$ in K/bar) for CO_2 and for ammonia.

This difference is particularly large at lower temperatures. Therefore, CO_2 is especially suitable for medium- and low-temperature applications. Strictly speaking, it comes down to the fact that R744 systems have higher pressure drops than with ammonia (absolute values) under the same

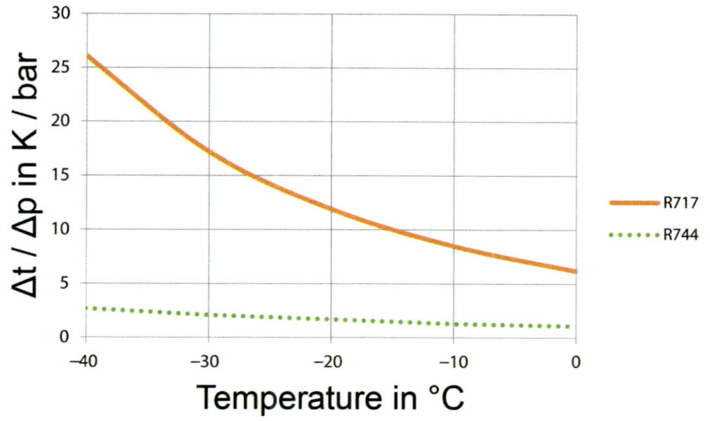

Figure 5.10
Temperature loss (Δt) resultant from pressure loss (Δp) for CO_2 and NH_3 at various saturation temperatures

conditions. However, the advantage shown in Figure 5.10 is still large enough to compensate for this.

Smaller diameter pipes for gas and two-phase flow in R744 systems can be selected, and at the same time the evaporator and the condenser or gas cooler can work more efficiently because of the lower sensitivity to pressure losses. Suction pipe diameters can sometimes be reduced by up to 50 %. For example, a suction gas line for a low-temperature application at –40 °C (–40 °F) that has to be dimensioned for ammonia with 150 mm (5.9") could for CO_2 be manufactured in 80 mm (3.15") pipe size. The corresponding difference also arises in commercial refrigeration applications when comparing CO_2 with R404A.

Evaporators and condensers / gas coolers can be dimensioned for large (mass) flow velocities, which enable more effective convective heat transport and better heat transfer. In addition, the number of parallel lines can be reduced, which is particularly advantageous in the evaporator since this can improve the liquid distribution. In this way, evaporators can also better withstand overloads without "choking", as is usually the case at the start of the freezing processes.

The advantage of the smaller changes in the saturation temperature due to the pressure drop is a major reason why the freezing temperature in plate freezers can be reduced from the usual –40 °C (–40 °F) to below –50 °C (–58 °F) when CO_2 is applied as the refrigerant. This results in an increase in freezing capacity of 25–40 %.

5.3.5 Importance of good heat transfer properties

Liquid CO_2 has a relatively good thermal conductivity, a low viscosity and a high specific heat capacity. These properties, in particular the good thermal conductivity, are favourable for convective heat transfer in heat exchangers. The low viscosity also minimises friction pressure loss.

The low Δt/Δp ratio of CO_2 means that gas bubbles will form and expand even within the liquid with very little additional superheating. The low molecular surface tension also contributes to this. The heat transfer is very effective due to the nucleate boiling, the dominant form of evaporation with CO_2, and is much more effective than with other refrigerants.

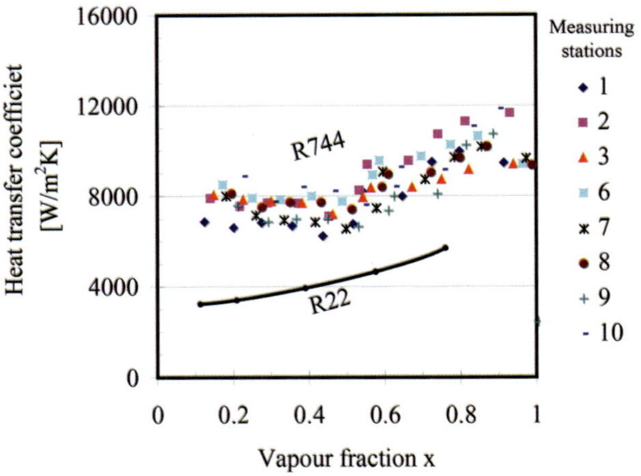

Figure 5.11
Measured heat transfer coefficient versus vapour fraction for oil-free CO_2 measured in a 7 mm round tube. t = −10 °C, M=200 kg/m²s, q=6000 W/m². Solid line is for R22 at approximately the same conditions from a different investigation [5-5]

Usually CO_2 evaporators are significantly more efficient than HFC evaporators; see Figure 5.11. The figure shows measured heat transfer coefficients in W/(m²·K) in heat exchanger tubes, depending on the vapour fraction. For comparison, the values for R22 are shown under almost the same conditions.

The heat transfer coefficient of CO_2 is almost twice that of R22. In addition, the temperature drop is lower due to the reduced pressure loss in the evaporator. Especially at low temperatures, this phenomenon can be just as important – or even more important – than efficient heat transfer. Correctly designed CO_2 evaporators can therefore be very efficient and compact.

5.3.6 Provision for the high coefficient of thermal expansion of the liquid

Liquids cannot (practically) be compressed. If the liquid in a container (pipe or component) heats up without a gas vent, the pressure rises uncontrollably until the weakest system component fails. This applies to all refrigerants and is a potential cause of pipe breaks.

In the critical point region, however, the fluid has a certain compressibility, such as for CO_2 at ambient temperature 20–30 °C (68–86 °F). On the other hand, liquid CO_2 has a high coefficient of thermal expansion. Thus, CO_2 expands by 25 % when heated from -10 °C to 20 °C (14–68 °F). With other refrigerants, the thermal expansion is just 5–7 %. There is a serious risk of bursting parts from trapped liquid CO_2 refrigerant.

Large volume expansion means that collectors, receivers and other liquid containers must be dimensioned with a low liquid-fill quantity at the prevailing operating temperatures. It is also necessary to check whether the container has enough volume for expansion when the highest operational temperature is reached. There should be sufficient gas volume above the liquid under conditions when the safety relief valve on the container opens. Draining liquid could increase the risk of a clogged valve or drainpipes, due to the formation of dry ice.

It is also important to avoid locking supercooled refrigerant liquids in pipes and system parts, preferably by providing overflow valves, by good maintenance procedures, or by at least having good warning signs, especially on hand-operated valves.

It should also be noted that in the case of R744 systems that are not designed and approved for the stand-still pressure, it must generally be assumed that all components, system parts or systems containing liquid refrigerant are to be classified as if supercooled refrigerant can be locked inside. Therefore, these systems must be fitted with safety relief valves.

5.4 Transcritical refrigeration / heat pump process

As CO_2 was "rediscovered" for car air-conditioning systems in the late 1980s, the so-called transcritical process with its unfavourable temperature glide during heat rejection was not desirable. This had to be overcome if the other favourable properties of CO_2 were to be used.

The temperature "gliding" can be very useful and thermodynamically advantageous in other applications, such as when producing domestic hot water to a high temperature level. This led to the development of a transcritical heat pump process at the NTNU-SINTEF Institute, Norway, in the late 1980s, including domestic hot water heat pumps. This was developed into a commercially significant product, particularly in Japan, where more than five million "EcoCute" CO_2 hot water heat pumps were sold before 2020. The system is also available on the European market and has a COP of 4 or higher.

Further development of this process has shown that refrigeration systems and heat pumps with a transcritical CO_2 process can be competitive in many areas, especially if the temperature glide on the warm side can also be utilised at the same time when cooling is provided. Current applications are explained in Chapter 5.10.

5.4.1 The process on the pressure-enthalpy diagram

Figure 5.12 shows the basic component layout of a transcritical CO_2 refrigeration system, as well as the thermodynamic process on a pressure-enthalpy diagram.

The refrigeration system consists of a compressor, gas cooler, suction gas heat exchanger, expansion device (high-pressure control valve), evaporator and low-pressure liquid receiver. The low-pressure receiver is used to store the working fluid to compensate the charge variations on the high-pressure side at various operation conditions, similar to a conventional system. This is particularly important regarding the active regulation of the high pressure in the system. The low-pressure receiver absorbs refrigerant when the system high-side pressure drops. Refrigerant may be shifted from the low-pressure receiver to the high-pressure side by means of the high-pressure control valve and the compressor if higher system pressures are desired.

The suction gas heat exchanger, also called the internal heat exchanger, is used to provide sufficient superheat upstream of the compressor suction port and also boil the remaining refrigerant from the oil / CO_2 mixture leaving the low-pressure receiver. Thereby, the oil return can be guaranteed and at the same time the compressor is protected against liquid hammer.

The pressure-enthalpy diagram shows three processes with different gas cooler pressures, however, with the same refrigerant outlet temperature from the gas cooler (40 °C / 104 °F isotherm). The dashed lines show isotherms, 120 °C (248 °F) on the far right and 40 °C (104 °F) on the far left. The letters used in the system sketch are used accordingly in the diagram. The gas cooler

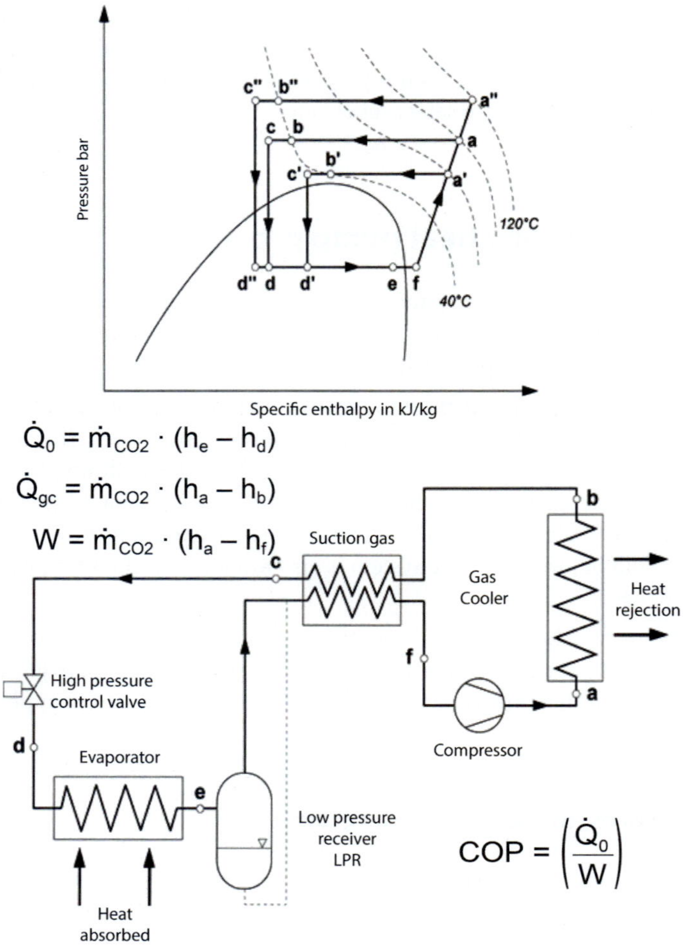

Figure 5.12
Typical transcritical CO_2 installation and pressure-enthalpy diagram

and evaporator changes of state are shown by the lines **a-b** and **d-e**. The change of state in the internal heat exchanger is **b-c** (on the high-pressure side) and **e-f** (on the low-pressure side).

The diagram shows:

- Temperature glide in the gas cooler, that is, the cooling of the CO_2 gases can sometimes be considerable.
- Cooling capacity ($\dot{Q}_0$) and heat output ($\dot{Q}_{gc}$) increase with increasing gas cooler pressure.
- If the exit condition from the gas cooler (point b) is close to the critical point, the process, due to a low cooling capacity and low COP, becomes less efficient.

The practical importance of these findings will be discussed in the next chapters.

5.4.2 Temperature curve in the gas cooler

5.4.2.1 Temperature curves on the temperature-enthalpy diagram (T-h diagram)

The temperature-enthalpy diagram (T-h diagram) is suitable for examining the temperature curve in the gas cooler. The diagram for CO_2 is shown in Figure 5.13. The cooling follows an isobar (constant pressure) from the inlet temperature to the outlet temperature, e.g., along the 100 bar (1450 psi) curve from 100 °C (212 °F) to 25 °C (77 °F), depending on the temperature level of the heat sink.

The specific heat capacity (c_p value) of the medium varies with temperature. This is the reason why the temperature does not decrease in proportion to the change in enthalpy (or in proportion to the amount of heat given off). The cooling curve has a more or less pronounced S shape. The deviation from a straight line increases with decreasing gas cooler pressure, especially near the critical point. The curves flatten out here, and there is a continuous transition into the two-phase area. Here the pressure curves run horizontally.

The specific heat capacity expresses – by definition – the change in enthalpy with the change in temperature ($c_p = \Delta h/\Delta t$). The slope of the curves in the diagram corresponds to the reciprocal ($= \Delta t/\Delta h$). When the temperature curves flatten out, the c_p value becomes infinite.

If heat is absorbed by air or liquid, the temperature of the heat-absorbing medium rises in proportion to the heat absorption since the specific heat capacity is almost constant and independent of the temperature. The heating up process of the cooling medium will therefore be linear.

The shape of the temperature curves of CO_2 is of great importance for the design and dimensioning of the gas cooler. If there is a lack of understanding of the thermal characteristics of the CO_2 gas cooler, false conclusions about the cooling capacity and the gas outlet temperature will be the result. However, since the performance and efficiency of the process strongly depend on the fact that this exit temperature is reached in the real process, the design of the gas coolers can be more difficult than the corresponding dimensioning of condensers in conventional refrigeration systems.

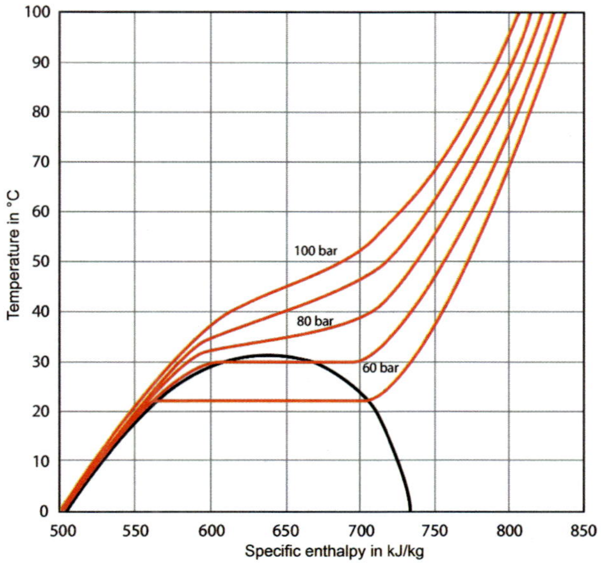

Figure 5.13 Temperature-enthalpy diagram for CO_2

5.4.2.2 Pinch point in gas coolers

A parallel temperature curve is the most efficient for heat transfer. In this case, the temperature difference can be reduced by increasing the physical size of the heat exchanger. If there is a deviation from parallelism, as with a linear and non-linear temperature curves, the temperature difference can only be reduced until both temperature curves touch. This contact point, or in a practical case the point with the smallest temperature difference between the two fluid flows, is called the pinch point.

The pinch point can either be at the inlet of the gas cooler (typically at low pressures) or at the outlet of the gas cooler (typically at high pressure). The relationship is shown in principle in Figure 5.14. In addition to the CO_2 temperature curves at 100 bar (1450 psi) and 75 bar (1088 psi), the curve for the maximum heating of the refrigerant in the gas cooler is shown.

The abscissa expresses the heat output transferred (enthalpy difference multiplied by the mass flow), which makes it possible to show the curves for CO_2 and the cooling medium on the same diagram.

The inlet temperature of the cooling medium is 20 °C (68 °F) (= 5 K lower than the outlet temperature of the CO_2). At 100 bar (1450 psi) it is possible to heat water to over 75 °C (167 °F). At 75 bar (1087.7 psi), this is not possible due to the pinch point being near the critical point, even if the gas inlet temperature is higher. In this case, a temperature of just under 50 °C (122 °F) can be reached.

Note that the graph does not show the temperature curve within the gas cooler itself. The water side therefore also shows a non-linear pattern.

In both cases, the pinch point is inside the gas cooler – both at 100 bar (1075 psi) and at 75 bar (1087.7 psi). Ideally, the pinch point should be at the cold end of the gas cooler (see Chapter 5.4.2.3). The temperature difference at the cold end can be selected according to predetermined criteria, such as the optimal size of the gas cooler, and is not dependent on the conditions inside the gas cooler.

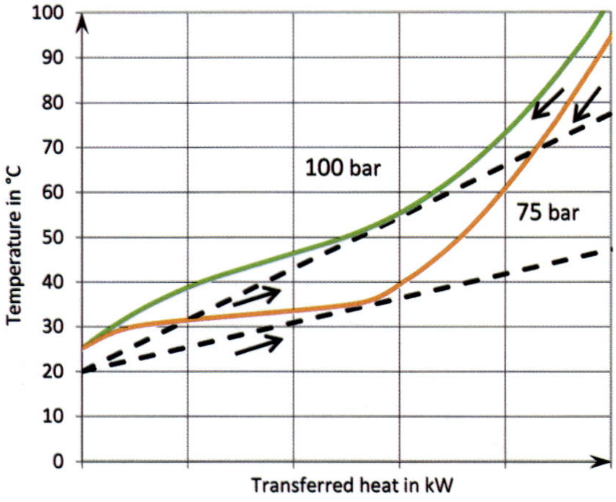

Figure 5.14 Temperature curve with pinch effect in a gas cooler

The placement of the pinch point can also be influenced by the cooling medium mass flow. The larger the mass flow, the less the slope of the heating curve and the greater the probability that the pinch point will be at the end of the gas cooler. This can be checked for a counter-flow heat exchanger by constructing the temperature curves as shown in Figure 5.14.

5.4.2.3 Temperature approach approximation

As with all classic subcritical processes, it is essential for the effectiveness of the process that the refrigerant be at the lowest possible temperature before throttling in the expansion device. In reality, this is, for the transcritical CO_2 process, even more important, due to the throttle losses already mentioned. Because of the temperature glide, it is also easier to enable the CO_2 gas cooler outlet temperature to reach the inlet temperature of the cooling medium, if the process (high-side pressure) and the gas cooler (sizing) are a correct match.

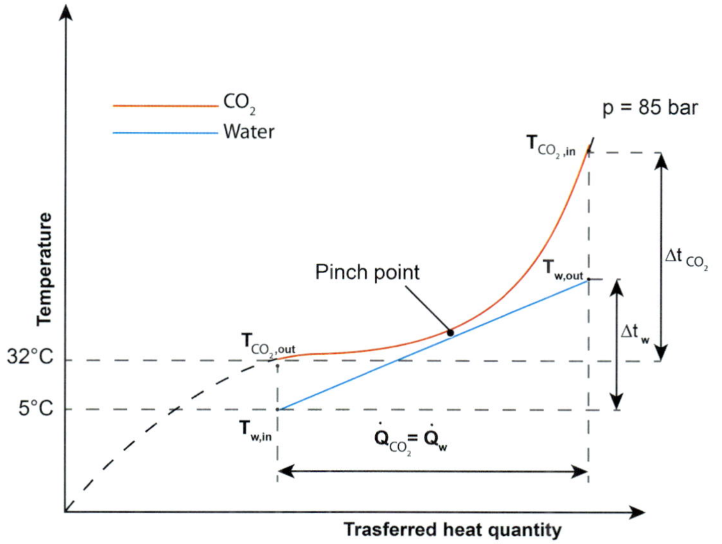

Figure 5.15
Low-pressure (85 bar / 1233 psi) gas cooler for water heating: temperature profile. Pinch point within heat exchanger

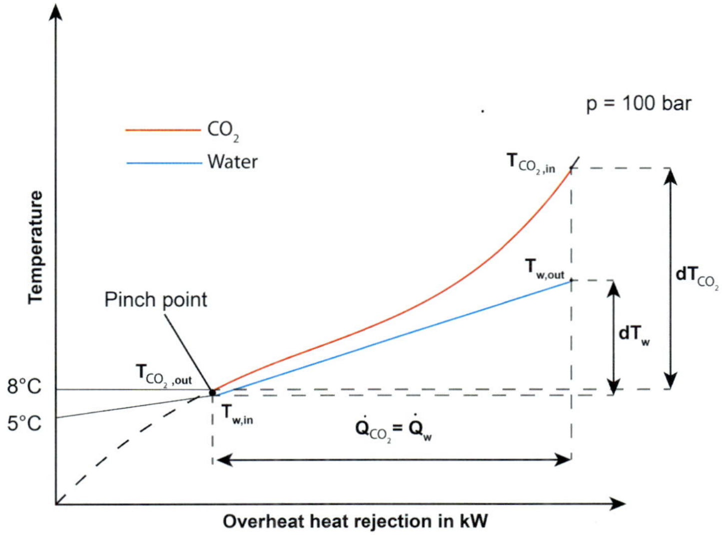

Figure 5.16
CO_2-gas cooler (at 100 bar / 1450 psi) with pinch point at end of heat exchanger

The temperature difference between departing CO_2 and incoming cooling medium at the cold end of the gas cooler is the designated temperature approach (Δt_A).

The temperature approach is a measure of how well the system is adapted to the transcritical operation. With correctly dimensioned gas coolers and a suitable gas pressure, the approach value can drop to 1–4 K. In the introductory example in Figure 5.14, the temperature approximation was assumed to be 5 K. The importance of correctly selecting the gas cooler pressure is shown in more detail in Figure 5.15 and Figure 5.16.

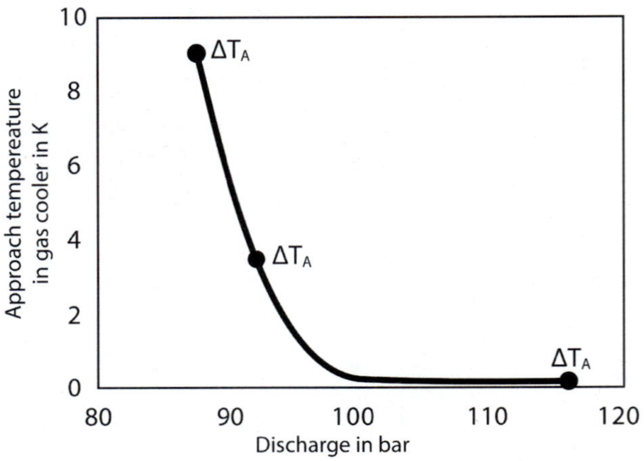

Figure 5.17
Measured approach temperature for gas cooler dependent on gas cooler pressure

Figure 5.15 shows an example of a calculated temperature curve in the gas cooler of a CO_2 heat pump for hot water generation, where the gas cooler pressure is relatively low (85 bar / 1233 psi). The average temperature difference is limited due to the pinch point being inside the heat exchanger. The gaseous CO_2 is cooled down to 32 °C (90 °F) only, despite the cooling water inlet temperature of 5 °C (41 °F). The temperature approach at the CO_2 outlet (Δt_A) is high (27 K). The installation will have a large throttle loss, a low specific cooling capacity and consequently a low COP.

By increasing the pressure in the gas cooler to, for example, 100 bar (1450 psi), the conditions become significantly better, as shown in Figure 5.16. The heat rejection curve of the CO_2 gas is less curved, and the pinch point has moved to the cold end of the gas cooler. With such a temperature profile, the temperature approach at the CO_2 outlet (Δt_A) can reach a small value (3 K in the figure). This significantly increases the system's COP.

Measurements taken on a prototype water heat pump illustrate the importance of gas pressure for the temperature approach, as shown in Figure 5.17. If the pressure falls below a certain level, below 95 bar (1378 psi), Δt_A increases significantly.

5.4.2.4 Temperature adaptation for heating needs at multiple temperature levels

If a heat pump heating circuit demands different temperature levels, for example by a combination of space heating and water heating, a transcritical CO_2 process can efficiently achieve various temperature levels, as shown in Figure 5.18.

This example describes a CO_2 heat pump for single-family houses. The heat pump has three separate gas coolers on the CO_2 circuit. These are located in series. Tap water is preheated to approximately 30 °C (86 °F) in gas cooler A and then heated to a final temperature (70 °C/ 158 °F) in gas cooler C. In gas cooler B the circulating water of the hydronic heating system is heated from 30 °C (86 °F) to 35 °C (95 °F).

This example shows the conditions at a specific design point. A particular challenge is to design the system in such a way that it also achieves a high level of efficiency when the different heating needs do not peak at the same time or the distribution between them varies. Suitable water storage tanks on the hot water side can be an advantage here and may extend the combined heating cycle times accordingly.

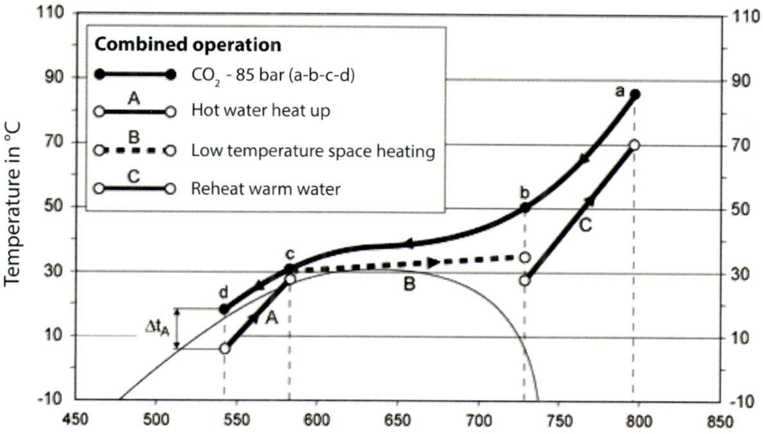

Figure 5.18 Tripartite CO_2 gas cooler with heat rejection at three temperature levels [5-6]

5.4.2.5 Calculation methods

Due to the change in the specific heat capacity of the supercritical CO_2, the mean temperature difference in the heat exchanger cannot be calculated as the logarithmic mean of the inlet and outlet temperatures. For precise calculations, the gas cooler must be divided into several sections, which are then calculated separately. In practice, this requires the use of computers.

An approximated analysis of the gas cooler to check the pinch point is possible with the help of the pressure enthalpy diagrams, refrigerant programs (e.g., CoolPack) or by simple manual calculations.

5.4.3 Significance of the gas cooler pressure on cooling capacity

When looking at the pressure-enthalpy diagram in Figure 5.12, one can see that the cooling capacity $\dot{Q}_0$ scales up with increasing pressure in the gas cooler. The gas cooler exit point (b) moves to the left. The cooling capacity increase is particularly striking in the area above the critical point.

In the supercritical area, there is no clear relationship between pressure and temperature. In addition, the amount of CO_2 in the gas cooler relative to the volume (the relationship between pressure, density and temperature of the gas) is of crucial importance.

The pressure in the gas cooler for transcritical operation must therefore be actively controlled by adjusting the refrigerant charge quantity in the high-pressure part of the system accordingly. In practice, a throttle control valve, called high-pressure control valve, is often used here.

Figure 5.19 shows how the cooling capacity changes at different pressures and outlet temperature values downstream of the gas cooler. The illustration shows the theoretical cooling capacity for a 10 m³/h (353 ft³/h) suction volume flow rate, dependent on the CO_2 temperature, before throttling. The gas cooler pressure varies from 75 to 110 bar (1088 to 1595 psi), while the evaporation temperature is −10 °C (14 °F) in all cases. The cooling capacity of a subcritical process with liquefaction is shown, for comparison, by the dashed curve. The following conclusions can be drawn from the figure:

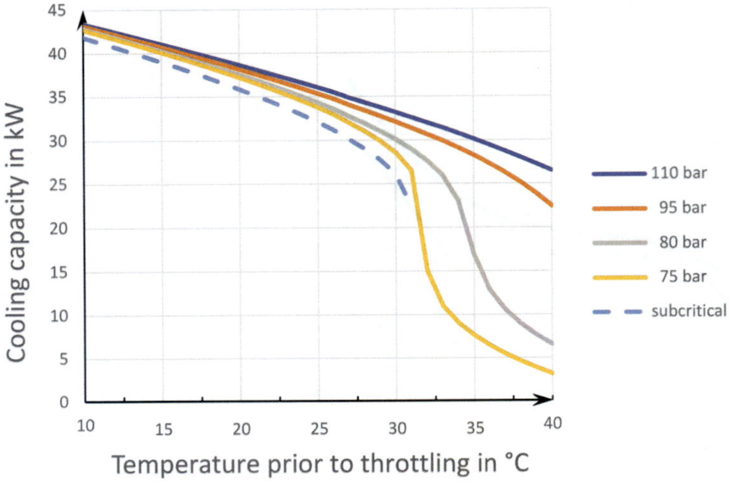

Figure 5.19
Cooling capacity per 10 m³/h (353 ft³/h) suction volume flow rate for transcritical CO_2 process, dependent on CO_2 temperature prior to throttling with −10 °C (14 °F) evaporating temperature

- The transcritical CO_2 process has a slightly higher cooling capacity than a subcritical process with the same refrigerant temperature before throttling.
- If the gas cooler CO_2 outlet temperature is above the critical temperature, a transcritical process with a "too low" high-pressure side level will lose a large part of the cooling capacity.
- The cooling capacity drops rapidly with increasing CO_2 temperature leaving the gas cooler. This can only be partially compensated for by increasing the pressure.

The curves in Figure 5.19 apply to a theoretical CO_2 process. In practice, the increase in cooling capacity resulting from the increase in the gas cooler pressure is compensated for by the decrease in the volumetric efficiency of the compressor. As a result, the cooling capacity remains unchanged, or may even decrease.

The above findings provide the basis for the following practical information:

- With air-cooled gas coolers, the maximum working pressure on the high-pressure side must be relatively high in order to be able to release heat on (extremely) hot days.
- The actual extreme values, i.e. the maximum inlet temperatures of the cooling medium (air or water), must be taken into account when selecting the compressor. The compressors should not be designed for an outside temperature of 30 °C (86 °F) if the temperature on the hottest day of the year could reach 40 °C (104 °F).
- With air-cooled gas coolers, it may be necessary and smart to distribute the required compressor output over several compressors in order to be able to achieve good partial load efficiency, capacity control, and redundancy with the system.

5.4.4 The importance of gas cooler pressure for the coefficient of performance

In transcritical operation, both the refrigeration capacity and the motor power requirement of the compressor increase when the gas cooler pressure is increased; see Figure 5.12. Initially, the cooling capacity and the COP increase. As the temperature curve becomes steeper, the increase in cooling capacity decreases, while the increase in compressor power remains almost constant. This change causes a reduction in the COP from a certain pressure onwards. Ultimately, for the transcritical CO_2 process, there is an optimal pressure in the gas cooler, at which stage the COP, for a given gas cooler outlet temperature, reaches the maximum value.

This relationship is shown in Figure 5.20. It shows the COP for a theoretical, transcritical cooling process depending on the pressure in the gas cooler. The CO_2 temperature after the gas cooler is a parameter and the evaporation temperature is considered constant at −10 °C (14 °F).

The gas cooler will have a theoretical optimal high pressure at 40 °C (104°F) and an evaporation temperature of −10 °C (14 °F), which will be at approximately 100 bar (1450 psi). At 30 °C (86 °F) it drops to 75 bar (1088 psi). If one takes into account that the compressor efficiency decreases with a reducing pressure ratio, the optimal values will occur at somewhat lower pressures. The temperature approximation of the gas cooler (Δt_A) must be taken into account as well; see Figures 5.15 to 5.18. Under certain circumstances, it is also neither sensible nor possible to lower the pressure ratio to such an extent that the optimum performance state is reached if, for example, the heat recovery unit demands increasing heat outputs.

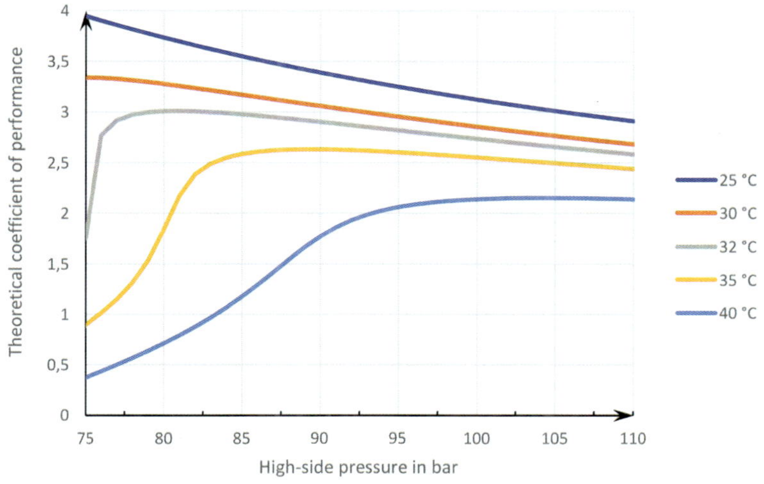

Figure 5.20
Theoretical COP value as function of the high-side pressure with varying CO_2 gas cooler outlet temperature. Evaporating temperature at −10 °C (14 °F)

5.4.5 Optimal high-pressure control of CO_2 processes

5.4.5.1 Optimisation basics

The theoretical COP, regardless of whether the system is operated transcritically or subcritically, depends on the pressure and on the CO_2 temperature after the gas cooler or the condenser. Figure 5.21 illustrates this relationship. The theoretical COP at an evaporation temperature of

−10°C (14°F) is represented as a function of the temperature before throttling (the temperature after the gas cooler / condenser) with the pressure as a parameter. The dashed line in the subcritical pressure range (approximately 70 bar / 1015 psi and lower) represents the refrigeration capacity for condensation without subcooling the refrigerant (condensation temperature = temperature upstream of the throttle). The solid lines in the range show the effect of subcooling.

The figure clearly shows that it may be advantageous to combine a higher pressure with increased subcooling instead of operating the process with a lower pressure and no subcooling. For example, the theoretical COP at 65 bar (943 psi) and 10 K subcooling is 5,4 which is higher than the COP 5,2 at 60 bar (870 psi) without subcooling.

As long as low refrigerant temperatures can be reached after the gas cooler, the refrigeration process is effective even when the high-pressure side is close to the critical pressure. This is the reason and the prerequisite for being able to implement continuous pressure control and "switching" between supercritical and subcritical process control without sacrificing efficiency and performance.

In the case of supercritical operation, the pressure control must be carried out as described in the previous section and shown in Figure 5.20. Figure 5.21 shows this relationship. If possible and sensible to liquefy the CO_2, the pressure will be reduced. In this area, active pressure control is generally not necessary. In practice, however, it is advisable to actually control the pressure in order to obtain a uniform control strategy in the entire application area.

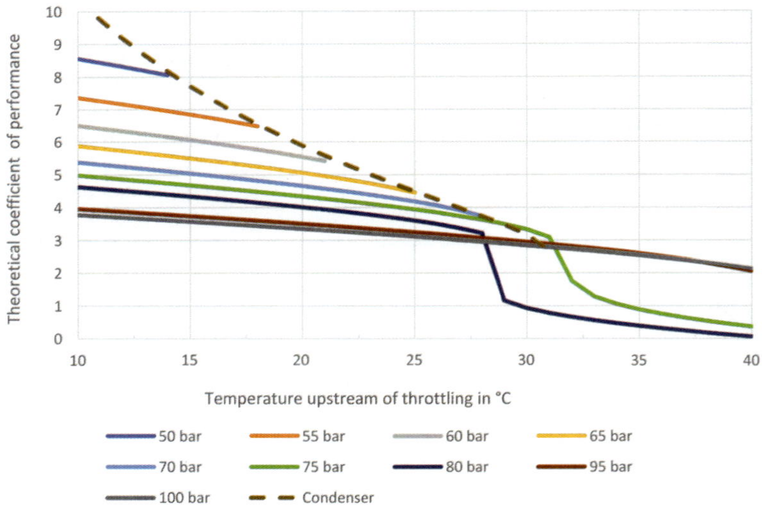

Figure 5.21 Theoretical COP for cooling process as function of CO_2 temperature, after gas cooler/condenser with varying pressures. Evaporating temperature at −10 °C (14 °F)

The optimal pressure depends on the operating conditions and the performance of the gas cooler / condenser. Under otherwise identical conditions, a lower pressure can be selected if less cooling air or water is to be heated in the gas cooler, see Figure 5.14. A properly designed gas cooler with a pinch point at the end of the heat exchanger is the prerequisite for good and satisfactory results. An example of a general control strategy is explained in the next section.

5.4.5.2 Control strategy

Controlling the pressure in the gas cooler should primarily aim at an energy-efficient operation. This is achieved by using a pressure control valve downstream of the gas cooler.

At stable temperatures of the cooling medium, such as heat pumps for warm water heating and some commercial refrigeration systems, the high-pressure side can be set to a fixed pressure. Mechanical and electronic controllers can be used here. With varying cooling medium temperatures, the pressure is dependent on the CO_2 temperature of the gas cooler, which is regulated according to a predetermined pressure-temperature curve. This requires a control strategy, control logic, electronic controller and a motor-operated throttle device. An example of such a pressure control strategy's characteristic is shown in Figure 5.22.

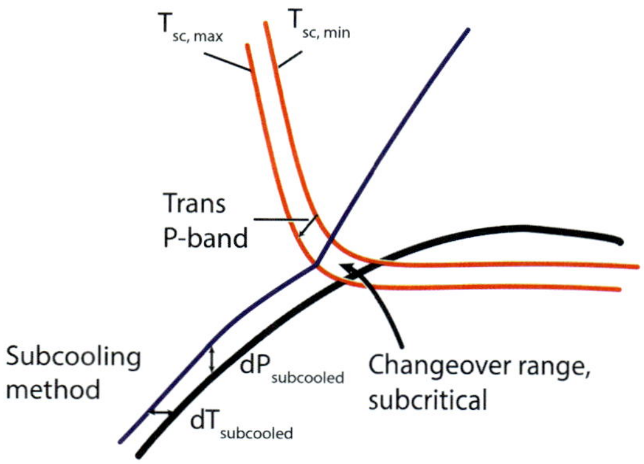

Figure 5.22 Controller characteristic for optimal gas cooler pressure

In the subcritical range, the given subcooling normally controls the pressure, which is usually set to 2–3 K. In the supercritical range, the pressure is controlled by a curve that (approximately) reflects the optimal settings, see Figure 5.20. The two curves are linked below the critical pressure and temperature in a transition area (possibly by means of a separate controller function). In this area, the subcooling is increased with the pressure. There are also other control strategies.

On particularly warm days, it can make sense to regulate the system to maximum cooling capacity (higher pressure), see Figure 5.19. This regulation overwrites the optimal energy regulation.

In the case of heat recovery, it may also be necessary to increase the high-side pressure to reach a certain specified temperature in the heat recovery process.

The control of the gas cooler fan can also be part of the control strategy. For example, the fan speed reduction can be more effective if, for example, less cooling power is required (if the increased workload of the compressor motor is less than the fan work saved) or to prevent the high pressure in the system from becoming too low (winter operation). Airflow control can also be relevant for heat recovery.

Controlling the pressure according to the function shown in Figure 5.22 is useful with respect to energy saving, if the heat is released to the environment without recovery or, if the heat recovery does not affect the process, e.g., as in a traditional refrigeration circuit. Efficient heat recovery, utilising high temperature lifts on the heated fluid and low CO_2 temperatures downstream of the gas coolers is one of the most important properties of CO_2, when the process is transcritical. In heat pumps or systems with heat recovery at high temperature levels, the pressure is controlled via the signals from the heating system itself. This requires that the heat recovery system is properly designed and dimensioned so that operation is as energy efficient as possible. Heat recovery options are explained in more detail in Chapter 5.7.

5.5 Technical solutions with CO_2

5.5.1 CO_2 as an evaporating coolant

The number of indirect refrigeration systems[7)] has increased in recent years – primarily to reduce the amount of environmentally harmful and expensive refrigerants (HFCs), but also for safety reasons (ammonia, hydrocarbons, other flammable refrigerants). In the case of indirect refrigeration systems with harmful refrigerants, the entire system is usually located in a machine room or outdoors. This ensures that only authorised persons get access to these systems. In large industrial installations, the refrigerant charge quantity can be reduced by a factor of up to 50 if the classic refrigeration cycle (usually ammonia) is replaced by an indirect system (usually brine).

Typically, fluids are used, i.e. coolants with a low freezing point. Examples of coolants are salt solutions, a mixture of water and glycol or water and alcohol in solution. The heat is absorbed and released in the form of sensible heat only, which leads to the associated rise or fall of its temperature through the heat exchangers.

A disadvantage of indirect systems is the larger energy demand due to the lower evaporation temperature of the refrigeration system and the operation of the coolant pump. This disadvantage can be greatly reduced by using an evaporating and condensing fluid as the secondary coolant, i.e., so that the latent heat of the fluid can be utilised as well. CO_2 is suitable for such an application in many industrial sites. Figure 5.23 shows an indirect cooling installation, using CO_2 as the evaporating coolant.

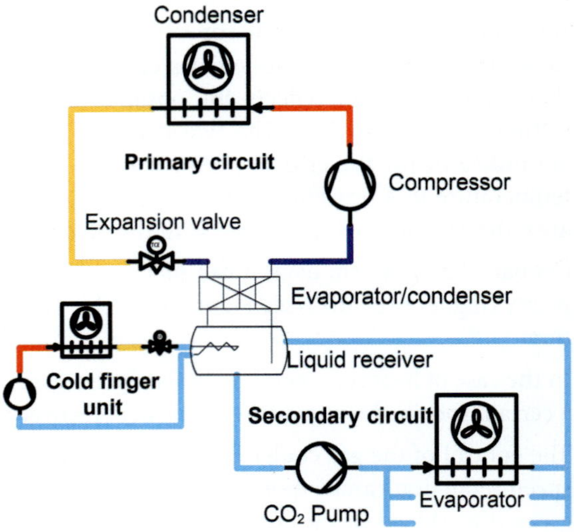

Figure 5.23 Schematic for indirect cooling installation with CO_2 as the evaporating coolant

For industrial and large commercial installations, the secondary circuit usually is designed and operated at working pressures / saturation pressures far below the pressure of CO_2 at ambient temperatures. In order to avoid the loss of refrigerant through the safety relief valve during (longer) standstill periods, the system is usually equipped with a small refrigeration system, a so-called "cold finger" unit. During standstill, CO_2 evaporates due to the heat input from the surroundings. The CO_2 is therefore liquefied by the cold surface of the cooled finger unit or another external evaporator, thus limiting the pressure rise in the secondary circuit. For smaller systems, it is advisable to dimension the refrigerant circuit to be able to cope with the pressure levels at ambient temperatures. This ensures it can withstand the temperatures even on

7) Systems with circulation of a secondary refrigerant between the evaporator and the medium to be cooled or frozen.

Table 5.3 Kinematic viscosity for CO_2 and other secondary coolants at −40 °C (−40 °F)

Medium	Ethylene Glycol/water	Ethanol/ water	Calcium chloride/water	Potassium/ water	CO_2
Viscosity in mm²/s or cSt	109	86	23	15	0,15
Viscosity in in²/s	0.17	0.13	0.036	0.023	0.00023

the hottest days. Liquid CO_2 wets surfaces very well due to its low viscosity, which has a favourable effect on the heat transfer in the heat exchangers. A low viscosity also contributes to a low pressure drop and thus results in a low demand of pump work compared to conventional secondary fluids. Table 5.3 shows a comparison of the kinematic viscosity for liquid CO_2 and other common coolants at −40 °C (−40 °F).

Furthermore, the heat transfer during a phase transition is much higher (latent heat) than heating or cooling with a liquid (sensible heat) only. As a result, the necessary temperature difference can be reduced both in the secondary fluid cooler (evaporator) and at the cooling point. Due to the evaporation, the heat absorption per supplied mass flow rate of CO_2 is many times higher than that of other coolants. This means that the coolant charge can be reduced to a fraction. For example, 1 litre (0.2 in³) of evaporating CO_2 absorbs about 13 times the heat of ethylene glycol / water mixtures changing the temperature by 5 K. Even if the evaporator operates with a circulation rate of 2–3, the pump work with CO_2 is still 90–95 % less than with conventional brine solutions. The energy consumption can be reduced by more than 20 % for the entire system. In many cases, these kinds of systems can compete with direct expansion systems in terms of energy demand. The small amount of circulating CO_2 required and the small influence of the pressure loss in the two-phase mass flow also means that the dimensions of pipes, valves etc. can be significantly reduced – in some cases by over 50 %. This simplifies the piping, saves space and reduces costs. That is why systems with CO_2 as an evaporating coolant (especially for an industrial application), when compared to conventional indirect systems have been proven to be very cost effective.

After the "rediscovery" of CO_2 as a refrigerant towards the end of the 1980s, the first industrial CO_2 refrigeration systems used CO_2 as an evaporating coolant. Since then, many systems have been manufactured for a variety of applications, initially for supermarkets, industrial applications and ice rinks. However, the most common applications can be found in low-temperature applications down to −50 °C (−58 °F). Cascade systems with CO_2 are suitable for freezing and other low-temperature applications. They are particularly relevant at these lower temperature levels.

Since the CO_2 flows as an evaporating coolant in the secondary circuit without a compressor unit and circulates with appropriate pumps, these circuits can be operated without any lubricant. Problems with oil return, negative effects on the heat exchangers and an increased pressure loss due to the oil are avoided. However, it may still be advisable to introduce some oil into the circuit in order to lubricate the automatic valves, where this is still necessary.

In general, the cooling coils are electrically defrosted. However, a separate heating cycle for deicing is a good alternative, for example, by using circulating warm glycol, which defrosts the heat exchangers. Hot gas defrosting is also conceivable, in which the liquid is pumped up to a correspondingly high pressure and then evaporated with waste heat in order to then defrost the cooling coils. In order to achieve a sufficiently high pressure, which corresponds to a saturation

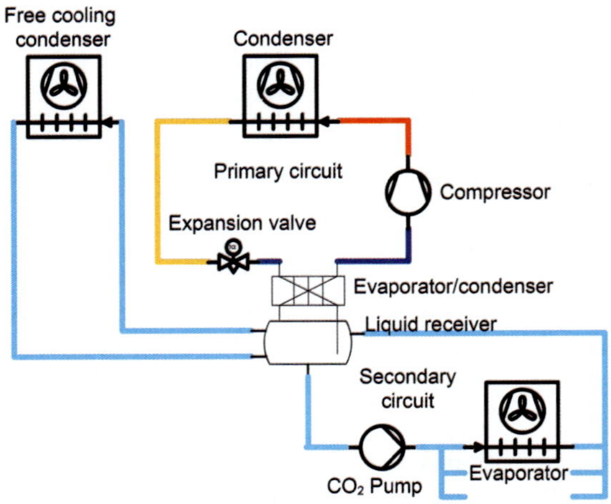

Figure 5.24 Indirect system with evaporating CO_2 as coolant. Low-temperature application

temperature of 10 °C (50 °F), the pump must generate a pressure of 45 bar (653 psi). The CO_2 circuit must then be designed for 53 bar (769 psi).

Indirect systems with evaporating CO_2 can easily be combined with low-temperature systems, as shown in principle in Figure 5.24. For such an application, the system design pressure calculations should be based on the standstill pressure, which can occur if the entire system is at the maximum ambient temperature of the installation.

In case of sufficient low ambient temperatures, the free cooling condenser enables operation of the cooling system without the primary circuit.

5.5.2 CO_2 in the conventional refrigeration process – cascade system

If the condensing temperature can be kept below 27 °C (80.6 °F) and there is no need for additional heat recovery, the CO_2 system is normally operated subcritically. In this case, the operation of CO_2 systems does not differ fundamentally from the operation of other refrigeration systems.

Year-round subcritical operation of the system with condensing compressed CO_2 is also possible if there is access to cold cooling water, for example, from sea-, ground- or river water. If it is not possible to keep the temperature constant, as with an air-cooled condenser, it is necessary that the system is able to switch between supercritical and subcritical operation accordingly. Transcritical operation is explained in the next chapter. A conventional subcritical process using CO_2 is always possible if the high-pressure side is cooled with another cooling system (cascade arrangement), or if the condenser heat is used as the heat source of a heat pump.

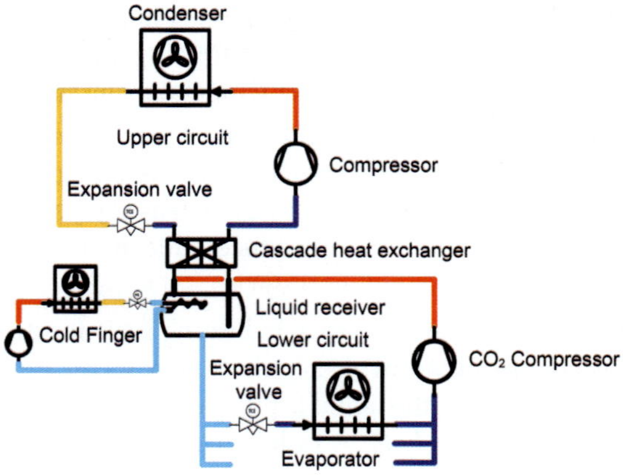

Figure 5.25 Basic cascade system

A cascade system consists of two separate refrigeration circuits. A lower and an upper stage are connected by a heat exchanger (cascade heat exchanger). The heat of condensation in the lower stage is transferred to the evaporator in the upper stage, which in turn releases the heat after compression to the air, cooling water or in case of a heat pump, to the heating system. The principle of a cascade system is shown in Figure 5.25.

The cascade solution enables a combination of refrigerants with favourable properties to perform efficient heat rejection in the upper stage and for efficient low-temperature evaporation in the lower stage. Because of the high pressure (and the other features mentioned above), CO_2 is particularly suitable for low-temperature applications in the lower temperature range, whereas ammonia and hydrocarbons are more suitable for moderate and high temperatures and hence the upper stage of the cascade system. Cascade refrigeration systems usually cover the same applications as systems with CO_2 as an evaporating coolant but are the preferred solution especially for evaporator temperatures below –35 °C (–31 °F). The reason is that the benefits of a CO_2 compressor rise with falling evaporating temperatures. A system with CO_2 in the lower stage and ammonia in the upper stage is a typical industrial solution. The early supermarket refrigeration systems consisted of a CO_2 process with CO_2 forming the lower stage of a cascade. By that time, HFC was mostly used in the upper stage; however, even some ammonia units have been delivered. Propane was used in Denmark. In Northern Europe, including Norway, transcritical systems have gradually taken over the retail market, while after 2020, cascade systems have not been further implemented in the other (warmer) parts of the world anymore, as transcritical systems have become more energy efficient, as shown in the next chapters.

In the past, it was common to connect a separate heat transfer circuit between the condenser in the lower stage and the evaporator in the upper stage. A typical argument for this was greater operational reliability and simpler control, because the cascade heat exchanger was often a weak point. The disadvantage of this intermediate link, however, is its significantly higher energy demand due to an additional heat transfer, temperature differences, and higher acquisition costs. This solution is used seldom today and has been overtaken by the further development of transcritical systems. Large CO_2 systems are usually protected against high pressures and the triggering of safety relief valves by a small additional refrigeration system. This is similar to systems with CO_2 as an evaporating coolant. There are also examples of low-temperature systems as pure (transcritical) CO_2 installations. These are used in supermarkets and are designed as a cascade with two temperature levels. Examples of this are explained in more detail in Chapter 5.5.3.5.

Cascade systems are also used to a certain extent in Europe for industrial refrigeration systems. Various technical design variants can be applied to achieve combined operation for medium- and low-temperature applications:

for medium temperature:

a) CO_2 as an evaporating coolant
b) dry expansion or self-circulation
c) circulation of the refrigerant in the upper stage as conventional refrigerants

for low temperature:

a) dry expansion or
b) self-circulation of CO_2 or
c) pump circulation

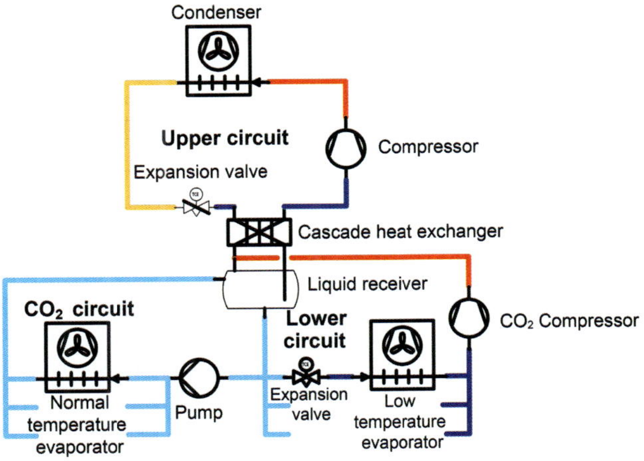

Figure 5.26 CO_2 cascade system with two cooling temperatures

The principle of a cascade system, using CO_2 as evaporating coolant for the medium-temperature and the dry expansion for the low-temperature application, is shown in Figure 5.26. The upper stage can work in small- to medium-sized installations with dry expansion, while in large industrial installations plate heat exchangers with self-circulation, tube and shell heat exchangers or special compact heat exchangers are used.

The refrigeration cycle in cascade systems with CO_2 is typically designed for 41 bar (595 psi) in industrial installations whilst 26 bar (377 psi) is sufficient for the low-temperature cycle. For the sake of simplicity and for safety reasons, it is sometimes advantageous as well to dimension the low-temperature circuit also for 41 bar (595 psi) in order to prevent the refrigerant from being vented out. If the compressor(s) are not designed for such a high working pressure, all of them must be equipped with their own safety valve.

For large industrial refrigeration systems, a conventional two-stage ammonia refrigeration system or preferably a cascade with NH_3/CO_2 can be selected. If the temperature requirements are –35 to –40 °C (–31 to –40 °F) or even lower, the cascade system with CO_2 is usually advantageous due to the operating costs. By using the good thermophysical properties of CO_2 at low temperatures and the dimensioning of freezers for temperatures of down to and below –50 °C (–58 °F), this can result in a significant increase of the cooling capacity, up to 25 % compared to conventional solutions.

In addition, the NH_3 / CO_2 cascade system brings some advantages when compared to a pure ammonia system:

- CO_2 exposure in the work area: low risk for employees who come into contact with the refrigerant
- no risk of damage to goods due to refrigerant leakage in the warehouse (reduced insurance premiums)
- no problems with air entry and moisture in the system (operating pressure in the CO_2 circuit is above atmospheric, even at the lowest temperatures)
- minimum space requirement (for compressors and pipes)

Contamination of the ammonia system with CO_2 is a potential problem in the event of a leak in the cascade heat exchanger. The substances react chemically and form a substantial powder (ammonium carbamate) which clogs the filter and can lead to malfunctions. This problem can be prevented by using a shell-and-tube heat exchanger with double-walled tubes where the intermediate space is connected to the outdoors. As an alternative, a spare cascade heat exchanger can be kept ready. The ammonia pump should definitely be equipped with a differential pressure switch to trigger an alarm as soon as such mixing of CO_2 and ammonia has occurred. The resulting powder can be removed with hot air or nitrogen at a temperature of at least 60 °C (140 °F). There have been such incidents of leakage, but this problem appears to decrease when only applying appropriately designed and tested heat exchangers.

There are similar procedures for the defrosting process as for the system with CO_2 as an evaporating coolant. In addition, hot gas defrosting from an additional high-pressure compressor[8)] can be carried out, as well as the evaporation of liquid at high pressure (as described above). The

[8)] Industrial low-temperature installations normally cannot apply sufficiently high pressure for the liquefaction of CO_2. High-pressure compressors from the commercial sector can be used for this purpose.

defrosting compressor must reach a discharge pressure of at least 45 bar (653 psi) to allow a defrosting temperature of 10 °C (50 °F).

There are also systems available utilising sensible heat of the gas for defrosting. The system compressors can be used without modification. This means that a lot of gas must be available for defrosting. In practice, this means that such systems must have several evaporators or cooling cabinets.

5.5.3 The transcritical process – basic solutions

If the transcritical process is to be integrated into an installation, some basic decisions have to be made:

- How should the high-side gas pressure be controlled?
- Should a low-pressure receiver be used?
- Is one-stage throttling efficient or is two-stage throttling required?
- Should the intermediate pressure be controlled or not (through two-stage throttling)?

The design decisions are to be based on the required efficiency level of the systems, the main operating conditions and the question of what is technically most suitable.

5.5.3.1 System with a low-pressure receiver and single-stage expansion

If a low-pressure receiver is applied, the fluid from the high-pressure side is normally expanded in one stage. A two-stage expansion is still possible and – due to the advantages offered by this solution – is increasingly used (see Chapter 5.5.3.4).

The basic sketch of a system with a low-pressure receiver is shown in Figure 5.12. It shows Professor Lorenzen's solution for a car air conditioning system. This principle is currently used primarily by smaller systems with only one evaporator, such as hot water heat pumps and small commercial systems for vending machines. It was also a basic solution for automotive air conditioning systems. Some automotive manufacturers came onto the market with this system around the year 2000. The latest generation of automotive air conditioning systems, such as the luxury Mercedes models from 2017, are also based on this principle, and since 2020 even other car manufacturers apply R744 in various models.

The expansion device controls both the high-side pressure of the refrigeration system as well as the refrigerant supply to the evaporator. A capillary tube, a constant pressure valve or an electronically controlled valve are standard here. An electronic expansion valve is the most flexible and enables optimum control of the gas cooler pressure (high-side pressure control). It also enables smooth transition between transcritical and subcritical operation.

The supply of liquid to the evaporator in systems with a low-pressure receiver is not regulated directly according to the requirements of the evaporator, as is the case with superheat-controlling expansion valves. The high-pressure control valve controls the desired amount of refrigerant in the high-pressure side so that the pressure level corresponds to the desired value of the control unit. This solution is simple, but requires careful adjustment of the control valve, the volumes on the high- and low-pressure side, and the refrigerant charge quantity.

A small amount of liquid (refrigerant and oil) is continuously drawn from the low-pressure receiver into the suction line. The refrigerant evaporates in the suction gas heat exchanger. This liquid flow leads to very effective (flooded / non-superheat) operation of the evaporator. At the same time, this ensures oil return to the compressor of the refrigeration system. Excess liquid is collected in the low-pressure receiver and stored in order to have enough refrigerant available for adjusting the high pressure. Compared to systems with thermal expansion valve control, the suction pressure in the compressor can now be increased by at least 5–7 K when the evaporator is in flooded operation. As this corresponds to a 5–7 bar pressure increase, significant efficiency improvements can be achieved.

In a properly dimensioned system, the liquid balance adjusts itself. If the evaporator is supplied with too much liquid, when more refrigerant is transferred from the high- to the low-pressure side than evaporates, the gas cooler pressure will drop. The throttle valve therefore closes and the pressure builds up again. In the meantime, the gas cooler passes on less refrigerant than is supplied by the compressor (= evaporated amount from the evaporator + liquid from the low-pressure receiver). More CO_2 is pumped to the high-pressure side than is discharged, which increases the pressure in the high-pressure side. The throttle valve is now opened, the evaporator gets more liquid through again and the initial state is reached again. If the parameters mentioned above are correctly adjusted, the variation in the liquid flow is small and the evaporator will never dry out completely.

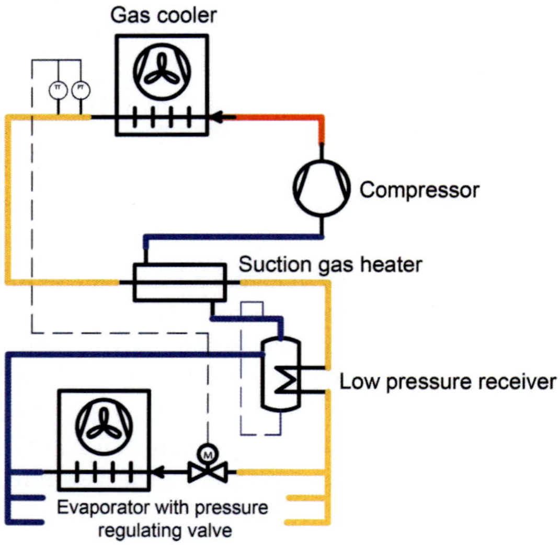

Figure 5.27 Transcritical installation using a subcooling heat exchanger in a low pressure-receiver

Under certain circumstances, it may make sense to overfill the liquid in excess of the evaporator requirements. This can be implemented by installing a subcooling heat exchanger in the low-pressure receiver, as can be seen in Figure 5.27. Heat contained in the high-pressure liquid refrigerant evaporates some liquid in the low-pressure receiver. This is compensated by a larger liquid supply to the evaporator. However, a certain amount of liquid must always be removed from the low-pressure receiver to ensure safe oil return.

5.5.3.2 System with intermediate-pressure collector, two-stage expansion

For applications with multiple evaporators, each at different temperature levels on the external fluid side, a solution with intermediate-pressure receivers may be most suitable [5-7]. Figure 5.28 shows a typical circuit.

With this solution, the refrigerant is first expanded into the receiver and then into the individual evaporators. There is no direct connection between the gas cooler pressure level and the inlet of the evaporator. The evaporator can be controlled in the usual way with an (electronic) expansion valve.

The pressure control of the gas cooler can be carried out as described above. However, since an intermediate-pressure receiver is only useful and applicable upward from a certain system size, it would still be necessary to optimally adapt the gas cooler pressure and regulate it (adaptively) based on the respective refrigerant outlet temperature. This is usually achieved by a motorised pressure control valve, as shown in Figure 5.28. The control is carried out using a gas cooler pressure algorithm, e.g., similar to the one in Figure 5.21.

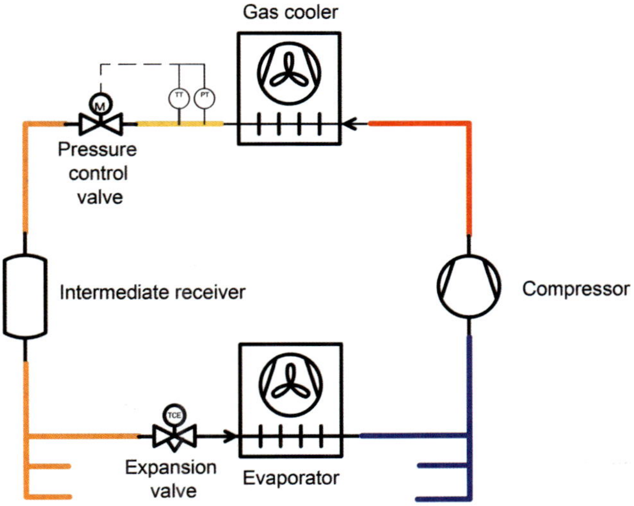

Figure 5.28 Transcritical installation with intermediate-pressure receiver. Gas cooler with pressure regulating valve

Due to the small density difference between liquid and vapour CO_2, there are often liquid droplets in the suction port of the compressor. To prevent this, such systems are preferably equipped with a suction gas heat exchanger to protect the compressor from liquid hammers. In theory, however, this heat exchanger causes a slight loss of performance and reduces the energy efficiency.

Another option is to use a differential pressure valve. With a constant pressure difference between the gas cooler and the receiver, the gas cooler pressure increases with increasing temperature and vice versa, provided the process runs almost optimally. This is a purely mechanical and therefore robust solution. If it is cold enough for subcritical operation of the evaporator, there must be an option to bypass this valve. A system diagram with this solution can be seen in Figure 5.29. However, this type of system is nowadays rarely built due to the electronic control valves becoming cheaper.

As shown in Figures 5.28 and 5.29, the receiver can be operated without pressure control or, as in Figure 5.30, have an active pressure control.

Without pressure control, the thermal state of the refrigerant in the receiver is automatically on the liquid saturation line because any transport of gas is prevented by the liquid in the receiver. Because the change of state at the throttle is isenthalpic

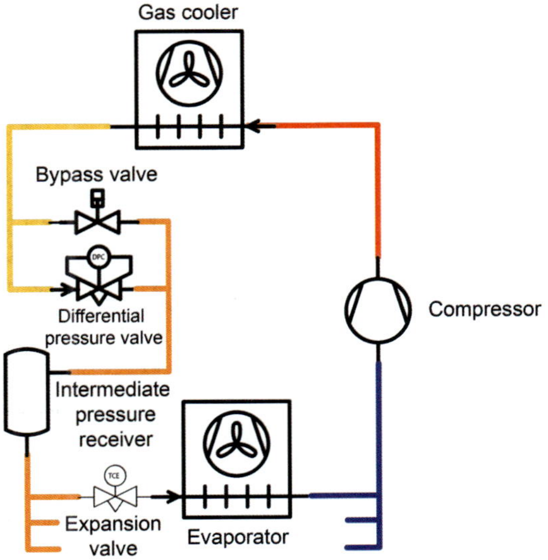

Figure 5.29 Transcritical installation with intermediate-pressure receiver. Gas cooled

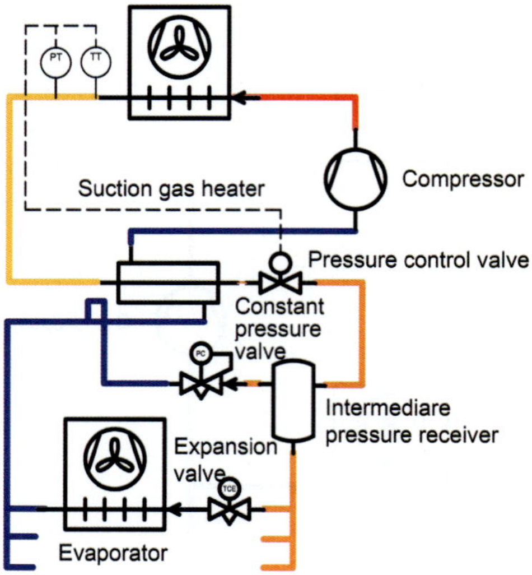

Figure 5.30 Transcritical installation with intermediate-pressure receiver. Dynamic regulation of receiver pressure dependent on suction superheat

(with constant enthalpy), pressure and temperature are determined by the conditions ahead of the throttle valve. The lower the temperature or the higher the pressure of the refrigerant in front of the throttle, the lower the pressure in the receiver, see Figure 5.31.

With an expansion from 80 bar (1160 psi) and 30 °C (86 °F) the pressure in the receiver is reduced to around 67 bar (972 psi), while with an expansion from 100 bar (1450 psi) and also 30 °C (86 °F) a pressure of around 63 bar (914 psi) is reached.

If the pressure is kept constant at 80 bar (1160 psi), but the temperature upstream of the throttle is reduced to 20 °C (68 °F), this will result in a receiver pressure of close to 54 bar (783 psi).

With active pressure control, the pressure is kept stable by means of the constant pressure

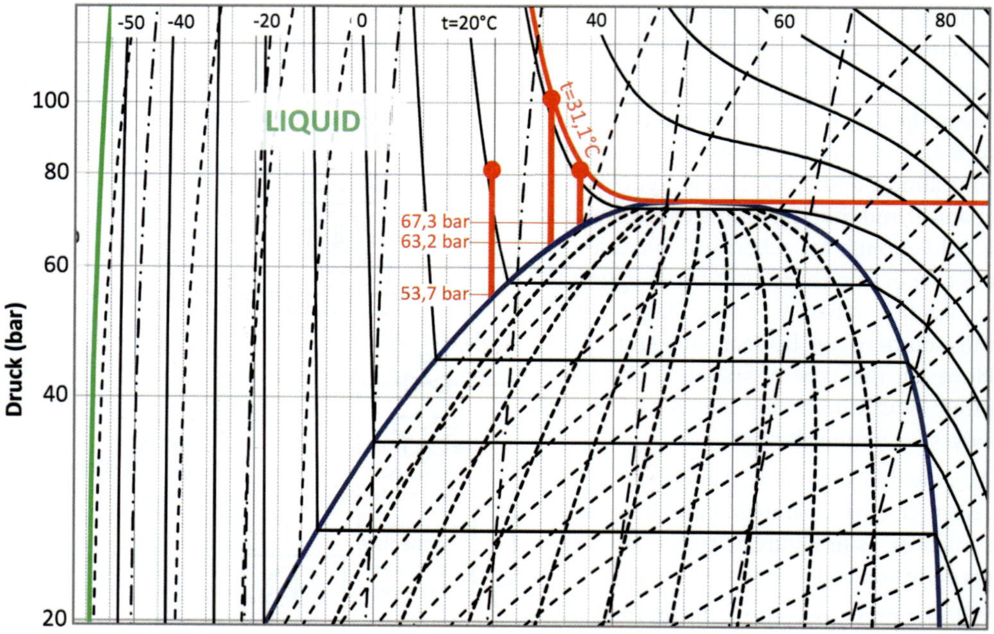

Figure 5.31 Pressure and temperature in receiver are determined by conditions ahead of throttle valve on pressure-enthalpy diagram

valve, also called a flash gas bypass valve. This valve is connected between the receiver and the outlet of the evaporators. The receiver pressure can be flexibly pre-selected or actively controlled. The process ensures a constant pressure level between the liquid receiver and the evaporator(s).

Since a proportion of gaseous refrigerant after the throttle (flash gas) is created regardless of the suction pressure, it is irrelevant for the COP which receiver pressure is selected. However, the feeding valves upstream of each evaporator requires a minimum amount of pressure differential, which must be maintained. Standard copper pipes and fittings can be used at pressures of less than 40 bar (580 psi). The need for using special materials such as $CuFe_2P$ (K65) has therefore decreased.

At pressures close to the critical point, throttling from the saturated vapour line ends in the two-phase region. Therefore, there will be a certain amount of liquid refrigerant after the throttle that still needs to be evaporated. This is another reason for the integration of a suction gas heat exchanger, which is usually installed in such systems.

5.5.3.3 Systems with both intermediate- and low-pressure receivers

The combination of low- and intermediate-pressure receivers will combine the advantages of both systems. The schematic diagram is shown in Figure 5.32. With an intelligent evaporator control, liquid will not escape from the evaporators, and the additional temperature difference that thermal valves require to operate is avoided.

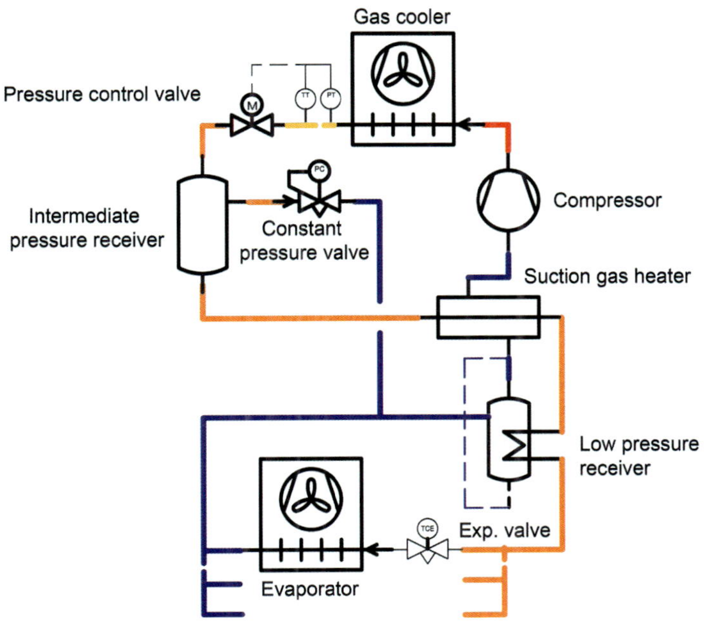

Figure 5.32
Transcritical installation with low- and intermediate-pressure receiver

5.5.3.4 Transcritical, two-stage process

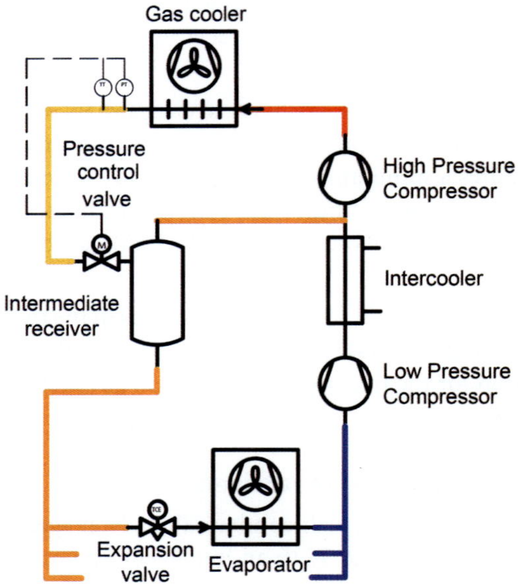

Figure 5.33 Two-stage transcritical process

Two-stage compression and throttling systems are usually employed in two-stage refrigeration systems. Small capacity two-stage CO_2 systems can also be throttled with just one expansion device. The compression process is optimised by intermediate cooling, a lower pressure ratio and a reduced pressure difference per stage. This solution is used, amongst others, in Japanese-made hot water heat pumps, where the high water-discharge temperature and high efficiency of the CO_2 compressors contribute to the success of these heat pumps.

The principle of a two-stage transcritical system with two-stage compression and evaporation is shown in Figure 5.33.

The receiver is kept at the intermediate-pressure level. In this case, an optimal intermediate pressure (and hence the receiver pressure) is available, just as with conventional two-stage systems. This optimal intermediate-pressure level can be calculated if all system components are specified. Normally, the intermediate-pressure level is set to the geometric mean (same pressure ratio) of the evaporator and high-side pressure.

If an additional cooling requirement is to be met at the intermediate-pressure level, it is necessary to adjust the pressure level and second stage compressor accordingly. This is also the case if the required pressure difference of the expansion valves must be kept constant.

CO_2 causes high discharge temperatures, which may require intercooling after the first compression stage. This can prevent the oil temperature and compressor components in the high-pressure part of the compressor from rising too high. This has a positive effect on the lifetime of the components as well as on the energy efficiency. These intercoolers are therefore regarded as a necessary component and hence are included in the schematic diagram.

Intercooling increases the performance and efficiency of the high-pressure compressor (HP compressor) and is a simple way to improve the energy efficiency of the system. The two-stage compressor systems mentioned above also have corresponding intercoolers for process improvement. Intercooling is explained in more detail in Chapter 5.6.1.

5.5.3.5 Solutions with cooling requirements at two temperature levels

Cooling in the food retail sector is an obvious example of meeting cooling demands at different temperature levels. It is also the industry in which all transcritical CO_2 refrigeration systems have become the preferred standard solution. Compared to the early systems, the concept has evolved impressively – with many measures to increase energy efficiency and competitiveness. Since 2018 it is therefore a realistic alternative, even in warmer climate regions.

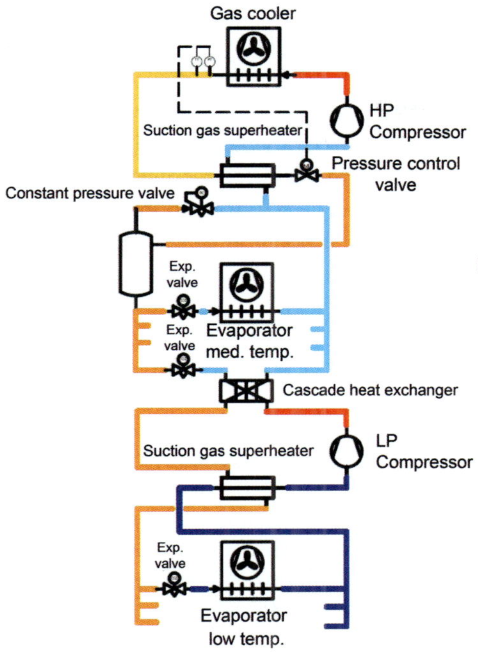

Figure 5.34 Transcritical cooling process using two temperature levels; left: Cascade CO_2 at both stages; right: Booster

The systems are based on two different principles. Either the two stages are connected indirectly by a cascade heat exchanger or there is a direct connection. The direct connection is usually referred to as a “booster arrangement”. The principles are shown in Figure 5.34.

With a cascade solution, possible problems with oil return are avoided. Another advantage is that oil with different viscosities can be used for the two evaporation temperature levels of the system. However, the implementation of booster systems has become the preferred option for commercial refrigeration. The maintaining of oil levels in all compressors is resolved. These various booster-type systems have the greater viability in terms of costs and energy efficiency.

5.6 Process improvement methods

5.6.1 Intercooling in two-stage systems

When compressing CO_2, relatively high discharge temperatures will arise. This requires active cooling, with cooling water or air, of the hot gas downstream of the low-pressure (LP) compressor in a two-stage / (booster) system. The principle of a two-stage system with an external cooler for intermediate pressure is shown in Figure 5.33. The intercooler may optionally be integrated into the CO_2 gas cooler with a separate refrigerant circuit [5-8].

A reduced inlet temperature into the high-pressure compressor increases its volumetric performance and reduces the required specific work for compression. In theory, both changes are proportional to the change in the absolute gas inlet temperature. The compressor capacity increases proportionally with the reduction in the absolute temperature, while the specific power consumption also decreases.

In reality, the performance improvements are even greater. Normal gas behaviour deviates from ideal gas behaviour and increases the effect. For example, the performance can increase by almost 0,5 %/K due to the suction temperature reduction, while the specific power demand and the efficiency are improved by a corresponding factor. Intercooling also improves operating conditions and increases the lifetime for the high-pressure (HP) compressor, which would become too hot if the suction gas is not cooled down properly, e.g., by liquid injection in emergency cases.

5.6.2 Parallel compression

In a system with an intermediate-pressure receiver and a controlled intermediate-pressure level (using a constant pressure valve or flash gas bypass valve), the process effectiveness is not affected by the pressure control, since the flash gas always reaches the suction gas pressure. On the other hand, the process is improved in terms of energy efficiency if the vapour volume flow after the first expansion is fed to a separate compressor, which maintains the pressure level in the receiver, instead of expanding the vapour to the low-pressure level. This solution is called parallel compression and is shown in Figure 5.35.

The flash gas is not expanded to the lowest pressure. It is compressed by the parallel compressor so that the suction volume flow of the main compressor is reduced accordingly. It is also compressed starting from a higher pressure, which reduces the energy demand to compress the vapour directly towards the high-pressure side by the parallel compressor if it is designed for this lower pressure ratio level. This system variant generally reduces suction volume flow requirements at the main compressor and improves the overall coefficient of performance (COP) by 10–15 %, which means that the additional costs are amortised in a relatively short timeframe.

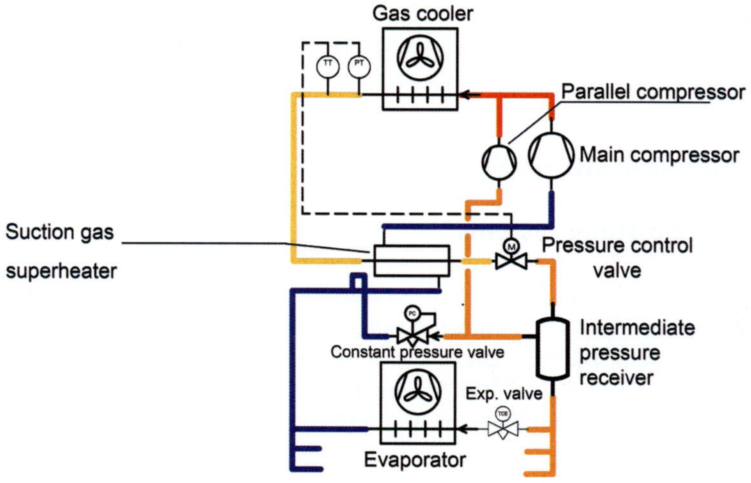

Figure 5.35
Transcritical CO_2 installation with parallel compression

It should be noted that in addition to the parallel compressor unit, a constant pressure valve / flash gas bypass valve is installed in the system. This valve is not absolutely necessary in normal operation of the parallel compressor; however, it can be useful as a replacement for the parallel compressor for very small amounts of vapour flow rates (e.g., for part-load and winter operation), and in a case of emergency. The valve also enables the main compressor to operate at normal system pressure during system start-up.

5.6.3 Cooling of the supercritical gas after the gas cooler

The easiest way to reduce the losses during expansion is to additionally cool the supercritical refrigerant downstream of the gas cooler. Subcooling increases the refrigeration capacity without affecting the compressor's power demand. The cooling capacity and COP of the system are thus improved as well.

The relationship is shown in Figure 5.36, which shows the relative increase in cooling capacity and the COP of a CO_2 installation per K subcooling. The applied evaporation temperature is −10 °C (14 °F) for this analysis. In the optimal range, one can achieve an improvement in the COP of 1,5–2 % per K. In the same way, the figure shows how large the losses are when the temperature downstream of the gas cooler is higher than expected; if for example the size of the gas cooler has been underestimated.

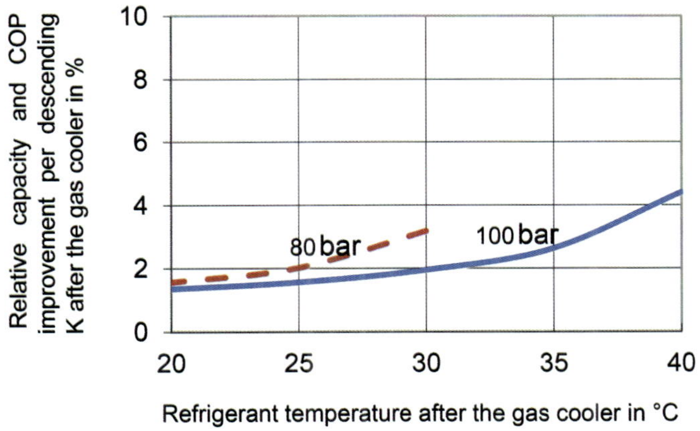

Figure 5.36
Estimated capacity and COP increase, per degree K subcooling prior to throttle, all as function of gas cooler outlet temperature t_{evap}= −10 °C (14 °F)

The maximum process improvement is achieved if subcooling is carried out naturally. This is possible, for example, in air-cooled systems that have access to additional cold cooling water in summer. Alternative additional heat sinks for this additional cooling of the transcritical CO_2 (second heat exchanger downstream of the gas cooler) could be geothermal probes that serve as a heat source in winter and are thereby regenerated during summer.

If there is no access to natural cold or heat sinks, an auxiliary cooling system can be used for subcooling (mechanical subcooling). A possible solution is shown in Figure 5.37. The cooling capacity increase of the CO_2 system is independent of whether the additional cooling comes from a natural source or from an auxiliary cooling system.

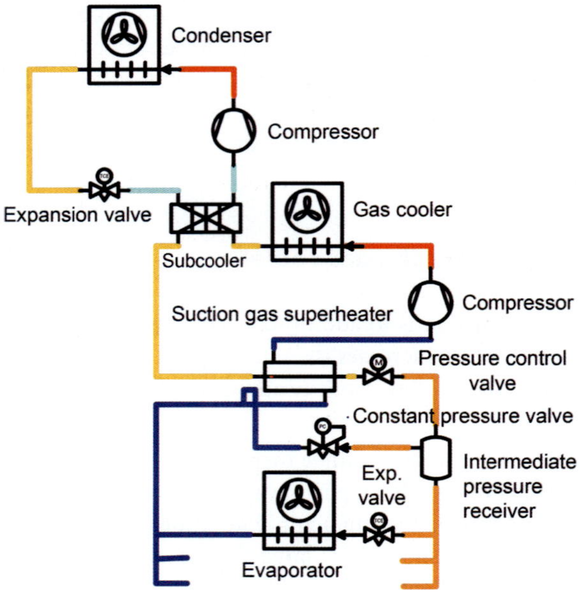

Figure 5.37 Transcritical cooling installation using subcooling via separate cooling system

For mechanical subcooling, the amount of additional cooling by the CO_2 system corresponds to the cooling capacity of the auxiliary cooling system. The overall energy yield is lower than shown in Figure 5.36, as the energy requirement of the entire system has increased. The overall improvement can nevertheless be significant, especially at high gas cooler outlet temperatures. With an efficient auxiliary cooling system, which should preferably contain a natural refrigerant, notable efficiency increase can be expected. These could correspond to about 50 % of the values in Figure 5.36 (additional drive power needed for the auxiliary cooling system). The amount of subcooling – by means of an auxiliary cooling system – is an optimisation problem, which has its technical limits. For CO_2 installations with an intermediate-pressure receiver, subcooling is limited by the lowest permissible pressure (saturation temperature) in the receiver. If the subcooling is too large, the pressure in the intermediate-pressure receiver cannot be maintained. The inlet refrigerant flow liquefies the vapour within the receiver. Hence the receiver's pressure drops below the required value to maintain liquid flow towards the feeding valves of the evaporators in the system.

With increasing subcooling needs, the size of the auxiliary system becomes more important. With an additional subcooling of 10 K, for example, the required cooling capacity of the auxiliary system can be 10–15 % of the entire CO_2 installation capacity.

5.6.4 Expansion work recovery

The refrigerant inside the gas cooler has a high pressure and a high energy content. Most of the energy or potential work is lost during expansion down to the evaporator pressure, i.e., it is converted into low-temperature heat. This loss of energy is commonly referred to as expansion loss.

Process energy can be recovered by (partially) converting the expansion loss into work. At the same time, the amount of flash gas decreases because less energy is converted into heat, and the specific cooling capacity increases. The expansion work can be recovered with all refrigerants, but this is particularly important with CO_2 due to the special thermodynamic properties of this high-pressure working fluid [5-9].

The work can be recovered using a turbine, an expander or an ejector. The output of the expansion machine can – under certain circumstances – be transferred directly to the drive shaft of the compressor or used to operate an auxiliary compressor.

The principle of energy recovery using a turbine / expander is shown in Figure 5.38.

At present, the use of turbines / expanders as standard in CO_2 refrigeration systems has not yet been fully developed, as their costs and maintenance effort are still relatively high.

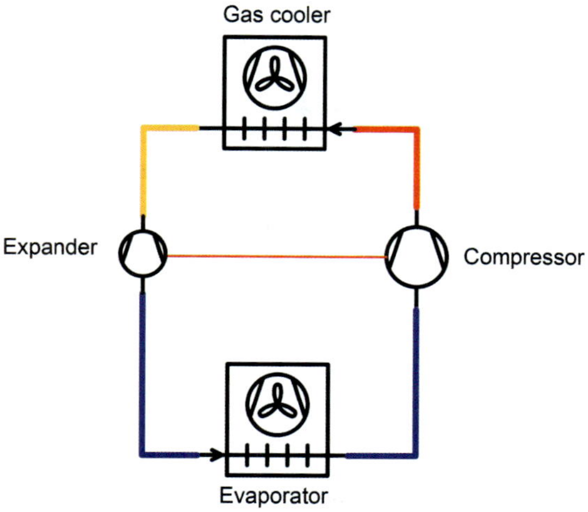

Figure 5.38 Transcritical CO_2 cycle using regenerative expansion turbine

On the other hand, an ejector can best be described as a gas-driven pump or compressor that can pre-compress the suction gas of the main compressor by means of the expansion work. The work input for the main compressor is reduced because its pressure ratio has now been reduced. An ejector can also be used to pump liquid. In this way, it ensures an excess of liquid at the exit of the evaporator and thus enables a much better heat transfer. The basic structure and pressure curve of an ejector is shown in Figure 5.39.

The working principle of the ejector is based on the relationship between pressure and velocity, which can be expressed by the Bernoulli equation. The high pressure (PH) of the supercritical refrigerant is converted into high speed in the motive nozzle (kinetic energy). Due to the very high speed, the pressure around the jet drops and a low-pressure level (PS) is created in the suction chamber prior to the mixing chamber. This causes the refrigerant to be drawn into the ejector at pressure level (PL). The motive flow and suction flow rate are now mixed in the mixing chamber and reach a common velocity. Slowing down the motive flow increases the pressure in the mixing chamber. A further increase in pressure occurs in the diffuser, where a steadily increasing diameter further reduces the flow velocity.

Since the year 2000, a lot of effort has been put into finding practical applications for both solutions. The ejector is much easier to implement than the turbine / expander and, moreover, easier to adapt to the mostly varying operating conditions. The development and integration of ejector-based expansion work recovery is at present the most advanced technology to improve the efficiency of CO_2 vapour compression units.

As shown in Figure 5.40, the ejector is connected to the outlet of the evaporator and compresses evaporated refrigerant to the pressure level of the receiver. Depending on the boundary conditions, i.e. the expansion work available, the pre-compression can be 5–10 bar (72.5–145 psi). The COP improvement reaches up to 17 % at the design point. The efficiency of the ejector, i.e. the ratio between the expansion work and the compression work, is approximately 30–35 % at the design point.

In the circuit shown in Figure 5.40, the ejector is solely used to maintain refrigerant circulation through the evaporator. The amount of refrigerant is controlled by an expansion valve in front of the evaporator. The low-pressure receiver enables the evaporator to operate in flooded condi-

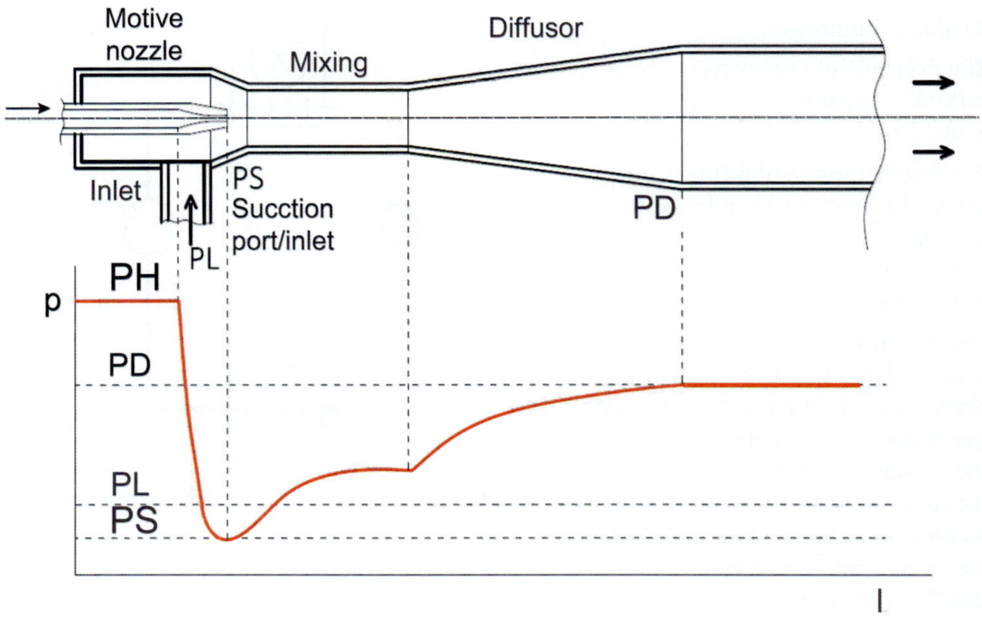

Figure 5.39 Typical ejector assembly showing pressure regain

tion. The pressure rise in the ejector is reduced if the suction gas contains liquid, but this is more than compensated for by a significant increase in evaporator efficiency (flooded operation) instead of the previously applied superheat settings.

The ejector is also able to work at low outside temperatures under subcritical conditions occurring and if the available expansion work is reduced. There will be less pressure increase though because the refrigerant flow through the ejector has a lower energy content, which means a significantly lower expansion work recovery potential. In addition, the ejector operates at a reduced efficiency because it is usually not designed for such operating conditions.

The ejector works best under design operating conditions. If the actual conditions deviate (strongly) from the design conditions, the efficiency of the ejector drops noticeably and thus the efficiency of the system is reduced. The system shown in Figure 5.40 is therefore best suited for relatively constant operating conditions.

Controlling the capacity of ejectors (e.g., flow rate control of the motive flow through the nozzle) is complex, especially if the efficiency of the ejector should be maintained at a high level. A multi-ejector circuit provides a compact solution for variable cooling needs. It can maintain set point conditions for different cooling capacities and for air-cooled gas coolers without significant performance losses. Ejectors used in various operating conditions will have different capacities and design points. Figure 5.41 shows a possible system diagram.

The system is equipped with two receivers (intermediate and low pressure) and parallel compressors. The ejector function is slightly different from the simple solution presented in Figure 5.40. Two ejectors draw vapour from the low-pressure receiver, thereby unloading the main compressors, and deliver the CO_2 to the intermediate-pressure receiver, from where it is compressed to the gas cooler pressure level by the parallel compressor. In contrast to the system in Figure 5.40, the ejectors have no influence on the intake pressure of the main compressor. A third ejector may pump liquid from the low-pressure receiver to the intermediate-pressure receiver, thereby ensuring that the evaporators can always be operated in flooded mode all year round [5-10].

The ejectors are switched on and off as required by solenoid valves (corresponding check valves are integrated in the ejector block, but not shown in Figure 5.41). When using multi-ejectors, there is no need to install an additional high-pressure control valve, because a multi-ejector has sufficient redundancy. The high-pressure control valve shown in Figure 5.41 has a safety function that is activated

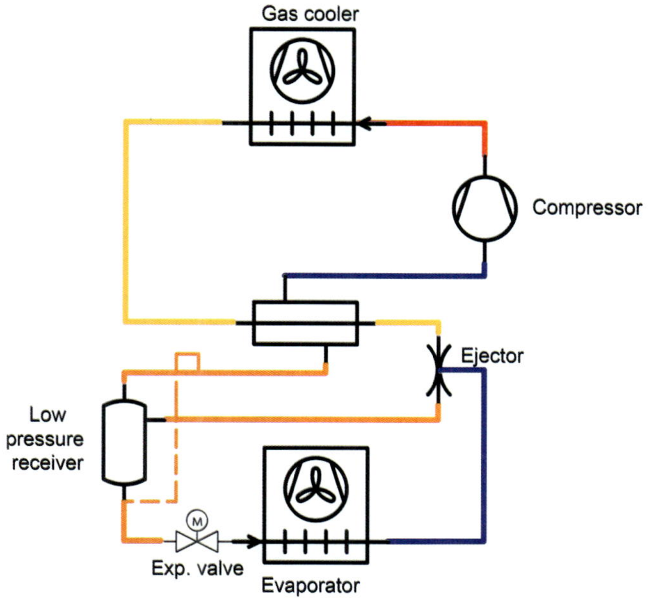

Figure 5.40 CO_2 cycle using regenerative expansion ejector

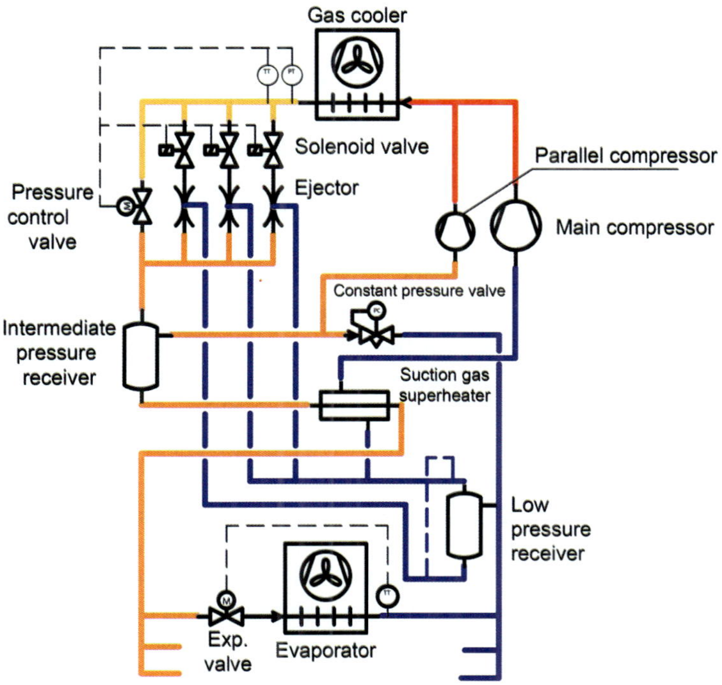

Figure 5.41
Schematic CO_2 installation with expansion work regain by multi-ejectors

when the pressure in the gas cooler rises too high or the pressure in the intermediate-pressure receiver falls below a certain value.

Similar to Figure 5.41 and also described in Chapter 5.10, ejector arrangements and alternative ejector integration concepts are used in cooling systems for supermarkets. The present integration of ejector technology in CO_2 refrigeration systems is just the beginning. Numerous manufacturers of this component are already able to offer them in mass production and integrate their operation into the installation's control system.

Further ejector applications will surely be seen in the future. These will make it possible to provide additional temperature levels in various parts of a refrigeration system. This means that the main part of the system can be operated at a higher compressor suction pressure level. An ejector can raise the refrigerant mass flow rate from auxiliary evaporators, with a limited cooling demand at reduced evaporation temperatures. These auxiliary evaporators can thereby be operated without an additional compressor stage. This increases the energy efficiency of the entire system. As an alternative, the refrigerant mass flow from these auxiliary units can be throttled to the next lower pressure level (e.g., the LT compressors) in order to avoid unnecessarily lowering the suction pressure of the main refrigeration system.

5.7 Heat recovery with CO_2 installations

The utilisation of waste heat from refrigeration systems is receiving increasing attention for both economic and ecological reasons. It is possible to engineer future integrated CO_2 systems for most types of buildings (supermarkets, hotels, etc.). These can fulfil both cooling and heating requirements. For example, if the energy system of an entire supermarket is optimised, the unit could also meet the heating and cooling requirements of surrounding buildings. In this case a central CO_2 system could represent the cost- and energy-efficient energy centre of the entire neighbourhood, e.g. the supermarket and other surrounding buildings.

The integration of heat recovery in an energy centre is particularly important for systems with CO_2 as a refrigerant. Such systems are compact and competitive. Due to the transcritical process, CO_2 systems can achieve very high temperature lifts. For example, hot water preparation from 10 °C to 70–90 °C (50 °F to 158–194 °F) can be achieved without auxiliary electric heaters, which would be necessary with conventional heat pumps.

The use of waste heat is also possible and can be realised if the system is operated in the subcritical mode. In this case, where the system operates near the critical point, the usable amount of waste heat in the heat exchanger downstream of the compressor is still larger compared to conventional refrigerants. For example, around 40 % of the waste heat for a CO_2 refrigerant circuit (between −10 °C and 25 °C /14 °F and 77 °F) is available at elevated temperatures in the superheated region. For conventional refrigerants this value is only 10 to 15 %.

A potential utilisation of this superheat depends above all on the final temperature after compression, and on the inlet temperature at the heat exchanger. CO_2 installations operate at significantly higher discharge temperatures than HFCs, but their temperature levels are still lower than for ammonia (NH_3). If the heat requirement and its temperature level can be achieved by direct heat transfer, without adjusting the high-pressure level, neither the power demand nor the energy efficiency (COP) are affected in a negative way. If the inlet temperature is too low to meet this requirement and therefore not achievable, this temperature can be increased when implementing a suction gas heat exchanger.

If the heating demand cannot be covered by the available superheat energy downstream of the compressor, it makes sense to adjust the high pressure of the CO_2 refrigeration system. This simple measure is part of standard control units and enables the usable amount of heat to be adjusted in accordance with the demand. To ensure that this pressure increase reduces the energy demand of the overall system, the proportion of additional work to provide the pressure increase must be less than the heat gained. This presupposes that the gained heat contains a minimum amount of useful heat. If all or major parts of this waste heat can be re-used, it always will be energy efficient by using realistic technical improvements to increase the high pressure of the system. However, the proportion of energy recovered can vary widely depending on the solution chosen. When the pressure is increased, it is likely that the critical pressure will be exceeded, and the process will become transcritical. With respect to the energy efficiency of the system, this is easily achievable as long as the temperature upstream of the pressure control valve is kept low (see Chapter 5.4.4). The next chapter describes how this can be implemented in practice.

If the process has to be carried out transcritically, more heat will become available at higher discharge pressures. Secondary fluids may thus be heated to a higher temperature level.

5.7.1 Heat recovery at one temperature level: a simple solution

The heat exchanger for this method of heat recovery is installed in the high-pressure line downstream of the compressor and upstream of the condenser or gas cooler. The technical solution is depicted in Figure 5.42. The circuit shown is used for hot water preparation and demonstrates one of many ways to use the waste heat.

This solution is suitable for both the recovery of heat content in superheat states (subcritical operation) and in transcritical operation for general heat recovery.

In subcritical operation it is possible that the refrigerant is already partially liquefied in the heat exchanger of the heat recovery unit. For fear of possible damage to the gas cooler (liquid hammer and erosion), some manufacturers warn against operation with two-phase flow into gas coolers. However, it is uncertain whether this is really a problem in practical installations.

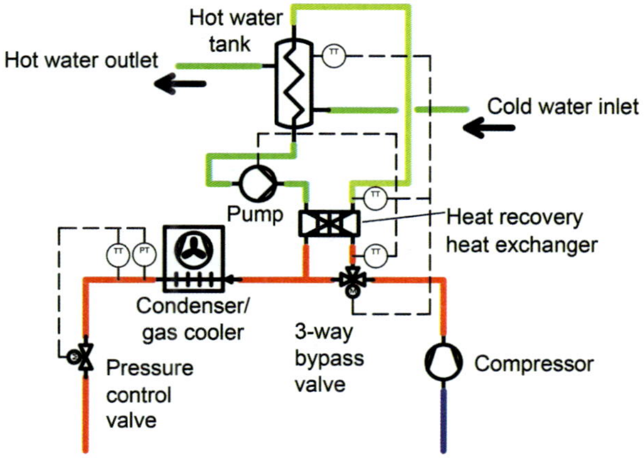

Figure 5.42 Heat regain at one temperature level, a typical hot water boiler circuit

The heat recovery system can be arranged in different ways. An intermediate circuit can be used – as in Figure 5.42 – or the water may be heated directly. The solution shown protects the drinking water circuit against refrigeration oil pollution and a significant pressure increase in the event of any leakage in the heat exchanger. In addition, temperature fluctuations are dampened more effectively by the irregular use of the hot water. With an indirect version, there are no practical challenges due to possible deposits in the heat exchanger if the drinking water is very calcareous.

The temperature level and temperature fluctuations of the high-pressure gas will vary depending on the operating conditions. Occasionally, temperatures above 100 °C (212 °F) can be reached, but sometimes only 80 °C (176 °F) is achieved. The thermal control of the water heating system must therefore be based on the operation of the CO_2 installation. A flexible solution would be a controllable pump for the heat transfer medium or a flow control valve. This then determines the flow rate in the intermediate cycle, depending on the CO_2 temperature and the supply temperature of the heat transfer medium. In addition, various safety measures are required to protect the heat exchangers and the heat transfer medium during low heat dissipation quantities and at high CO_2 entry temperatures.

- If the refrigeration system is switched off or the heat recovery is not in operation (the pump has been stopped), the CO_2 gas is then redirected through the bypass valve towards the next heat exchanger.
- The three-way valve is also brought into its bypass position when, e.g., a hot water tank is fully charged or the temperature in the heat exchanger exceeds a certain limit (safety mode).
- Complete temperature monitoring and a flow switch on the secondary loop side can provide additional protection against unintentional flow through the heat exchanger with hot CO_2, which could significantly shorten the lifetime of the heat exchanger and cause damage to the secondary loop system.

5.7.2 Heat recovery at different temperature levels

In many locations there is often a need for heat at various temperature levels. The waste heat from the cooling system can be used, for example, for water heating, space heating (radiation and ventilation) and snow/ice melting (in front of the building in winter) [5-11]. In such cases, it may make sense to install a separate heat recovery cycle for each purpose. Typical examples of this heat recovery chain can be found in supermarkets where CO_2 is used as a refrigerant and this advanced and integrated solution for heat recovery is used. In the Nordic regions in particular, this is a widely used solution. It is used in Europe and is starting to gain a foothold in the US. Figure 5.43 shows an example of heat recovery for three different applications.

The hot gas after the compressor is used for hot water generation, as described above.

In transcritical operation, CO_2 has a particularly high specific heat capacity around the critical point. It is important to cool the refrigerant down to temperatures of 30 °C (86 °F) or below. Heat in this temperature range can be used for air heating or low-temperature space heating. The diagram shown in Figure 5.43 assumes that the space is heated by warm air.

A de-icing or a snow melting system for the entrance area can be effectively provided with low-temperature heat. This heat exchanger should be implemented downstream of the heat recovery unit for the space heating system. To place this heat exchanger downstream of the air-cooled gas cooler can also be advantageous for other reasons. If necessary, the refrigerant can be

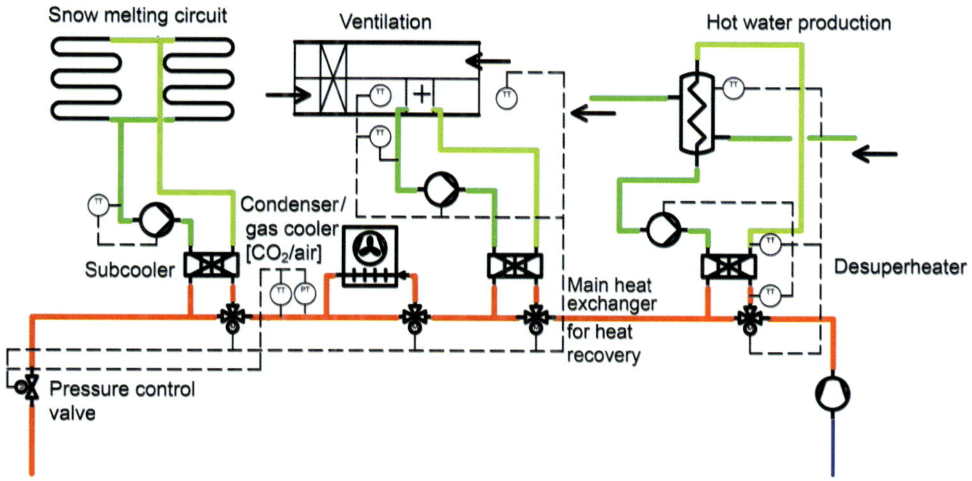

Figure 5.43 Typical layout for heat recovery at three temperature levels

cooled further on warm summer days (see "Summer operation" section) the CO_2 temperature is kept low all year round after flowing through all aforementioned heat sinks. (In some early cases, fatigue fractures have occurred in heat exchangers located downstream of the compressor. The heated fluid with its varying and relatively low temperature will meet the hot CO_2, thus causing thermal stresses. Nowadays the well-designed heat exchangers for heat recovery can cope with these fatigue issues.)

Summer operation

If heat is required neither for space heating nor for melting ice in the entrance area, the waste heat will then only be used for hot water preparation. The high pressure in the refrigeration system is thus not affected by the heat recovery.

High outside temperatures occur mostly for a few hours in the afternoon. Under such circumstances, the external circuit downstream of the gas cooler can help to subcool the refrigerant. The system control circuity detects whether it is possible to discharge heat through the "snow melting circuit". This depends on the level of the return temperature compared to the ambient air temperature. Energy wells or geothermal probes can also be implemented (for air conditioning or heat pump operation, see section "Supermarket cooling, integrated system concepts" in Chapter 5.10.2.2). These indirect circuits regenerate the energy well for winter operation and at the same time further subcool the refrigerant after the gas cooler on hot summer days. The CO_2 must always be routed past inactive heat exchangers to avoid unnecessary heat cycles. As a result, the lifetime of these components and the system can be significantly extended.

Operation in spring and autumn

As soon as space heating is required, the corresponding heat exchanger in the heat recovery chain is switched on and active within the refrigeration cycle. In the example shown, the temperature in the heat transfer circuit and the heat output are regulated by means of a speed-controlled pump.

If the demand for heat increases, the heat recovery may not cover the entire demand. To counteract this, the high pressure of the refrigeration system is increased by the high-pressure control device. The required increase in system pressure is managed by a signal from the heating system. If several heating systems are provided with waste heat, the heating system with the highest heat demand is prioritised, provided that this reference heating circuit covers the major part of the total requirement. (When designing, it should be noted that the energy of the heat recovery is not generated exclusively by additional compressor work.) The corresponding high pressure of the refrigeration system is then increased up to a specified maximum pressure, e.g., 80 bar (1160 psi). The upper level of the controller signal area, e.g., 8–10 V (volts), can be used to control fans and the bypass valve of the gas cooler (see below).

The provision of thermal storage (possibly with phase change materials) should be considered in order to dampen short-term fluctuations in heating requirements. This is the preferred option, instead of continuously adjusting the high-side pressure of the system.

Winter operation

All available heat can and should be used in winter. Transcritical operation is usually a prerequisite for using this heat. For this reason, it is important to keep the temperatures required for space heating (e.g., underfloor heating, etc.) as low as possible, to install the appropriate infrastructure, and to apply the appropriate control strategy.

In order to deliver the best result for the customer, it is important to carefully analyse the required temperature profiles of the various heating systems and heat sinks well in advance. Only then can the heating circuits and any heat storage be correctly dimensioned and integrated into the system architecture. Above all, care should be taken to ensure that a much larger temperature gradient on the heat transfer medium side must be achieved in the ventilation heat exchangers than is customary with conventional heating systems so that the return temperature to the gas cooler in the CO_2 circuit is as low as possible.

If there is a good temperature control upstream of the CO_2 high-pressure control valve, the critical pressure can be exceeded without having large losses in system efficiency. Without this regulation, and if the high-pressure refrigerant is not cooled sufficiently, control of the process near the critical point can become unmanageable.

If the gas cooler operates at full fan power, a low refrigerant temperature, close to ambient temperature, is ensured upstream of the high-pressure control device. However, at the same time, a significant amount of heat is released into the environment. This is not a problem as long as provision is made for the required system heat. If this is not the case, the performance of the gas cooler must be reduced. The first step is to reduce the fan speed, if necessary, to a standstill. It is controlled by the voltage signal from the heating system. For example, the voltage range of 8–9 V can be assigned to this function. If the heating requirement is still not covered, the gas cooler is bypassed using the bypass valve on the refrigerant side to prevent heat loss through wind and natural convection. The upper part of the voltage range, 9–10 V, is used for this control. If the snow-melting system is required, additional cooling of the refrigerant is achieved. If the demand is met without further increasing the high-side pressure, the heat for the snow melt is practically for free. At the same time, the specific cooling capacity increases, and the power consumption of the compressor decreases.

If more heat is required for the snow-melting system than is available in the previously described operating state, the system's high pressure can be increased further (use a separate 0–10 V signal). The heat is then released at the higher pressure level. However, it should be checked whether the pressure increase is still energy efficient. If not, it makes more sense to integrate an additional heat exchanger (heater) in the snow-melting circuit and combine it with other heat sources such as energy wells or sea water. This principle, as described above for maximising heat recovery, which was developed for the first time in 2009 by a Norwegian company, contributed significantly to the early introduction of CO_2 refrigeration systems in the supermarket sector in Scandinavia.

5.8 CO_2 heat pumps

5.8.1 General

Most space heating applications require temperatures that cannot utilise condensing CO_2 in the conventional refrigeration process (27–28 °C / 80.6–82.4 °F; see Chapter 5.3.2.1). Therefore, the CO_2 heat pumps are usually operated in transcritical mode. To achieve a high COP with such a system, this should be done as set out in Chapter 5.4. It is particularly important to note that the fluid to be heated by the heat pump must have the lowest possible temperature level [5-12].

The amount of supplied heat and the COP of the CO_2 heat pumps are mainly determined by the temperature glide in the gas cooler as well as by its operating pressure. The heating capacity increases when more refrigerant is cooled in the gas cooler (without increasing the electricity requirement of the compressors). However, energy efficiency has an optimal value (see Chapter 5.4.4) at which CO_2 heat pumps should generally be operated. In addition, it is important that the compressors have good performances both under design and at partial load condition [5-13].

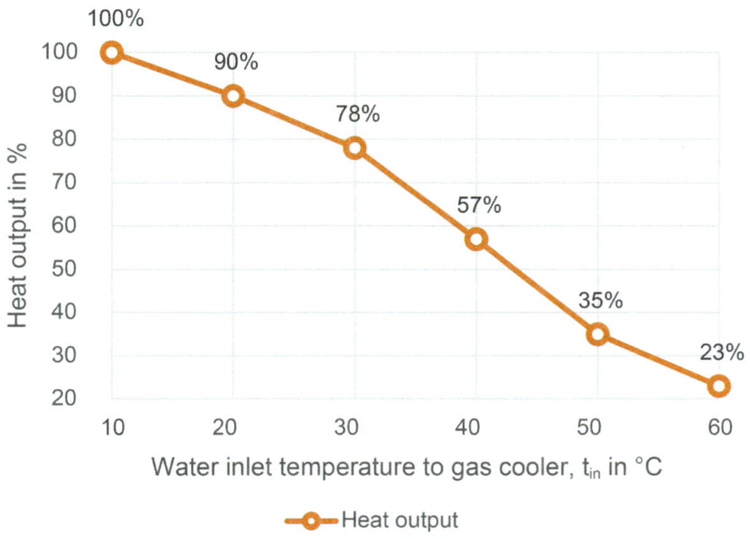

Figure 5.44 Percentage heat output of a single-stage CO_2 heat pump versus water inlet temperature into gas cooler

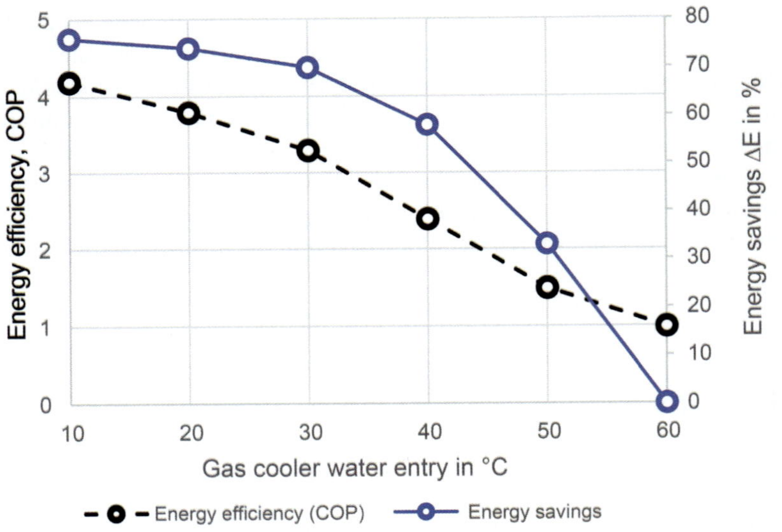

Figure 5.45
Typical COP and energy savings for single-stage CO_2 heat pump versus water inlet temperature into gas cooler

Figure 5.44 and 5.45 show the percentage heat output and the COP together with the percentage of energy savings with varying water inlet temperature into the gas cooler. The evaporation temperature is set to 0 °C (32 °F) and the gas cooler pressure to 100 bar (1450 psi). The isentropic efficiency of the compressor was assumed to be constant at 70 %. The heat output is set to 100 % at an inlet temperature of 10 °C (50 °F).

5.8.2 Production of industrial hot water

Hot water production is one of the best applications for CO_2 heat pumps, since water temperatures of 85–90 °C (185–194 °F) can be reached without the use of an additional auxiliary system [5-14]. In addition, a much higher COP is achieved than with conventional hot water heat pumps. As already mentioned, the CO_2 technology from SINTEF/NTNU was introduced onto the market for the first time in Japan ("EcoCute"); see Chapter 5.4. These heat pumps were ultimately exported to Europe. There are presently many European CO_2 heat pump manufacturers. CO_2 heat pump systems for multi-family housing, hotels and other industrial applications are built and delivered across Europe.

Hot water generation using CO_2 heat pumps achieves better performance figures and higher energy savings than conventional systems for the following reasons:

- lower average refrigerant temperature during heat rejection due to better temperature adaptation between the refrigerant and water side in the counterflow gas cooler
- due to the high CO_2 gas cooler inlet temperatures and the gliding temperature during heat rejection, tap water can be heated up to 60–90 °C (140–194 °F). There is also no need for a disinfection cycle
- high performance levels and efficiencies can be expected from the heat pump compressor due to the lower pressure ratio operations
- excellent heat transfer in the heat exchanger due to the favourable properties of CO_2

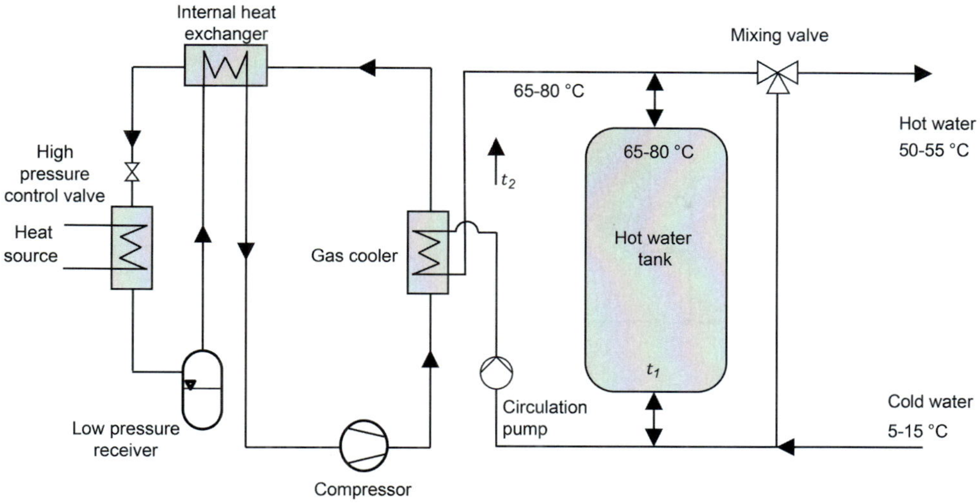

Figure 5.46 Principal circuit for a CO_2 heat pump connected to a basic hot water system

The conditions within the gas cooler change only slightly for hot water heat pumps with well-designed water storage devices. For a heat source with a relatively constant temperature all year round, e.g., exhaust air, the hot water systems can be operated with a constant gas cooler pressure [5-15]. To achieve this, the optimal pressure in the current temperature range is determined and then set. If conditions change, such as with outside air as a heat source, the gas cooler pressure set point is managed using an optimised pressure algorithm.

Figure 5.46 shows the basic structure of a hot water system connected to a CO_2 heat pump. A hot water storage tank, a rpm-controlled water pump in the gas cooler circuit and a mixing valve in the pipe system are also shown. The temperature level in the lower part of the tank (t_1) controls both compressor and water pump operation.

When hot water is drawn out, cold water flows into the tank from the inlet on the lower side, while the hot water flows into the building via the mixer. It is a must that the storage tank be equipped with a well-designed diffusor at the lower water inlet port, otherwise the stratified water will be mixed when warm water is required, and a jet of cold water enters the tank. The heat pump is most commonly switched on and off (intermittent operation).[9] If the water temperature in the lower range (t_1) falls below a certain value, e.g., 20 °C (68 °F),[10] the heat pump starts, and the rpm-controlled water pump circulates water through the gas cooler. The flow rate provided by the water pump is adjusted to achieve the desired storage tank temperature, e.g., 70 °C (168 °F) (t_2). If the water temperature t_1 reaches its highest design temperature in the tank, e.g., 30–40 °C (86–104 °F), the heat pump is switched off.

Alternatively, two temperature sensors, one in the tank and one in the water line to the gas cooler, can be used to achieve a slightly better storage capacity. Multiple tanks can be used, see Figure 5.47. The heat pump switches off when the temperature at the gas cooler outlet exceeds a

[9] A capacity-controlled compressor can also be used.
[10] Low temperatures are associated with a high coefficient of performance and a small storage capacity (and vice versa).

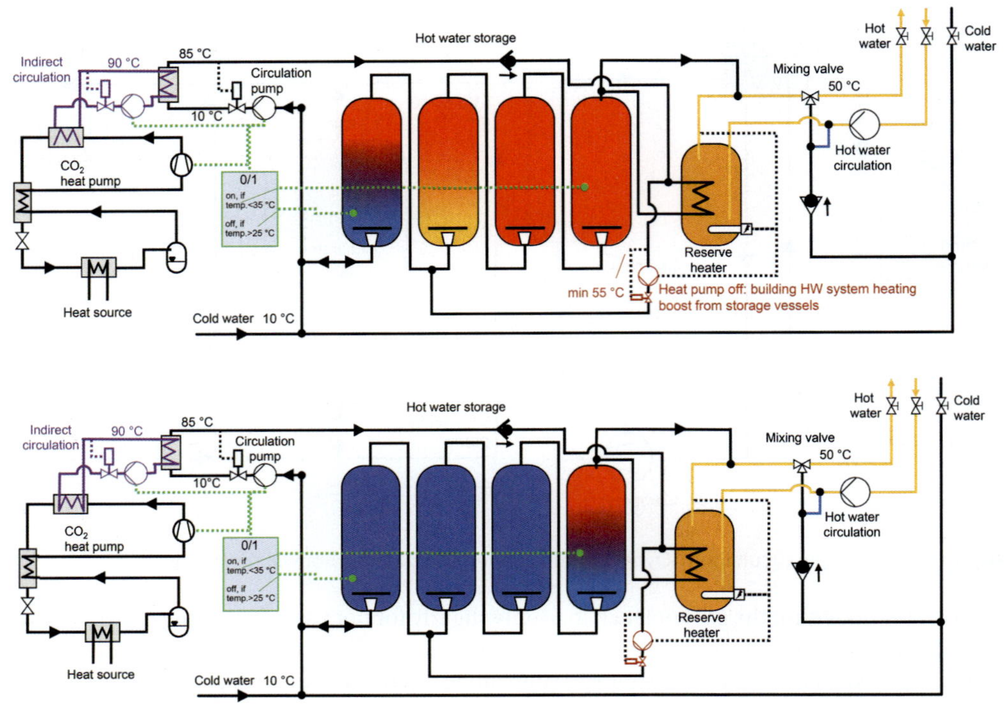

Figure 5.47 Two schematics for CO_2 heat pumps including the water storage devices. Upper: storage depleted, lower: storage full, heat pump off

certain temperature, for example, 20 °C (68 °F). In large systems with several tanks connected in series, a so-called reference tank can help to increase the effective storage capacity. The heat pump is switched off when the temperature in the reference tank reaches the specified temperature.

Important conditions for the design, installation and operation of CO_2 hot water heat pumps are:

- *Heating capacity:* The installed heating capacity of heat pumps should correspond to the average daily requirement + 20 % in reserve (that is a daily runtime of 20 hours and 4 hours on standby).
- *Gas cooler connection:* The heat exchanger must be operated in counterflow to achieve gas cooler outlet temperatures at the lowest possible CO_2 temperature and thus to achieve the best coefficient of performance.
- *Temperature adjustment:* In a properly dimensioned gas cooler for hot water production, the temperature difference between CO_2 gas cooler outlet temperature and water inlet temperature (Δt_A) should not exceed 3–5 K.
- *Heat pump control:* The compressor can be operated in on-off cycle mode. There is no need for frequency control; the storage devices should enable sufficient capacity to enable long operation intervals. The control system should facilitate reaching the optimum gas cooler pressure under all operating conditions.

- *Water flow rate control:* The pump in the gas cooler circuit should be rpm-controlled to maintain the water supply temperature accordingly (e.g., 70–90 °C/158–194 °F). A corresponding flow control valve can be used as an alternative.
- *Dimensioning of the storage tanks:* The total volume of the storage system is calculated according to the daily hot water demand. Good stratification between the warm and cold water is important. The internal heat transfer and especially mixing increases the average water inlet temperature in the gas cooler and thus lowers the efficiency of the system and the storage capacity. The tanks should therefore be slim and tall with a moderate cross-sectional area. If several storage tanks are required, they must be connected in series; very large systems may connect two series connections in parallel. In addition, the connections or pipes should be dimensioned large enough. Diffusers must be installed in the lower area of the storage tanks so that the water inlet velocity does not lead at all to an inhomogeneous distribution of the tank contents, particularly if larger quantities of hot water are drawn. The correct design of the storage system is the most important prerequisite for an energy-efficient CO_2 hot water heat pump system (see Figure 5.7).
- *Circulation pumps and reserve:* In large buildings, a buffer storage tank with a heat supply to reheat the return water from the recirculation pipe should be integrated into the hot water distribution pipework. Additional auxiliary heating elements (backup) can be dimensioned to have the same heating capacity as the heat pump.

Figure 5.47 shows an example of a hot water system with a CO_2 heat pump > 15 kW (4.3 TR). Outside air, an earth probe or surface water (river water, sea water, etc.) can serve as a heat source. The gas cooler is operated in counterflow, and the tanks are connected in series. The example shows an indirect system solution for the heat pump and the water circuit.

5.8.3 Combined space heating and hot water preparation

CO_2 heat pumps can provide space heating and hot water at the same time. However, in order to achieve the highest possible energy savings, it is important to keep the average temperature of the water in the gas cooler relatively low. The criteria for this are:

- The share of heat for hot water must make up a relatively large part of the total heat requirement. Homes, nursing homes, residential complexes and apartment buildings with low heating load or built to passive house standards generally have a low space heating requirement. In these buildings, hot water treatment typically requires 50–80 % of the total annual heat, depending on the standard of living, the type of building and the climate zone.
- The heat distribution system for space heating and any heated ventilation air from heat recovery should have a medium flow temperature and a low return temperature. Large heat exchanger surfaces in the heating system of the ventilation system, combined with low-temperature radiators (50–40 °C / 122–104 °F) and / or floor heating (35–30 °C / 95–86 °F) are ideal. The flow temperatures of such systems can then be controlled based on the outside temperature.
- The hot water system with the arrangement of the storage tanks, the operation of the pumps and the gas cooler etc. is identical in construction to the pure hot water preparation with a CO_2 heat pump.

- The hot water system and the heat distribution system are connected via heat exchangers. The system is used to achieve the lowest possible temperature in the gas cooler. An optimal system then has a minimal primary energy requirement for the simultaneous provision of warm water, space heating and, if necessary, space cooling.

5.9 Safety when using CO_2

Even if CO_2 is not acutely toxic in low concentrations, a number of safety measures must be taken into account when used in refrigeration systems and heat pumps. The measures relate, on the one hand, to physiological effects, and on the other hand, to technical conditions such as pressure, temperature and the possible formation of dry ice.

5.9.1 Pressure and temperature – safety conditions

The pressure levels in CO_2 refrigeration systems are significantly higher than in systems with other refrigerants, which initially pose additional safety-related challenges. The current permissible high pressures (HP) are given in Table 5.2 in Chapter 5.3.4.2. The system pressure will increase when liquid CO_2 is present when the unit is at standstill (idle). This pressure might be higher than the permissible operating or working pressure of system components as well as the entire system. Therefore, it may be necessary to:

- Cool down all or part of the system during standstill (idle) periods with a separate cooling system ("cold finger") or a controlled, partial CO_2 "blowing off" initiated at a pressure below the safety valve opening pressure
- Secure all the pipes and components that may contain trapped liquid by equipping them with an automatic internal blow-off valve or an external safety valve
- In the case of manual shut-off valves, to point out a potential danger of liquid inclusion. If necessary, install safety valves on "dead legs"
- Define the shutdown and discharging processes in the service manual, e.g., the sequence for manual shutdown and how to operate the various valves
- The entire system up to and including the valve on the filler neck is to be dimensioned for a pressure of 90 bar (1305 psi)

The compressor casing of CO_2 compressors used for cascade systems can be selected for a significantly lower maximum pressure than standstill pressure, for example, 26 bar (377 psi). If these cascade systems are designed to operate at higher pressures, the low pressure (26 bar / 377 psi) compressors must be switched off automatically and disconnected from the main system. In addition, the compressor casing should be protected with its own safety valve. It should also be noted that the pressure rise is unusually high when heating pure CO_2 (due to the high initial pressure level). This should be taken into account when estimating the installation's standstill pressure. When using CO_2 in cascade systems, particularly low temperatures are achieved. The risk of frost damage to the skin from touching cold parts increases accordingly.

When the system is discharged, the risk of direct contact with cold components or refrigerant increases. Dry ice, formed when the liquid is blown out, has a temperature of −78 °C (−108.4 °F). Even short-term contact can cause frost damage. Pipes and components made of unalloyed steel become brittle and are prone to breakage at low temperatures. Pipes that are exposed to particularly low temperatures, as when discharging a system, should be made of stainless steel or copper.

5.9.2 Dry ice – safety and operating conditions

The formation of dry ice when the system pressure falls below 5.18 bar (75.13 psi) requires various measures, both in operation and in terms of safety. There is a great danger that, if the discharge / drain valve becomes blocked by dry ice near the liquid outlet, it can be erroneously concluded that the system is empty. It is therefore extremely important to develop and follow good work processes when maintaining CO_2 facilities. Some important elements of such work processes are:

- connecting a pressure gauge to the part of the system that is being emptied
- draining liquid-containing system parts only from the gas phase and through a pressure control valve; this ensures that the pressure in the system is always kept above approximately 6 bar (87 psi) so that dry ice cannot form in the drainpipe
- no direct draining of liquid into the environment, liquid refrigerant should be poured into a pressure-tight container at a pressure above 6 bar (87 psi)
- taking a manometer reading to confirm that it no longer contains any liquid, then releasing CO_2 into a pressure-tight container (pressure level above 6 bar / 87 psi)
- relieving blocked valves with a gas flame or similar is not permitted
- charging installations with gaseous CO_2 only until a system pressure of approximately 6 bar (87 psi) is reached

Dry ice can be a particular challenge in connection with safety valves and discharge / drain lines for safety valves. Under certain conditions, dry ice forms due to pressure relief; see Figure 5.4 in Chapter 5.3.3. View state point "a" at t = 20 °C (68 °F). Upon pressure relief, from 60 bar to 4 bar (870 psi to 58 psi), line "a" drops vertically into dry ice phase of CO_2 chart. The amount of dry ice that is formed is greatest when the liquid is pressurised beforehand, but as the illustration shows, dry ice can be formed even when a system is emptied from the gas phase. In the worst case, dry ice can block the safety valve or drainpipe, causing an uncontrolled rise in pressure in the refrigeration system. Measures that best prevent this are:

- a safety valve or liquid overflow valve installed internally in the system
- situating the vapour / gas safety valve far away from any liquid part of the system
- situating a safety valve at the end of the blow-out pipe. Alternatively, the drainpipe after the safety valve must be very short and without bends. The pressure drop between the tank or system part and the safety valve must be less than 3 % of the opening pressure (general rule)

The last topic concerns, among other things, intermediate-pressure receivers, where the setpoint for the safety valve is usually 80 to 90 bar (1160 psi to 1305 psi).

At pressure levels below the triple point pressure, dry ice is formed when the temperature is below 50 °C / 80 bar (122 °F / 1160 psi) or 62 °C / 90 bar (144 °F / 1305 psi). It must also be pointed out that the discharge temperature from the compressor should be above 80 °C (176 °F) to avoid the formation of dry ice. At lower temperatures, dry ice will occur if the safety valve blows off at 120 bar (1740 psi).

The applicable safety-related conditions for the use of CO_2, and the meaning they have, will vary from system to system. It is therefore important that a full risk analysis of the installation design, which is required by regulations, is carried out on every individual installation system.

5.10 Examples for CO_2 refrigeration system applications

CO_2 is principally suitable as a refrigerant and coolant (secondary circuits) for all applications. Due to unique process properties, CO_2 is better suited for certain applications. In some areas, it is challenging to compete with conventional systems in terms of energy efficiency.

The technical and practical reasons why the use of CO_2 is increasing for new applications can be attributed to the fact that CO_2 is neither flammable nor toxic. In addition to the large temperature glide during heat rejection and the good process properties due to the high-pressure levels, the excellent heat transfer properties are very advantageous.

The temperature glide in the gas cooler is very beneficial in heat pumps when a high temperature increase is required, for example in hot water generation. CO_2 is increasingly used in systems with heat recovery from industrial refrigeration systems, in areas that require a lot of hot water, such as in the processing of milk and other foodstuffs. Heat recovery in supermarket refrigeration systems, taking advantage (at least partially) of the temperature glide, is standard today. Other areas of application like heat pump chillers will no doubt follow soon and are described below. The high evaporation pressure is advantageous in all CO_2 systems because the dimensions of the compressor, the valves and pipes, as well as the pressure / temperature losses due to friction in the pipes and fittings are very small. CO_2 refrigeration systems can be operated very effectively at low temperatures (down to –50 °C/–58 °F) without facing the challenge of conventional systems to operate below 1 bar (14.51 psi) (ambient) on the low-pressure side, as e.g., with ammonia systems.

The good heat transfer characteristics contribute to the fact that the heat exchangers can be manufactured more inexpensively and thus the CO_2 systems, viewed holistically, are economically competitive.

5.10.1 Cascade refrigeration installations using CO_2 as a refrigerant

5.10.1.1 Industrial installations

There are many cascade refrigeration systems operating with CO_2 and ammonia. CO_2 as a refrigerant for low temperatures has many important advantages that most end users have not yet considered and applied in all areas. The advantages are described in more detail in Chapters 5.10.1.4 and 5.10.2.9 [5-16].

Cascade systems are slightly more expensive to build than conventional ammonia refrigeration systems. However, the operators of such systems are very satisfied with the systems and appreciate their advantages [5-17].

However, cascade systems are very common in Europe and the USA, since those countries have strict safety requirements for ammonia refrigeration systems. This is due to the usually high mass of the refrigerant charge. NH_3 / CO_2 cascade systems now enable the use of ammonia in new and unusual applications. CO_2 covers the refrigeration area with a classic subcritical compressor circuit, where the CO_2 is used as an evaporating refrigerant. The pressure rating of a normal CO_2 system is 41 bar (594.7 psi) on the low-pressure side and 53 bar (768.7 psi) on the high-pressure side. This is very advantageous because this pressure range is covered by common reciprocating piston and screw compressors. The use of CO_2 as a refrigerant in industrial systems started in cascade installations. Ammonia is used as the main refrigerant in the upper / warm stage. This system solution is used in many refrigeration installations for safety reasons, as direct product contact with ammonia can be excluded. Today refrigeration systems with CO_2 / ammonia in smaller refrigeration capacity ranges and very low gas / liquid charge quantities can be provided.

In larger cold stores, CO_2 is used as an evaporating refrigerant in cascade systems – both to cover the cooling requirements for medium and low temperature. The main refrigerant is mostly ammonia, but occasionally also hydrocarbons or still HFCs. For small- or medium-sized refrigerated warehouse buildings, pure CO_2 is now becoming the more common refrigerant (transcritical booster design) instead of a CO_2 cascade. An advantage of CO_2 as a refrigerant, compared to ammonia, is that a leak will not damage valuable goods. Food that has come into contact with ammonia is unusable and can no longer be sold. By using CO_2, the insurance premium for the goods becomes cheaper.

5.10.1.2 Artificial ice rinks

CO_2 was first used in artificial ice rinks in 1999 [5-18]. CO_2 was used as an evaporating refrigerant for cooling ice rinks (especially ice hockey rinks) in Austria. Over time, more and more systems for artificial ice rinks were built, using CO_2 as a direct evaporating refrigerant. On the one hand, new installations were built. On the other hand, old installations, in which ammonia was used as a direct evaporating refrigerant, were converted to CO_2 for safety reasons. Nowadays there are more than 100 artificial ice rinks worldwide which use CO_2 as a refrigerant [5-19]. The first artificial ice rink with a transcritical CO_2 system was built in Canada in 2011. This technology is now also used in other countries.

Compared to conventional systems with an ammonia refrigeration system that cools the usual coolant (calcium chloride), the energy requirement of a comparable CO_2 refrigeration system with direct evaporation under the ice is reduced by 25 %. This can largely be attributed to the higher evaporation temperature and the significantly lower pump power (reduced by 90 %). The first CO_2 ice rink in Canada achieved a 33 % lower energy requirement in its first season compared to the previously used R22 rink. Conventional copper pipes (ø 12 x 1 mm / ø 0.47 x 0.04 in) are often used as evaporator tubes for CO_2 systems. Steel pipes (ø 21,3 x 2 mm / ø 8.39 x 0.079 in) were often installed in the earlier direct-evaporation installations. Plastic-based pipes (ø 33,7 x 2 mm / ø 1.33 x 0.079 in) are often used for conventional coolant systems (mainly $CaCl_2$). In comparison to systems with circulating calcium chloride brines, CO_2 systems can be installed using much longer pipes. The distance between the pipes is usually 100 mm (3.94 in). The design pressure for the pipes is typically 41 bar (594.7 psi). The experiences with evaporating CO_2 in pipes for artificial ice rinks are summarised below:

- pump power is very low (approximately 2 kW (6828.3 Btu/h) for the entire ice hockey and curling track)
- evaporation temperature can be increased by 2–3 K compared to conventional refrigerants (high heat transfer coefficient on the inside of the pipe, lower heat transfer resistance in the metal pipe wall, pipes can be fixed closer to the ice surface)
- much more stable temperatures on the ice surface (Δt = 0,5 K compared to 2–3 K with $CaCl_2$ or glycol)
- faster adjustment of the ice temperature is achievable

5.10.1.3 Retail installations

The first retail refrigeration systems with CO_2 as refrigerant were warehouses. CO_2 was used as an evaporating refrigerant in conventional ammonia cascade systems. The aim was to reduce the amount of ammonia and avoid the use of ammonia in the warehouse. There are still a few warehouses using ammonia (or HFC) / CO_2 cascade systems. Nowadays, however, most developers seem to realise that transcritical CO_2 installations are particularly suitable to Central and Northern Europe, while modern CO_2 installations or well-designed cascades with ammonia / CO_2 represent an ideal and energy- efficient combination for warehouses in warm climates.

The food retail sector was one of the first activity areas where CO_2 was used as a refrigerant (initially mainly in cascade systems). HFC was often used in the upper cascade [5-20].

In the early years prior to the millennium, cascade systems upward from a certain system size were the standard solution in supermarkets, particularly in Scandinavia. There were two installation concepts: one used CO_2 only for the low-temperature area, while medium-temperature cooling was covered by an indirect system with conventional coolants. In the other system concept, CO_2 was also used in the medium-temperature cooling area, but transported to the respective evaporators by pumps. In some cases, conventional cascade heat exchangers were used, or the CO_2 was liquefied by using a secondary circuit.

Nowadays mainly transcritical systems have taken over the market in Scandinavia as well as the rest of Northern Europe. CO_2 cascade systems are no longer considered viable options for supermarkets in Southern Europe, the United States and Asia. A change in attitude occurred prior to 2020, as energy-efficient transcritical CO_2 systems for warmer climates [5-21] became available and are now considered state of the art. The former CO_2 "equator" no longer exists.[11)]

5.10.1.4 Fishing boats

The first Norwegian industrial cascade system with CO_2 / ammonia was installed in the MS Kvannøy, a fishing vessel, in 2002 [5-2]. The goal was to increase the freezing capacity by lowering the evaporation temperature without using additional ammonia in the cooling system. The CO_2 refrigeration system cools eleven plate freezers and nine rooms (empty RSW[12)] tanks)

11) The CO_2 equator was the name of a boundary defined by annual mean temperatures. South of this limit, conventional CO_2 booster installations had a higher annual energy requirement compared to conventional installations.

12) **R**efrigerated **S**ea **W**ater.

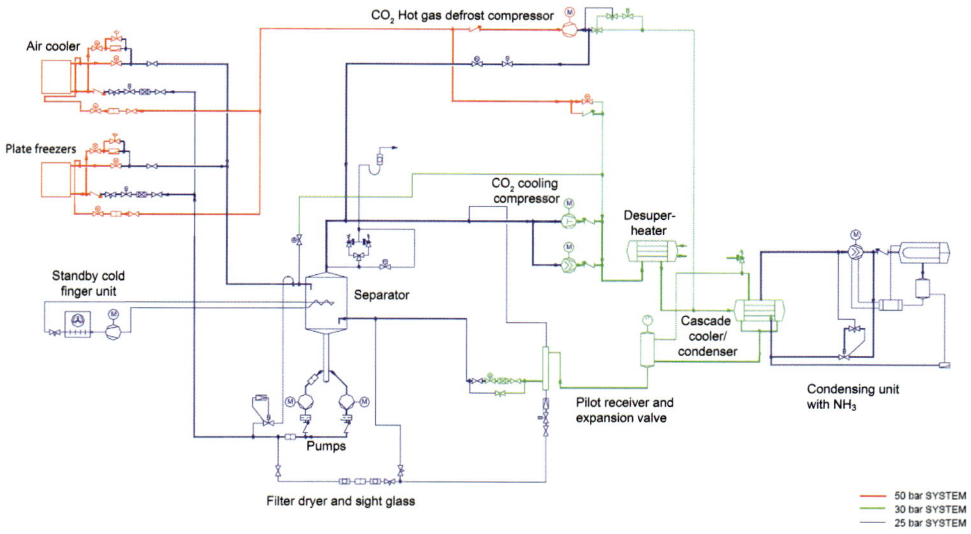

Figure 5.48 Installation scheme of the CO_2 ammonia cascade system on board the MS Kvannøy

when used as a deep-freeze storage hold and an ice maker. The cooling capacity is 1359 kW (386.4 TR) at an evaporation temperature of –48 °C (–54.4 °F). RSW is processed in a separate RSW system, which applies ammonia as the refrigerant [5-22]. Figure 5.48 shows a detailed installation diagram for the cascade system.

Compressor selection: The system diagram in Figure 5.48 depicts a concept study, in which only one piston compressor and one screw compressor were shown. The installation was actually equipped with six reciprocating compressors for CO_2 and two screw compressors for ammonia. In addition, a specially manufactured reciprocating compressor was installed that compresses gas up to 50 bar (725.2 psi), which is used for hot gas defrosting. The defrost compressor sucks CO_2 gas from the high-pressure side of the ordinary CO_2 compressor downstream of the desuperheater. This compressor could also be used for normal operation, however, only outside the programmed defrost phases.

Chiller oil: A POE (polyol ester) was used.

Cold finger: The CO_2 circuit is secured with a so-called "cold finger" so that the temperature of the system does not rise above –25 °C (–13 °F) when the refrigeration system is stopped. Should the system pressure nevertheless rise and approach the discharge pressure of the safety valve, the CO_2 quantity is reduced (blown-off) with the help of a solenoid valve and vent line.

Challenges and experiences:

A very important finding from the first CO_2 / NH_3 facility was that the interaction of CO_2 and NH_3 compressors needs careful analysis for normal operation and defrosting. Due to the high volumetric cooling capacity of CO_2, there is an immediate change in the load on the ammonia compressor when the CO_2 compressor pistons switch on or off. Screw compressors react sluggishly to such changes, which complicates keeping the pressure in the CO_2 receiver within the desired pressure limits. For this reason, advanced control algorithms must be ensured. With the help of efficient frequency converters, which are now available, the adjustment of the cooling capacities can be implemented in a straightforward way.

In this case, the greatest challenge was the operation of the defrost compressor, as it worked close to permissible operating range and required a relatively stable pressure on the suction side. The problem was solved by a hot gas bypass for the defrost compressor, which was controlled with a fast-acting control valve.

Another important factor for installations that are operating close to the triple point of CO_2 is that the minimum cooling capacity of the CO_2 compressor must be carefully analysed and chosen. If this minimum capacity is too large, compressors must be switched off in order to avoid dry ice formation in the liquid separator and the evaporators, at low load. If the system then requests cooling power again after a short time, this cannot always be provided immediately, since the specified number of compressor starts / stops is limited in certain periods.

The experience with the refrigeration system on the MS Kvannøy was positive. Compared to conventional units providing –40 °C (–40 °F), the freezing time was reduced by 40 % and the freezing capacity increased. The reliability of the system was also very satisfactory. This was recorded in an interview with the chief engineer in 2005. The feedback from the operator of a sister ship that was similarly equipped was also very positive. In the meantime, these types of fishing vessels were mostly fitted with ammonia in combination with CO_2 or with CO_2-only systems [5-23]. In addition to the high freezing capacity, the effective and gentle defrosting of the plate freezers is of great practical value as the fish quality remains high.

5.10.2 Transcritical systems

5.10.2.1 Small systems for retail

This section deals with bottle coolers, vending machines, freezers and other self-contained refrigerators for the storage, sale and transportation of chilled food, etc.

The starting point for the development of CO_2-chilled vending machines were campaigns by non-governmental organisations (NGOs) in the run-up to the 2004 Olympic Games in Athens. These drew attention to the enormous number of beverage cooling machines worldwide. A leading global lemonade manufacturer then decided that all of its vending machines (over 10 million units) should be replaced with CO_2-cooled devices at the end of their lifespan. As of January 2014, 1 million CO_2 beverage cooling machines were already in operation. The cooling system is integrated in an easily exchangeable cassette, which simplifies service even for people without any special technical knowledge.

Systems with even smaller cooling capacities from other manufacturers were also converted to CO_2, especially in Japan. In conventional beverage vending machines, which contain both warm and cold drinks, the usual electrical heating has been retained because the CO_2 system could effectively provide hot drinks and chilled drinks at the same time.

Stand-alone refrigerated shelves for retailers have also been developed. These are obviously in competition with similar hydrocarbon systems that can also be operated in the long term and have now become state of the art. These types of refrigerated shelves are very attractive for very small markets (kiosks, petrol stations etc.) because the costs for installation and maintenance are lower compared to central refrigeration systems. However, the corresponding annual energy requirement is usually higher than that of central installations. The type of heat dissipation from the cooling units plays a crucial role. If the waste heat can be released into the public market, it may have to be released outside in summer, using an air conditioning system. If a water-cooling cycle (waterloop) is installed, the waste heat can be used specifically to also heat other parts of the building, or it can be released into the environment via dry coolers.

With these waterloop systems, it is particularly important that all the market's cooling points be adequately supplied with the cooled secondary fluid, e.g., water for the cooling shelves. Only then will the condensing temperature of the conventional circuit operate at a normal level. If this is not the case, this will lead to a significantly increased energy demand. The basic advantage of CO_2 systems of this type is that they achieve a higher temperature rise at reduced secondary fluid flow rates (similar to hot water production with CO_2 heat pumps). This allows for extensive use of the waste heat and may reduce the challenge of the uneven circulation rates through the cooling points and shelves.

A Japanese manufacturer has shown that single CO_2 cooling shelves could save more than 40 % in energy compared to the previously common R404A systems. The major advantage in this case was a more efficient compressor and the implementation of a frequency converter for the compressor, as well as appropriately adapted evaporator temperatures.

5.10.2.2 Large capacity central systems for retail and industrial refrigeration

The first CO_2 supermarket systems had separate circuits for medium-temperature (MT) cooling and low-temperature (LT) operation. The first pilot installation was installed in Belluno in Italy in 1999, and the first installation in Scandinavia was installed in Trondheim in 2004.

Figure 5.49 shows the example of such a system with a separate medium-temperature and low-temperature circuit. The waste heat from the different circuits can either be discharged directly to the air using air-cooled gas coolers or – as shown here – via a secondary circuit with a common dry cooler. A simplified system connection diagram shows the installation of a small supermarket in northern Italy, which was built in 2003. The cooling requirement is 120 kW (34 TR) for medium-temperature cooling (evaporation temperature –10 °C / 14 °F) and 25 kW (7.1 TR) for low-temperature cooling (evaporation temperature –35 °C / –31 °F).

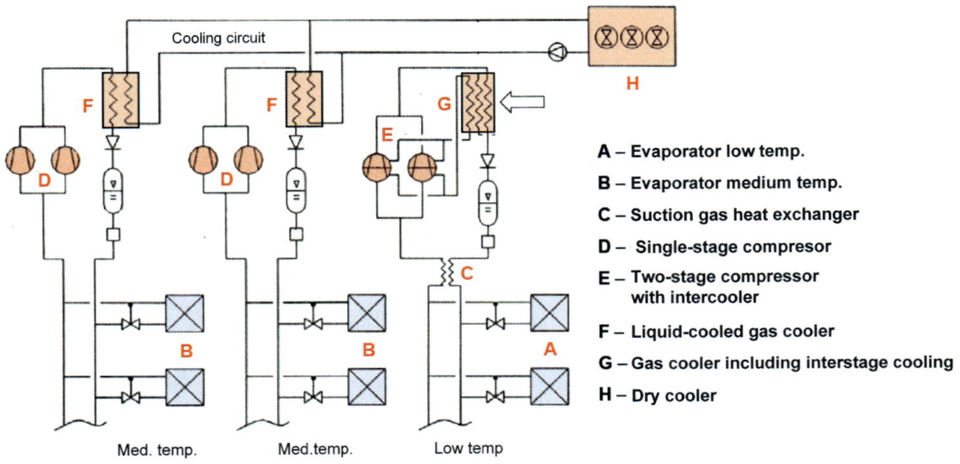

Figure 5.49 Schematic for transcritical CO_2 installation with separate medium- and low-temperature circuits

It should be noted that the cooling circuit at that time still consisted of two separate units for reasons of redundancy.

In transcritical operation, the pressure is reduced in two stages. A differential pressure valve with a fixed setting is used for throttling between the gas cooler and the liquid receiver. The evaporators are fed by electronic expansion valves.

The intermediate-pressure receiver is not pressure-controlled and remains at the pressure level of the corresponding saturation line (liquid at the boiling point). The pressure depends on the gas pressure and the temperature before the first throttling; see Chapter 5.5.3.2, and is usually between 50 and 70 bar (725.2 and 1015.3 psi). If the outside temperature permits subcritical process control, an overflow channel (bypass) in the differential pressure throttle valve is opened so that the same pressure prevails in the condenser and the downstream liquid receiver.

The detailed scheme for a freezer is shown in Figure 5.50. A suction gas heat exchanger is integrated to protect the compressor from liquid hammer. To limit the compression end temperature and the oil temperature, the refrigerant is intercooled after the first compression stage. At the same time, the compressors have their own oil coolers, and oil from the central oil separator is cooled before being returned to the suction line in a controlled manner. The Main components are: 01 compressor with integrated second compression stage; 02 subcooler; 03 liquid receiver; 04 oil separator; 05 oil cooler; 06 external oil cooling of the compressor; 07 liquid subcooler, suction gas superheating; 08 suction side receiver (protection of the compressor against liquid hammer); 09 suction side collecting manifold (first stage); 10 suction-side collecting manifold (second stage); 11 collecting manifold after the 1st compressor stage; 12 collecting manifold after the 2nd compressor stage; 28 differential pressure throttle valve with overflow valve; 29 intercooler (1–2 compression stage).

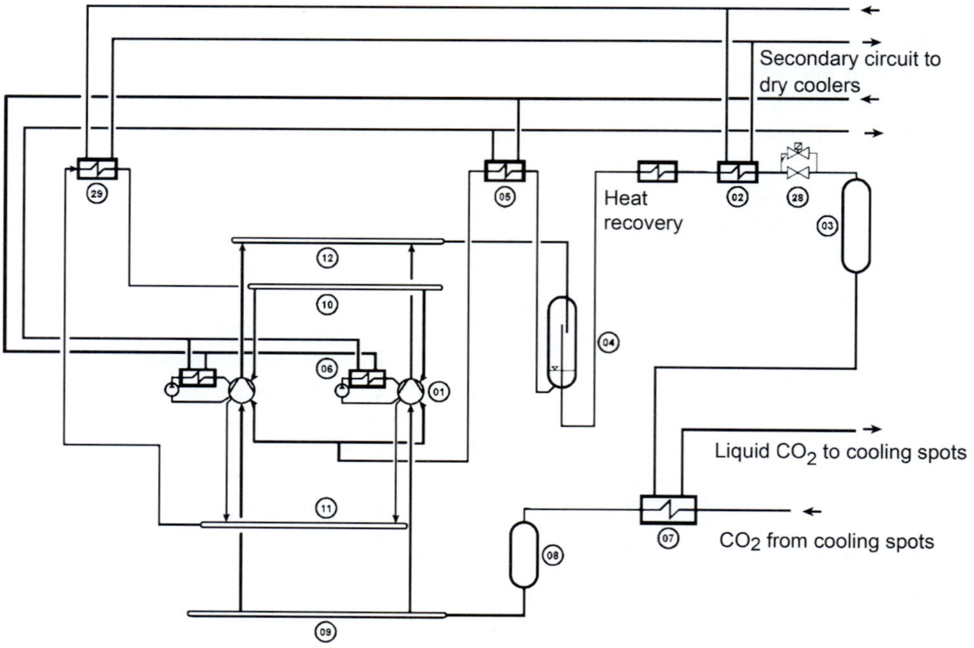

Figure 5.50 Schematic diagram of a two-stage transcritical CO_2 installation

Supermarket cooling, standard solution ("Booster")

Supermarket refrigeration has been the major application for transcritical CO_2 refrigeration systems [5-9]. Most supermarket refrigeration systems are using standard refrigeration systems based on the booster principle, which means that both medium- and low-temperature cooling are provided with the same refrigeration circuit. The technical solution is in principle as shown in Figure 5.34 (right-hand side example). The standardisation of these systems and components has resulted in very reliable operation, provided that they are carefully installed and maintained. Figure 5.51 shows the first example of a CO_2 booster installation diagram (Linde / Carrier-Design, 2004) [5-24]. The freezer still had its own CO_2 circuit that was integrated with a cascade connection. However, all systems today are equipped with a direct connection between medium- and low-temperature cooling, as shown in Figure 5.34 (right-hand side example). Two independent circuits were installed in the first system in 2004. This offered operational safety and redundancy for the end user, because each refrigeration cycle had its own oil balance. Today, however, oil-return can be handled properly in booster systems. There are three pressure levels in this system with appropriate design and protection of the components: 116 bar / 1682.4 psi (high pressure), 41 bar / 594.7 psi (medium pressure) and 26 bar / 377 psi (low pressure). The pressure level of the medium-pressure and low-pressure stages was deliberately set to normal operating limit values; this ensured the use of refrigeration components that were standard at the time. In the start-up phase of these systems, the maximum working pressure in the pipes and components that were used in the area of the shop open to customers was set at 41 bar (594.7 psi). The pressure level in the refrigerant receiver after the high-pressure control valve was measured using a constant pressure valve (flash gas bypass valve).

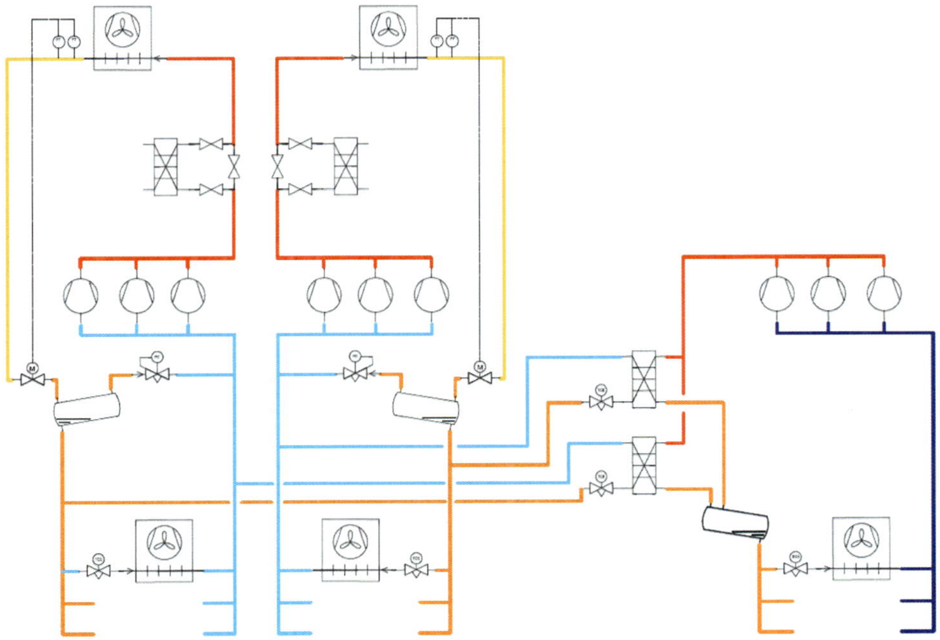

Figure 5.51 First-generation booster system in a supermarket in Switzerland

This valve was connected to the suction line of the medium-temperature cooling compressor. Heat recovery was integrated by transferring heat between the medium-temperature cooling compressors and the gas cooler.

Today, standard components are available, e.g. pipes, fittings and heat exchangers, all of which are designed for higher pressures. In order to extend the time until the refrigerant is blown off in the event of an installation shutdown, the design pressure of the medium-pressure collector and the entire medium-temperature cooling circuit are selected at 61 bar (884.8 psi). Compressors manufactured to a design pressure of 101 bar (1464.8 psi) for the low-temperature circuit in the system are now also available.

When laid out for 61 bar (884.8 psi) or more, this can be directly connected to the medium-temperature circuit. This now makes the cascade arrangement unnecessary.

These booster systems are usually built with heat recovery. As described in Chapter 5.7.1, the amount of useful heat or the heating capacity of the system can be adapted to the need by regulating the high pressure. These CO_2 booster systems are more energy efficient than those previously used with HFC 404A. As a rule, all HFC systems in Scandinavian supermarkets have now been replaced with CO_2 booster systems. This trend can also be seen in other regions such as in Central and even Southern Europe.

Supermarket cooling, integrated system concepts

Heat recovery in CO_2 refrigeration systems is regarded as standard nowadays. A further development of the previously shown systems are the so-called integrated systems or energy centres [5-25]. These units provide cooling and heating for the entire building. There are countless practical examples of such systems. Figure 5.52 shows a system for a supermarket in Trondheim [5-26].

The heating systems are designed for low supply temperatures. If there is an increased demand for heat in the building, which cannot be adequately covered by the available waste heat from the cooling racks, the heat shortage is balanced by cooling the brine from a geothermal probe. The soil around the geothermal probe was previously charged with waste heat resulting directly from the operation of the supermarket's refrigeration system.

With this concept, three gas coolers are connected in series downstream from the medium-temperature compressors. The heat is recovered from the CO_2 booster system at three temperature levels. The first gas cooler (CO_2 / water) is designed for a supply temperature of 60 °C (140 °F) and loads the warm water storage tanks, which are typically connected in series. After that heat recovery, the warm water is used as peak load to heat the air flowing to the shop area within the ventilation system. The air curtains for the entrance areas are heated with the warm water as well – instead of direct heating using electricity.

In order to reduce the heat required for heating supply air in the ventilation system (HVAC), it usually operates in recirculation mode until the CO_2 sensor in the shop area determines that the ratio of outside air must be increased. If outside air is supplied and it needs to be heated, it is first preheated using the internal heat recovery coil in the HVAC system. If necessary, it can be reheated using the supply air heat exchanger in peak demand situations. This is fed with hot water, i.e. the waste heat from the refrigeration system. If there is no need for heat due to the underfloor heating in the shop area and there is no need for cooling the supply air due to low ambient temperatures, both the fresh air and the exhaust air are bypassing the heat exchangers in the

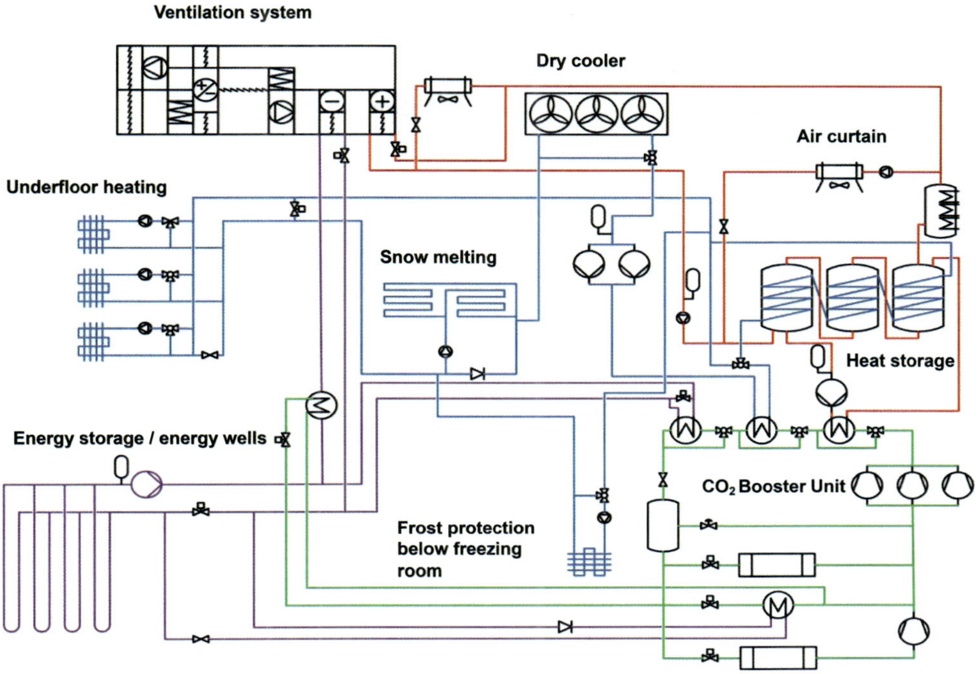

Figure 5.52 Integrated cooling and heating system in a Norwegian supermarket

HVAC system. This leads to a substantial reduction in fan work. The two fans in this example have very high energy efficiency levels because they are equipped with EC motors.

In the second gas cooler, a heat transfer fluid (water glycol) with a supply temperature of around 30 °C (86 °F) is provided for the floor heating in the entire shop area. If necessary, the floor heating circuit can also be supplied with additional heat from the hot water storage tanks. After the fluid has left the floor heating area, it circulates through the defrost circuits in front of the public entrance areas of the market. These are thus kept free of snow and ice by using the waste heat from the refrigeration system. This used to be achieved with heat energy directly from electric heaters. The basement under the low-temperature warehouse is fitted with slab heating, to prevent the formation of frost in the ground. Except in the winter months, the soil surrounding the geothermal probe is loaded with waste heat from the third gas cooler. A secondary fluid (water-alcohol) is used for this circuit, the return temperature from the well remains below 15 °C (59 °F) all year round. Furthermore, the soil around the geothermal probe is directly charged with waste heat from the HVAC system directly in summer. This means that the market premises can be air-conditioned for many weeks without active cooling (i.e. using electricity) by the refrigeration system. If there is a need for additional cooling during the summer period, the additional evaporator within the circuit provides the required cooling capacity.

Figure 5.53 shows a simplified installation diagram of a CO_2 pilot refrigeration system in a building complex in Trondheim. The CO_2 booster system with parallel compression and ejector support is the central energy installation of the whole building with its indoor supermarket and several office floors. The medium-temperature (MT) cooling parts in the market are operated at

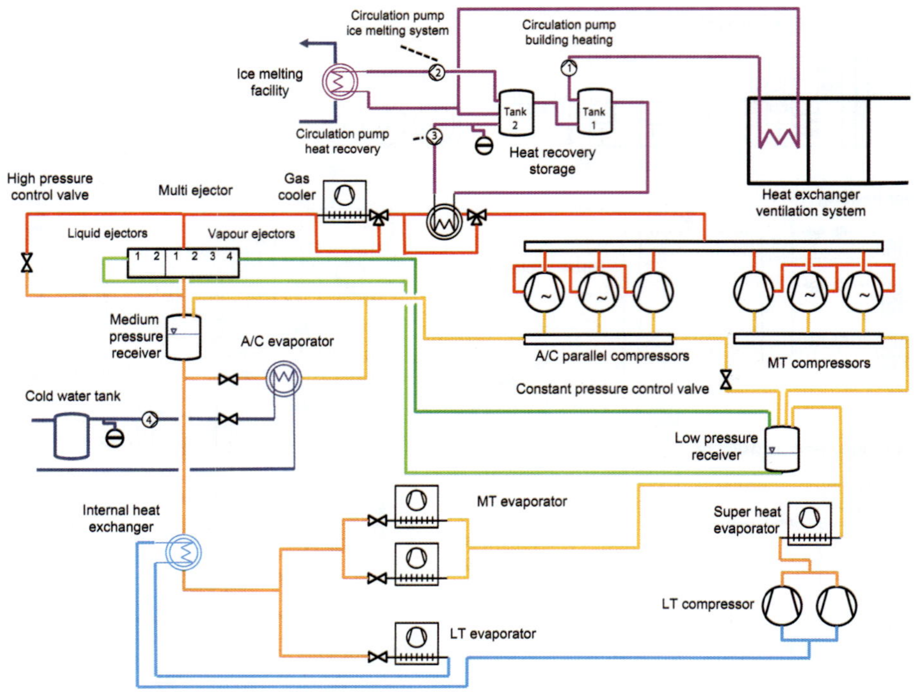

Figure 5.53 Integrated cooling/heating system in a combined office building with supermarket CO_2-booster system with parallel compressors and ejector support (Trondheim) [5-27]

setpoint temperature (without overheating), at an evaporation temperature of –2 °C (28 °F). Up to 250 kW (71.1 TR) cooling capacity can be provided for cooling the circulating chilled water for the AC systems within the building. The waste heat is used to heat the offices in the building via the air conditioning systems, on the one hand, and to keep the entrance areas and walkways free of ice in winter, on the other hand.

Supermarket cooling for warmer climates, parallel compression, ejectors

Methods to improve installation efficiency, such as parallel compression, flooded evaporators and the use of ejectors to recover expansion work, are alternatives for use in CO_2 refrigeration systems. These alternatives will improve overall energy effectiveness, even in warmer climates [5-28].

Figure 5.54 shows a simplified basic installation layout of a CO_2 Booster System with parallel compressors and ejector support (right-hand side example). Evaporators for cooling and air conditioning the building are also integrated. The example shows how modern CO_2 booster systems can cover the entire cooling and heating requirements of a building. This leads to significantly reduced operating costs and less cost for service and maintenance.

If the booster system is equipped with parallel compressors, as in the example on the left in Figure 5.54, these compressors can unload the normal cooling compressors. This additional suction pressure level regulates the pressure level in the intermediate-pressure receiver. In nor-

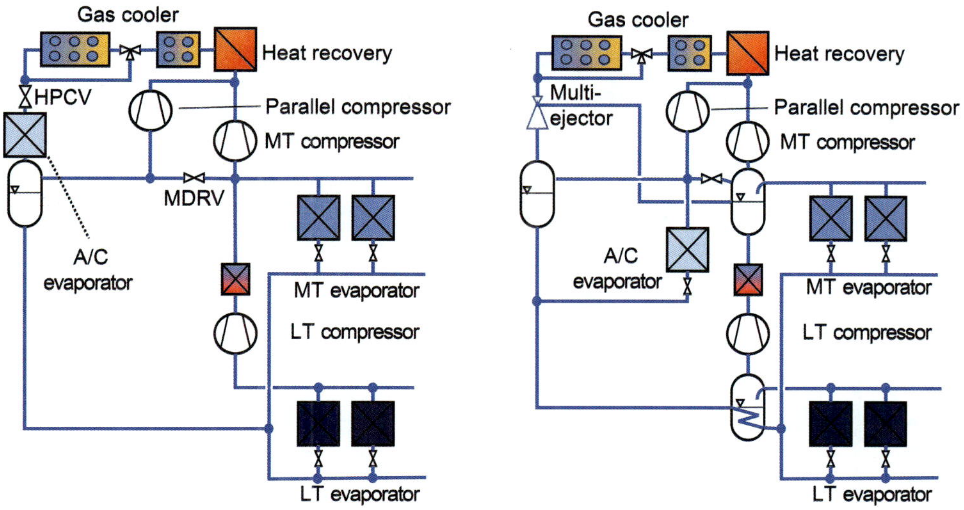

Figure 5.54 CO_2 booster system with parallel compressor and AC evaporator before the intermediate collector (left). Setup with ejector and AC evaporator gravity fed with liquid out of receiver (right)

mal booster systems this pressure control can only be achieved in using a constant pressure valve (flash gas bypass valve, FGBV).

This reduces the power requirement of the entire system, since the suction pressure of the parallel compressor is much higher than that of the medium-temperature cooling compressor. However, the parallel compressors must be dimensioned for this operation so that the optimum efficiency can be achieved at reduced pressure ratio. Since the constant pressure valve is only used in this type of system when the parallel compressors are not in operation due to low suction mass flow rates, the risk of liquid hammer in the medium-temperature cooling compressors is significantly reduced. Thanks to the third compressor stage of the booster system, it is possible to integrate the air conditioning into the system. The cooling for the air conditioning evaporator is provided by the parallel compressor. The evaporation temperatures can be adapted to the cooling requirements or set points of the air conditioning system. The evaporators can be integrated either upstream or downstream of the intermediate-pressure receiver, as shown in Figure 5.54.

The integration of ejectors (see example on the right-hand side in Figure 5.54), which can continuously deliver liquid and gaseous refrigerant from the low-pressure receiver, enables the medium-temperature cooling evaporators to be operated without superheating (i.e. flooded). This makes the heat exchangers much more effective, so that the evaporation temperature in the refrigerated cabinets and cold rooms can be increased significantly. With appropriately designed MT cabinets with doors, the evaporation temperature can be raised to –2 °C (28 °F), which can lead to a reduction in the power consumption of the system by up to 15 %. At these high evaporation temperatures, the defrosting requirement of the heat exchangers is also significantly lower. Defrosting can be reduced from several times a day to possibly once a week.

It is also possible to operate low-temperature (LT) evaporator without superheat, as shown on the left in Figure 5.54. As a result, the suction pressure of the LT compressors can be increased significantly, which in turn leads to a reduced power consumption of the overall system.

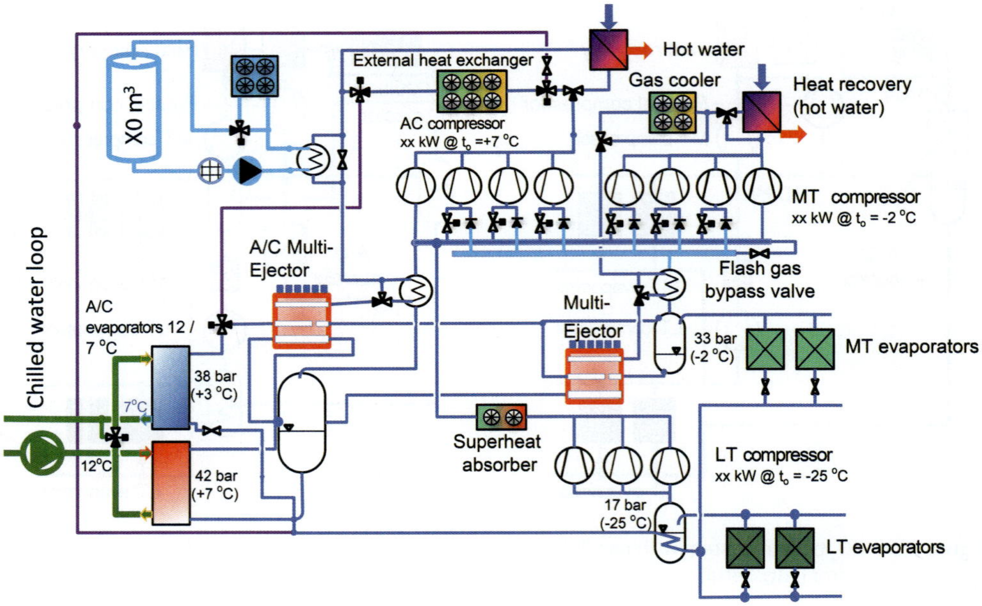

Figure 5.55 Example of a future integrated circuit for supermarkets

Supermarket cooling, future options for integrated systems

Modern supermarket systems are considered to be integrated systems in the sense that both the hot and the cold sides of the refrigeration system are utilised and that all the essential thermal systems of the building are connected to the central refrigeration / heat pump unit. In the previous sections, some examples were shown of how air conditioning can be provided, even on hot summer days. However, integrated solutions are constantly being developed.

The development of such systems generally focuses on reducing their unit costs while maintaining equal or even improving the energy efficiency. Nowadays, end users have the option to implement standardised, energy-efficient CO_2 refrigeration systems with integrated solutions for air conditioning and for heat pump functions from different manufacturers. The energy requirement of appropriately modified, integrated CO_2 refrigeration systems will be reduced by up to 30 %, compared to standard booster systems with separate HFC air conditioning systems. Figure 5.55 shows a circuit of an advanced CO_2 refrigeration system.

Compressor allocation:

The allocation of compressors to the respective suction group is to be flexible with respect to demand and to reduce the number of installed compressors for the future. The different compressors can be assigned either to the medium-temperature cooling suction manifold or to the parallel compressors, as required. The central operational control determines which compressors maintain the pressure level in the low-pressure receiver (MT standard cooling) and in the intermediate-pressure receiver (flash gas, refrigerant supplied by the ejector from the low-pressure receiver of the MT side and air conditioning) most effectively at the corresponding target pressure level.

Integration of the air conditioning:

Chilled water, for the air conditioning of the building, can be provided in a very energy-efficient manner with an ejector-supported refrigeration unit. The chiller evaporator consists of two heat exchangers that are connected in series on the water side. The evaporator, which is connected to the return water line from the air conditioning system, is gravity fed with refrigerant from the intermediate-pressure receiver. In the example shown, the saturation temperature in the intermediate-pressure receiver will be 7 °C/42 bar (45 °F/609.2 psi) when the water returns from the air conditioning system at 12 °C (54 °F). Depending on the cooling demand of the building, the supply temperature of the water can (and should) be adjusted. If lower temperatures are required in the warm season due to an increased cooling demand, the water can be further cooled using the second evaporator. The evaporated refrigerant from the second evaporator is then fully drawn in by the ejector. This ensures a pressure increase of up to 4 bar, if appropriately designed. In this case, the evaporation temperature in the second heat exchanger is lowered to 3 °C/38 bar (37 °F/ psi). Therefore, a chilled-water supply temperature of 7 °C (45 °F) is possible with a suction pressure of 42 bar (corresponding to +7 °C (45 °F) saturation temperature of the CO_2) at the compressor.

External cold absorber:

In climate zones with very high ambient temperatures during the warmest hours of the day and reduced temperatures during the night, it may be economical to include an existing external cold thermal storage unit. An example of such a cold storage device could be the water contained in the building's fire extinguishing water tank. This water can be used as an additional heat sink during the warmest time of day. As a result, the refrigerant temperature downstream of the gas cooler can be further reduced to values below the ambient temperature. During the night, the absorbed heat in the water stored in the fire extinguishing water tank can dissipate into the surroundings via heat exchangers. If available, existing heat exchangers from the previous refrigerating system, which is supposed to be replaced, could be applied for the purpose of night time heat rejection.

Supermarkets with fresh food counters and limited low-temperature demands

In climates such as those found in Central and Northern Europe, the use of ejectors to increase efficiency during the relatively few operation hours at high outside temperatures is often not economically viable. For certain applications, the ejector technology can nevertheless be used advantageously. For example, in markets with open meat and fish counters, it is often difficult to work with the same evaporation temperatures as are required for the closed food counters in the same market. Figure 5.56 shows how these open-display counters can be specifically operated with slightly reduced evaporator temperatures by using an ejector (EJ_{MT_Minus}) without the entire MT suction group's having to work with reduced suction pressures. This leads to a significant energy saving for the market operator, which amortises the additional effort (separate suction line and ejector) in a short time.

In some market configurations, most of the medium- and low-temperature counters are self-contained units that either reject the waste heat to a water cycle or directly into the market's indoor air. In future, instead of using LT compressors for a limited number of evaporators, the integration of these individual LT evaporators in a booster system could be achieved with the

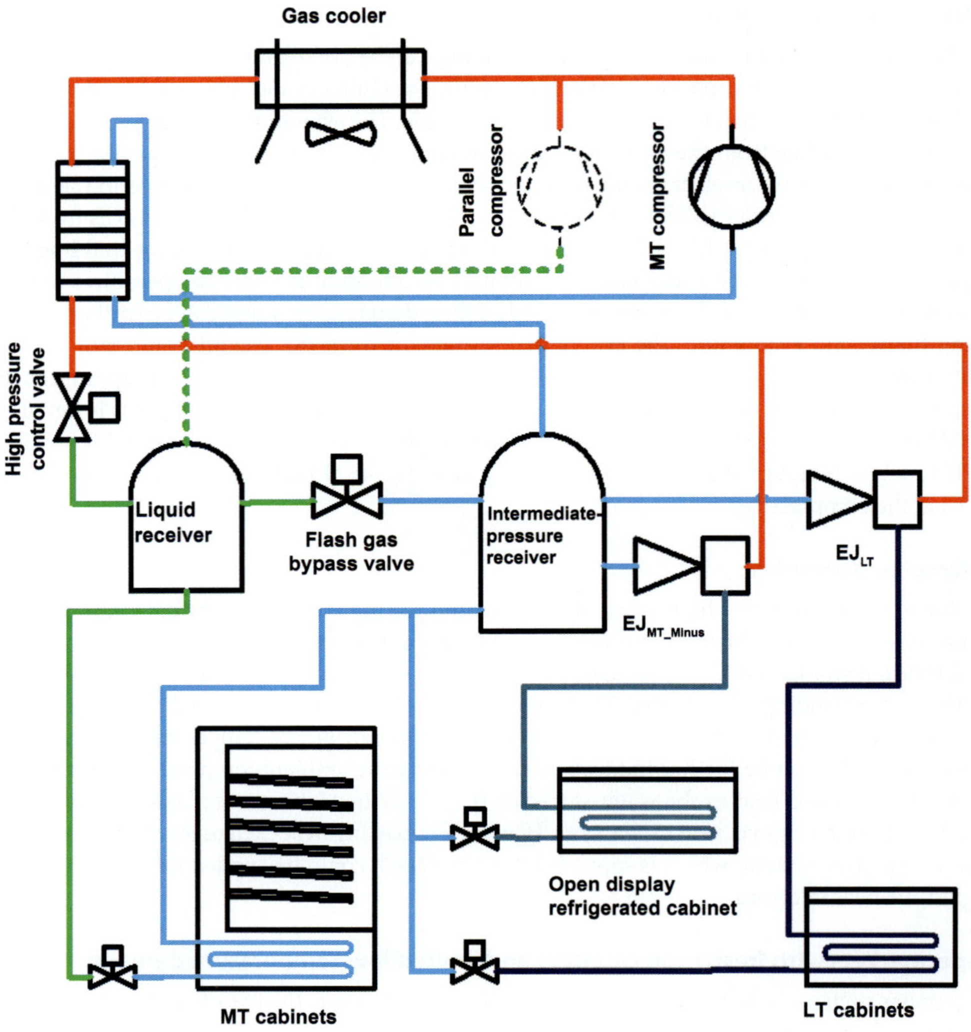

Figure 5.56 Refrigeration concepts with open counters and a few low-temperature cooling points

help of appropriately designed ejectors (providing the low-temperature level within the evaporators (EJ_{LT}) in Figure 5.56).

The serial arrangement, as shown in Figure 5.56, is a further development of the conventional parallel compressor system of medium-temperature-cooling and low-temperature units with liquid refrigerant from the intermediate-pressure receivers. The MT cooling units are always consciously operated with a certain excess of liquid. This enables the year-round, efficient operation of the MT evaporators without superheating. The overflowing liquid refrigerant is stored in the low-pressure receiver according to the requirement and operation of the refrigeration counters with lower evaporation temperatures. This concept can also be used in CO_2 booster systems with ejector support (see Figures 5.53 to 5.55).

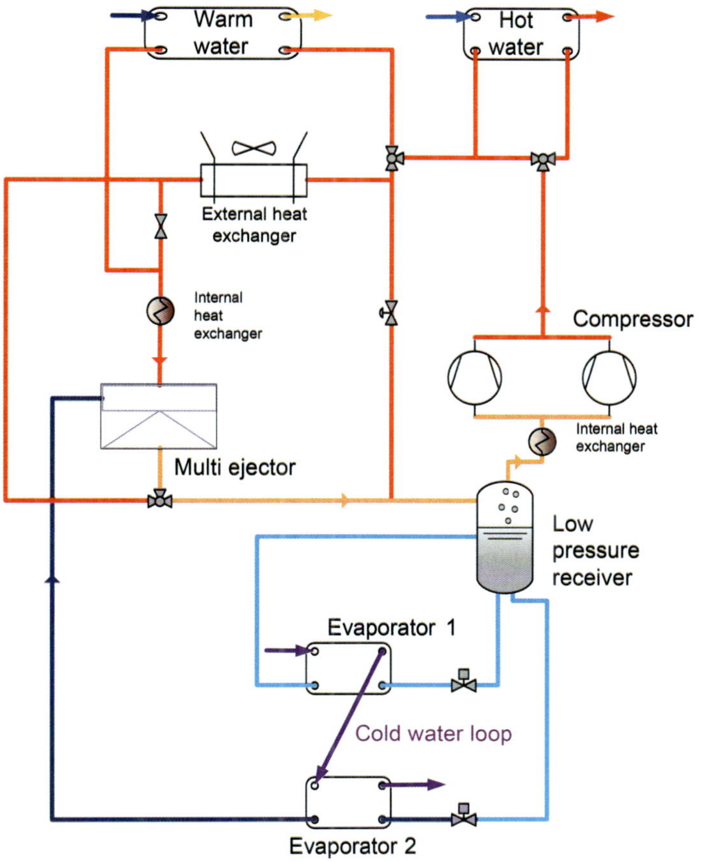

Figure 5.57
MultiPACK-CO_2 chiller and heat pump for building cooling, heating and hot water requirements

Industrial cooling / low-temperature systems

After the successful introduction of transcritical systems in supermarkets, this type of system is now also being adopted in other industrial sectors, especially for small- and medium-sized medium- / low-temperature (MT/LT) applications. Today, transcritical CO_2 units can handle large parts of the former HFC market and CO_2 cascade systems. This is not surprising since the systems are based on standardised compressors, heat exchangers and control units, similar to equipment already used in modern supermarket systems.

5.10.2.3 Hot water heat pumps

The successful entry of CO_2 hot water heat pumps (EcoCute) in Japan was described in Chapter 5.4. The following examples show that CO_2 heat pumps can also be used successfully in other countries and regions. Above all, this is the case with energy systems for buildings with simultaneous hot water and cooling requirements, such as hotels, sports facilities, hospitals, etc.

Figure 5.57 shows a compact CO_2 heat pump system with integrated cooling circuit for air conditioning, developed by Enex in the EU-funded MultiPACK project.

This system can provide hot water and space heating water, using either the outside air or the cold-water circuit as a heat source. If only the cold-water circuit has to be cooled, the waste heat is released by the outside air heat exchanger, which then operates as a gas cooler. By using the two evaporators, which are connected in series on the cold-water side, this "chiller" can work very efficiently by using the available ejector technologies (see also Figure 5.55).

Example of a residential cooperative building in Oslo:

The building complex consists of three identical apartment blocks, each with 13 floors and 819 apartments. The apartment blocks were extensively renovated in 2011. In each apartment block, a heat pump for space heating with low-temperature radiators and a 100 kW (341413 Btu) CO_2 domestic hot water heat pump replaced the previous oil heating system [5-29].

As a result, the annual energy requirement could be more than halved. The annual energy requirement for hot water for each block is approximately 600000 kWh/a (2047680000 Btu/a). Each hot water system consists of 13 x 400 litres (88 gallons) storage tanks arranged in series (see also Figure 5.47). The exhaust air from two common exhaust air ducts in each apartment block serves as the heat source for the heat pumps. The heat transfer from the exhaust air to the heat pump evaporator takes place in each case via an air / liquid heat exchanger using a frost-protected water-based heat transfer fluid. The CO_2 heat pump is equipped with a plate heat exchanger as an evaporator, two series-connected plate heat exchangers as a gas cooler and a reciprocating compressor. The heat pump heats the cold water to approximately 73 °C (163 °F) and is operated at a constant gas cooler pressure of 100 bar (1450.4 psi). Field measurements have shown that the CO_2 heat pump for an apartment block reached an average COP over the year (SCOP) of 4. If one also takes into account the power requirements for pump operation, this corresponds to an annual primary energy savings of approximately 70 % compared to the previous electrical heating system.

Example from the University Hospital in Tromsø:

Europe's most powerful CO_2 heat pump at the time of commissioning was put into operation in the autumn of 2014. It provides hot water at the University Hospital in Tromsø (UNN). A thorough pre-analysis showed that it was worthwhile to have a CO_2 heat pump installed that could cover the hospital's entire hot water demand. The hospital's cooling system (chilled water circuit) is used as a heat source; the CO_2 system thus covers the basic requirements for process cooling. This unit unloads the existing cooling machinery and reduced its energy demand to provide chilled water cooling. At the same time, this unit provides additional cooling capacity. The CO_2 heat pump unit consists of three parallel reciprocating compressors, plate heat exchangers as evaporators, gas coolers and an electronic expansion valve. The compressor capacity is regulated according to the cooling requirements of the chilled water circuit in the building. In order to ensure the most energy-efficient operation of the heat pump, the hot water subsystem of the building was also converted. Among other things, the hot water storage tanks were connected in series because this ensures the lowest possible return temperature to the gas cooler and thus guaranteed a maximum COP for the heat pump.

A serious challenge with the existing hot water system of the UNN buildings was that the poorly insulated hot water circulation pipes had significant heat losses. Preliminary field measurements show that the CO_2 heat pump in the UNN buildings has a COP of between 3,3 and 4,5. This means that the system delivers an average of around 3 kWh (0.85 TR) of usable process cooling and 4 kWh (1.1 TR) of usable heat, i.e. 7 kWh (1.99 TR) of usable thermal energy for each kWh (3414 Btu/h) of electricity provided.

Example of residential cooperative building Romolslia:

In 2014, the first water / CO_2 heat pump for hot water preparation was installed in Norway. This was for a block of flats with 56 apartments in Romolslia (Trondheim). The CO_2 heat pump replaced an older R22 heat pump system that was used to preheat the tap water to 45 °C (113 °F). The heating system is designed in a similar way to the example in Oslo, i.e., the exhaust air from the ventilation systems serves as a heat source. A secondary circuit connects the heat source with the heat pump [5-30].

Figure 5.58 CO_2 heat pump system from Cadio AS

The CO_2 heat pump has a nominal heating capacity of 30 kW (8.5 TR); see Figure 5.58. The unit consists of plate heat exchangers as evaporators, gas coolers and suction gas heat exchangers, a reciprocating compressor with on / off control and an electronic high-pressure control valve. The system heats tap water to 70 °C (158 °F), which is stored in series-connected water tanks with a total volume of 3000 litres (793 gallons). The annual heat requirement for hot water for the building block is in the order of 220000 kWh (750816000 Btu), i.e. 4000 kWh (13651200 Btu) per apartment. The CO_2 heat pump has, among other things, integrated measuring devices for heat energy and for electricity supplied. The measured average COP for the CO_2 heat pump, including the operation of the hot water circulation pump and the pumps in the frost protection circuit, was approximately 3.8. This corresponds to approximately 70 % energy savings compared to a hot water system using electrical heating only.

5.10.2.4 Air conditioners

Comfort cooling

Chillers for the provision of process cooling and, above all, comfort cooling have not been the most preferred area of application for CO_2 until recently.

However, the situation has changed with the introduction of flammable refrigerants as the only alternative in future installations, as in some cases this type of refrigerant cannot be used. If, as shown in Figures 5.55 and 5.57, the ejector technology is used skilfully, CO_2 water chillers are built that meet the performance requirements of the EU Ecodesign Directive.

Figure 5.59 shows a simplified system circuit for such a CO_2 chiller which can, at the same time, provide hot process water. The biggest challenge for the CO_2 chiller is the price advantage of commercial mass-produced systems with conventional refrigerants. However, an efficient and

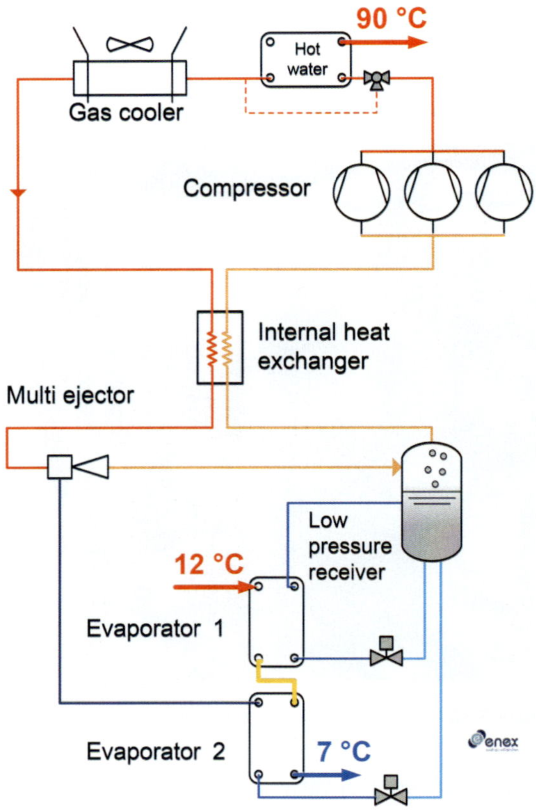

Figure 5.59 CO_2 chiller from Enex; further development from EU project MultiPACK

future-proof alternative is now available for cases in which a non-flammable refrigerant is required.

Cooling for data centres

CO_2 is suitable as a refrigerant for the direct cooling of data centres and computer rooms in which many PCs are used [5-31]. The evaporation temperature typically is 14–15 °C (57–59 °F) (or sometimes higher) at a high cooling load demand. The systems can be designed for normal direct expansion, or a circulation pump can simplify the cooling and especially facilitate free cooling. In such a pump circuit, CO_2 can serve as a refrigerant, as shown in Figure 5.24.

The use of flooded evaporators with self-circulation is also an alternative. For technical reasons – among other things due to the oil return – it can make sense to use pure (oil-free) CO_2 systems, i.e. one primary and one secondary circuit with CO_2, as shown in Figure 5.60. This type of installation has already been built in England, Japan and Hong Kong.

5.10.2.5 Transport

Due to the varying ambient temperatures and the very hot climate in some parts of the world, CO_2 was not the preferred refrigerant for medium- (MT) and low-temperature (LT) containers. Nevertheless, Carrier, one of the largest manufacturers of medium- and low-temperature installations for containers, has developed CO_2 models, which are called NaturaLINE, as shown in Figure 5.61. These containers are currently being tested. The manufacturer advises that the energy requirement is comparable to that of optimised R134a systems if the conditions are the same.

5.10.2.6 Mobile air conditioners

Since the "re-discovery" of CO_2 as a refrigerant around 1988 (for hot water heat pumps), car air conditioning systems became the main focus of research and development groups in the 1990s. The pioneers were German automobile manufacturers, who had tried until 2007 to introduce CO_2 as the future refrigerant in air conditioning for this sector.

Daimler AG had already decided, before the EU MAC directive[13] came into force, for safety reasons, not to use R1234yf as an interim replacement refrigerant for R134a. CO_2 air condition-

[13] https://ec.europa.eu/growth/sectors/automotive/environment-protection/mobile-air-conditioning-systems_en

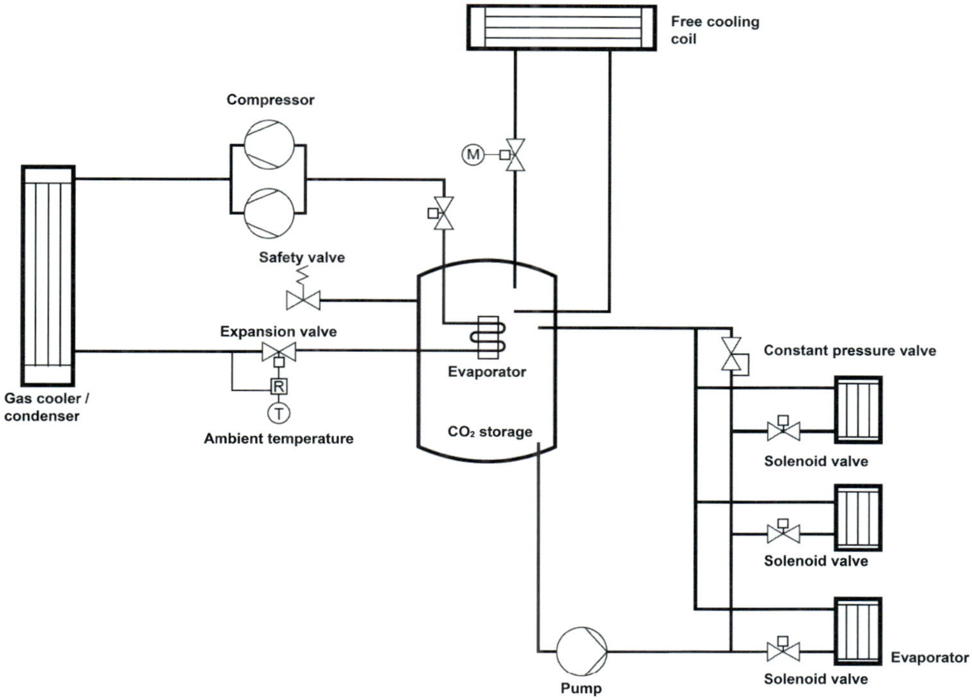

Figure 5.60 Cooling system for data centres with pump circulation of CO_2

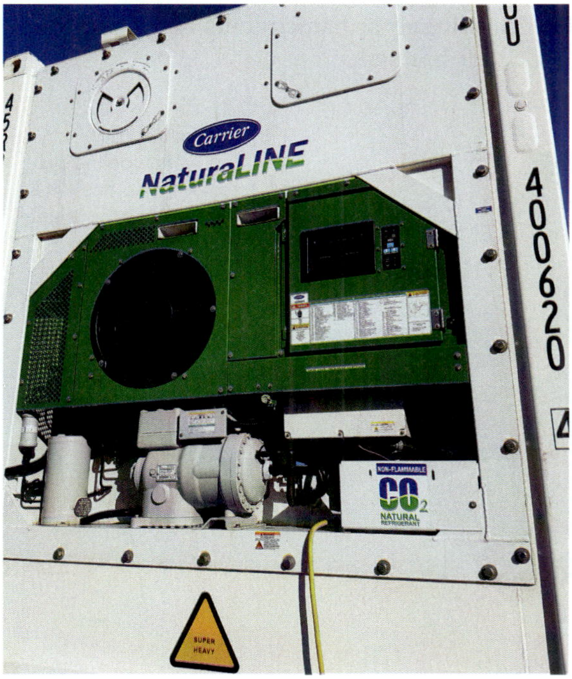

Figure 5.61
CO_2 container cooling unit (Photo: Armin Hafner)

er systems have been in production since autumn 2016 for selected vehicle models. In the meantime, other manufacturers also have certain models applying CO_2 as the refrigerant. Due to the major changes that are about to happen in the next few years in terms of the powertrains of vehicles, there will also be new requirements for vehicle energy management.

Especially for electric vehicles, there is considerable interest in efficient heating and cooling devices. The excellent heat pump characteristics of CO_2 help extend the driving range during wintertime compared to heat pumps using other refrigerants.

Since the vehicle's energy supply should be used primarily for motion, and because CO_2 is not flammable and enables these tasks with a very compact design, it is very likely that all major car manufacturers will apply CO_2 as in ID.3 models of VW (when choosing the R744 heat pump version).

Air conditioning systems for buses with CO_2 as a refrigerant have also been developed and tested by several bus manufacturers in Europe, e.g., since 2010 in a city bus in Berlin.

All manufacturers of railroad air conditioning systems already have the first CO_2 pilot installations in test operation in different regions of Europe. Due to the strict safety requirements in this sector, it can be assumed that CO_2 will prevail as a refrigerant on trains and in bus air conditioning systems once the necessary key components become available in small series.

5.10.2.7 Refrigerated sea water cooler for the fish industry

Refrigerated sea water (RSW) systems are designed to provide chilled sea water at –1 °C (34 °F) [5-32]. The fish catch is cooled in large tank systems by circulating the sea water before the fish is processed. The processing of the fish is done either directly on board the fishing vessel or onshore. The first RSW systems with CO_2 as refrigerants were developed in 2010. Figure 5.62 shows a system with a cooling capacity of 450 kW (128 TR), which is installed on board a fishing trawler. The RSW systems are operated either subcritically or transcritically, depending on the temperature level of the sea water, which is used as a heat sink.

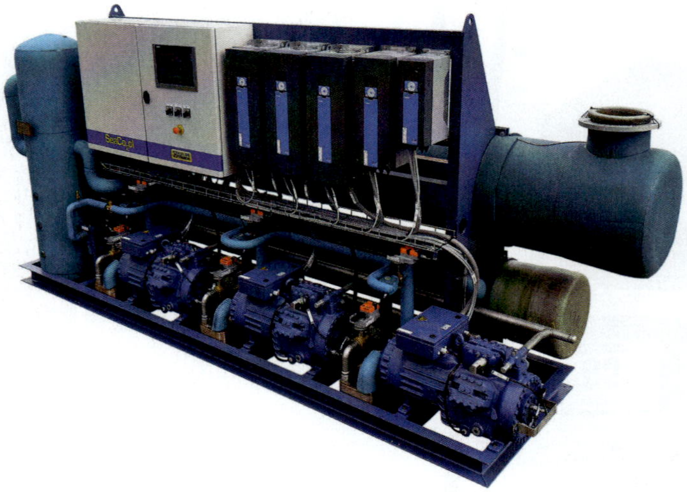

Figure 5.62
Sea water cooling system from Kuldeteknisk AS, supplied for fishing vessel Båragutt (2011) [5-33]

These sea water direct cooling systems meet the cooling needs of smaller fishing boats where the use of ammonia is difficult and expensive due to safety requirements. The aforementioned CO_2 system is equipped with three standard reciprocating compressors, designed for transcritical operation. The total stroke displacement volume is 225 m^3/h (7956 ft^3/h). A redundant-duty compressor was also integrated directly into the on-board refrigeration system. One of the compressors is frequency controlled for capacity control. The installation operator's experience is very positive. Nowadays CO_2 RSW systems are used on both new and refurbished smaller size fishing vessels.

5.10.2.8 On-board low-temperature installations

Experience from on-board low temperature systems and land-based refrigeration installations have shown that the energy demands of such installations can be reduced by applying CO_2 units for medium- or low-temperature cooling. The freezing capacity can be considerably increased. This is partly due to the properties of CO_2 and partly due to the possibility to lower the evaporation temperatures to approximately –50 °C (–8 °F). Conventional systems work with evaporation temperatures of approximately –40 °C (–40 °F).

In 2014, Kuldeteknisk AS installed a comprehensive pilot low-temperature system on board the fish trawler MS Roaldnes. The nominal cooling capacity was 250 kW (71 TR) at an evaporation temperature of –50 °C (–58 °F), which corresponds to freezing 36 metric tons (39.6 $tons_{US}$/ 35.4 $tons_{Imp}$) of fish daily. The refrigeration system is based on the same principle as the CO_2 booster systems in the commercial refrigeration units. However, the CO_2 liquid is pumped through the different evaporators with circulation pumps. Defrosting takes place via direct hot gas injection. A picture of such low-temperature systems is shown in Figure 5.63.

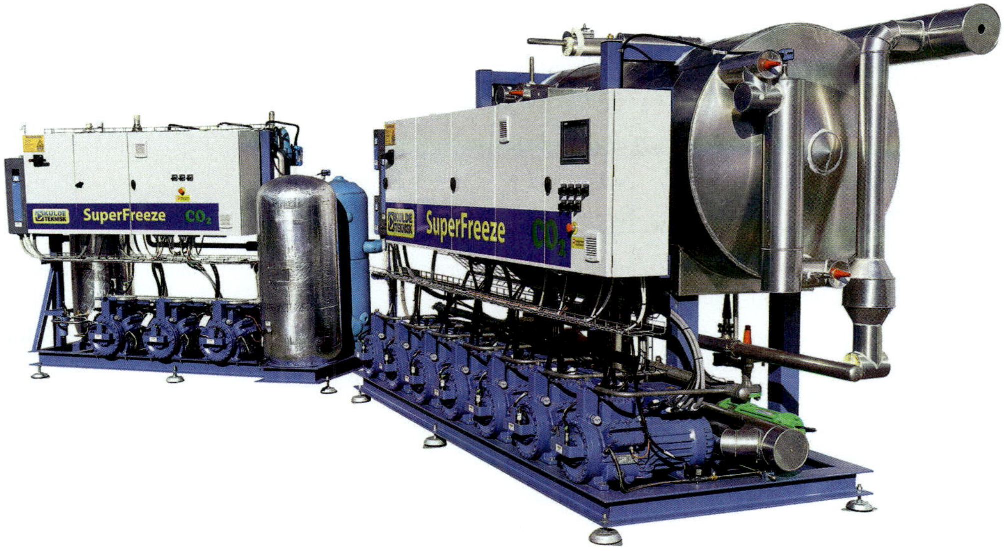

Figure 5.63 Low-temperature cooling installation from Kuldeteknisk AS

In summary, the results from the one-year test operation were very positive for the ship owner. It was shown that the freezer capacity was more than 30 % higher than with the conventional system (R22), which was previously installed in the same ship.

Defrosting of plate freezers with warm CO_2 is very fast, even and effective, so that there are no local discolorations on the fish. The faster freezing and the effective processing ensure a better quality of the frozen product, and therefore better prices can always be obtained for the catch. Other advantages of CO_2 installations are the compact design with significantly reduced space requirements than a conventional installation. Much of the assembly work for such units is not carried out on board or on site, but under controlled conditions in the assembly workshop / factory. This significantly reduces installation errors and reduces the period required to perform a refurbishment.

5.10.2.9 Other areas of application for CO_2 as a refrigerant

There are potential application areas for CO_2 refrigeration and heat pump systems which are currently still in the development or pilot testing phase, i.e., only small-scale series are available.

These developments and pilot phases include:

- Mobile heat pumps in the transport sector (passenger transport)
- Transport cooling of goods (containers, trucks, ship refrigeration systems)
- Combined heating and cooling / air conditioning of residential and commercial buildings
- Heat pumps for space heating and for combined space heating and hot water preparation
- Heat pump dryer (private and commercial)
- Power generation from waste heat (Rankine cycle)
- Air conditioning systems for residential buildings
- Chillers / heat pumps for warmer climates
- Industrial heat pumps

5.11 Literature

[5-1] Lorentzen, G.: *Revival of carbon dioxide as a refrigerant.* International Journal of Refrigeration, Vol. 17, No. 5 Elsevier, 1993

[5-2] Hafner, A., Nordtvedt, T.S., Gukelberger, E., Banasiak, K., Widell, K.: *Design of a R717 / R744 Cascade System for the Pelagic Fish Industry.* 11th IIR Gustav Lorentzen Conference on Natural Refrigerants, Hangzhou, China, 2014

[5-3] *Safety Code for Refrigerating Systems Utilizing Carbon Dioxide.* The Institute of Refrigeration. United Kingdom, 2007

[5-4] Hrnjak, P., Young Park, C.: *Evaporation at Low Temperatures.* C-Dig Meeting, March 16–17, 2006

[5-5] Bredesen, A.M., Hafner, A., Pettersen, J., Nekså, P., Aflekt, K.: *Heat transfer and pressure drop for in-tube evaporation of* CO_2, IIR Proceedings, 1997 – 5; *Heat transfer in natural refrigerants,* College Park (USA), November 6–7, 1997

[5-6] Stene, J.: Residential CO_2 Heat Pump System for Combined Space Heating and Hot Water Heating, Doctoral Thesis at NTNU 2004: 53

[5-7] Madsen, K.B.: *Transcritical CO_2 System in a Small Supermarket.* Danfoss seminar on CO_2 som Kulededium, Oslo, January 21, 2009

[5-8] Nekså, P., Hafner, A., Bredesen, A.M., Eikevik, T.M.: *CO_2 as Working Fluid - Technological Re-development on the Road to Sustainable Refrigeration.* 12th IIR Gustav Lorentzen Natural Working Fluids Conference, Edinburgh, 2016

[5-9] Nagata, H., Kakuda, M., Sekiya, S., Shimoji, M., Koda, T.: *Development of a Scroll Expander for the CO_2 Refrigeration Cycle.* 2010 International Compressor Conference, Purdue University

[5-10] Schönenberger, J., Hafner, A., Banasiak, K., Girotto, S.: *Experience with ejectors implemented in a R744 Booster System Operating in a Supermarket.* 11th IIR Gustav Lorentzen Conference on Natural Refrigeration 2015, August 16-22, Yokohama, Japan

[5-11] Bakken, K.: *10 års utvikling av CO_2 varmegjenvinningsløsning.* Norsk Kjøleteknisk Møte, March 14–15, 2013

[5-12] Stene, J.: Guidelines for Design and Operation of Compression Heat Pump, Air Conditioning and Refrigerating systems with natural working fluids. IEA Heat Pump Program, report no. HPP-AN22-4, 1988 ISBN 90-73-741-31-9

[5-13] Taira, S.: *Development of High Temperature Water Heating Using CO_2 Heat pump.* Air Conditioning Manufacturing Division, Daikin Industries, Ltd., 2008

[5-14] Nekså, P., Rekstad, H., Zakeri, G.R., Schiefloe P.A.: *CO_2 heat pump water heater, characteristics, system design and experimental results.* International Journal of Refrigeration, Vol. 21, No. 3, pp. 172–179, Elsevier, 1998

[5-15] Hu, B., He, Y., Wang, S., Cao, F., Xing, Z.: *Field Evaluation for Air-source Transcritical CO_2 Heat Pump Water Heater with Optimal Pressure Control.* International Refrigeration and Air Conditioning Conference. Purdue University. Paper 1483, 2014

[5-16] Pearson, S.F.: *Using CO_2 To Reduce Refrigerant Charge.* ASHRAE Journal Vol. 54, No. 3, October 2012

[5-17] Maratou, A., Masson, N. (shecco SPRL): *Examples of NH_3 / CO_2 Secondary Systems for Cold Store Operators.* https://www.issuu.com/shecco/docs/guide_nh3-final

[5-18] Berends, E.: *Natural working fluid in artificial skating rinks.* 7th IIR Gustav Lorentzen Conference on Natural Working Fluids, Trondheim, May 28–31, 2006

[5-19] Lunde, H.: *Erfari drift from CO_2 som kuldemedium og kuldebærer.* Isbane Seminar 17–18 March, 2014

[5-20] Christensen, O.: *Typical ammonia / CO_2* Cascade *System, Freezing and Cooling Application.* Technical Paper 1, 2006 IIAR Ammonia refrigeration Conference & Exhibition, Reno, Nevada, USA

[5-21] Hillphoenix Refrigeration Systems: *Second Nature CO_2 Capable of making a difference.* http://www.hillphoenix.com/refrigeration-systems/second-nature/

[5-22] Skærbek, P., Nielsen, Lund, T.: *"Introducing a New Ammonia / CO_2 Concept for large fishing vessels".* Technical Paper # 11, IIAR, 2003

[5-23] Hansen, H.: *CO_2 innfrysingsanlegg MS Roaldsnes.* Norsk Kjøleteknisk Møte, Kristiansand, 7–8 April 7–8, 2016

[5-24] Javerschek, O.: Commercial Refrigeration Systems with CO_2 as Refrigerant. IIR Bulletin, 2009-I

[5-25] Hafner, A.: *Kommersiell kjøling (butikk).* Norsk Kjøleteknisk Møte, Bergen, March 12–13, 2014

[5-26] Hafner, A., Claussen, I.C.:, F.Schmidt, Olsson, R., Fredslund, K., Eriksen, P.A., Madsen, K.B.: *Efficient and Integrated Energy Systems for Supermarkets.* 11th IIR Gustav Lorentzen Conference on Natural Refrigerants,Hangzhou, China, 2014

[5-27] Hafner, A.: *R744 ejektor butikk-kjøleanlegg. Driftserfaringer fra Norges første parallellkom-pres-sorkjøleanlegg with ejector support.* Norsk Kjøleteknisk Møte, Kristiansand, 7–8 April, 2016

[5-28] Hafner, A., Hemmingsen, A.K., Nekså, P.: *System Configurations for Supermarkets in Warm Climates Applying R744 Refrigeration Technologies. Case Studies of Selected Chinese Cities.* 11th IIR Gustav Lorentzen conference on Natural Refrigerants, Hangzhou, China, 2014

[5-29] Borge, M.: *Analysis of CO_2 varmepumper for vartvannsbedredning av boliblokker.* Masterop-pgave, NTNU, 2014

[5-30] Berntsen, M.R., Eikevik, T.M., Jenssen, S., Tolstorebrov, I.: *Optimised Operation Conditions for a CO_2 Heat Pump for Chilling of Ice Water / Heating of Hot Tap Water to 80 °C.* 11th Gustav Lorentzen Conference on Natural Refrigerants, Hangzhou, China, 2014

[5-31] Solemdal, Y., Eikevik, T.M., Tolstorebrov, I., Veiby, O.J.: *CO_2 as a Refrigerant for Cooling of Data-Centre: A case study.* 11th IIR Gustav Lorentzen Conference on Natural Refrigerants, Hanszhou, China, 2014

[5-32] Jensen, S.: *Ny type varmevekslere i RSW anlegg with CO_2 som kuldemedium.* Norsk Kjøleteknisk Møte, Kristiansand, 7–8 April 2016

[5-33] Ladam, Y.: *CO_2 RSW, drift og erfaringer med 250 kW anlegg om bord IM / S Båragutt.* FHF, Pelagisk Samling 2012, December 4, 2010

6 Hydrocarbons as refrigerant

Nina Schultz, Irmgard Bauer, Joachim Schadt, Matthias Koffler (Figures 6.3 to 6.11)

The authors of the revised version thank Daniel Colbourne and their colleagues at Secon GmbH, Messrs. M. Funes, J. Gregori, P. Schadt, M. Siegle and L. Siems, for sharing their technical knowledge.

6.1 Introduction

Use of propane (R290) and butane (R600) as natural refrigerants started at the beginning of the 20th century, around the same time as the development of the first domestic refrigerators. During the 1930s, however, halogenated hydrocarbons, where some hydrogen atoms are replaced by halogens such as chlorine or fluorine, found widespread use and were preferred over non-halogenated hydrocarbons. However, since it has transpired that halogenated hydrocarbons have ozone-depleting potential (ODP) and/or high global warming potential (GWP), they ought to be discontinued and it is widely acknowledged that the "old" natural refrigerants should be used instead [6-4].

Today, use of hydrocarbon refrigerants is once more of interest, as they have no ODP and negligible GWP. Their use in refrigeration technology is being reassessed; standards and guidelines are adapted to the progressing state of the art.

Hydrocarbon compounds differ primarily in the numbers and positions of their carbon atoms and are classified as saturated or unsaturated hydrocarbons. Saturated hydrocarbon molecules, have a basic atomic structure where atoms bond either with exactly one adjacent carbon atom

Table 6.1 Chemical formulas and R-numbers of hydrocarbons and some fluorocarbons (HFC)

R-number	Chemical formula	Refrigerant
R170	C_2H_6	ethane
R290	C_3H_8	propane
R600	C_4H_{10}	n-butane
R600a	C_4H_{10}	isobutane
R1150	C_2H_4	ethylene
R1270	C_3H_6	propene / propylene
R134a	CH_2FCF_3	HFC-134a 1,1,1,2-tetrafluoroethane
R404A	 CHF_2CF_3 CH_3CF_3 CH_2FCF_3	R-125/ 143a/ 134a (44 / 52 / 4 %): HFC-125 pentafluoroethane HFC-143a 1, 1, 1 trifluoroethane HFC-134a 1,1,1,2 tetrafluoroethane
R410A	 CH_2F_2 CH_4	R-32/ 50 (50 / 50 %): HFC-32 difluoromethane methane

* "Propene" is the formal IUPAC name

or the bond is occupied by a hydrogen atom. Unsaturated hydrocarbons include multiple bonds between carbon atoms; adjacent carbon atoms of the molecule can alternatively be connected to each other via double or triple bonds. Multiple bonds reduce the number of binding sites to be occupied by hydrogen. Hence, unsaturated hydrocarbons contain fewer hydrogen atoms than saturated hydrocarbons (for the same number of carbon atoms).

Single bonds are more stable than multiple bonds due to the geometric configuration of the molecule. Only sufficiently stable unsaturated hydrocarbons, e.g., R1270 can be used as refrigerants; hydrocarbon molecules with triple bonds are unsuitable. Saturated hydrocarbons, i.e. R170, R290, R600, are less reactive than unsaturated ones; they are more stable at higher temperatures and pressures and do not tend to polymerise. More hydrogen atoms normally result in higher flammability. HC refrigerants burn clean without toxic combustion products.

Hydrocarbons consisting of five or more carbon atoms (C_5 or higher) are heavier and have normal boiling points above room temperature. They are the main ingredients of petroleum and used as fuels and as raw materials for the production of chemicals and plastics. Aside from pentane (R60a) and isopentane (R601a) for use in centrifugal systems, e.g., turbo compressors, they are of little interest in refrigeration technology. Although, trace amounts are present as impurities in refrigerants.

Propane and butane are the main components of liquefied petroleum gas (LPG), although LPG often contains significant proportions of many other medium and light hydrocarbons, as well as moisture, sulfur and "aromatic" hydrocarbons (e.g., benzene). LPG also has an odourant, such as ethyl mercaptan (which gives the rotten, egg-like smell) added to it, which is not usually a component of refrigerant-grade hydrocarbons, due to concerns arising from the presence of sulfur. Refrigerant-grade hydrocarbons are usually much higher purity than LPG; this is not only to ensure chemical stability but also to avoid changes in system performance due to variations in thermodynamic properties arising from different compositions. Today, refrigerant-grade ethylene (R1150), ethane (R170), propene (R1270), propane (R290), isobutane (R600a) and n-butane (R600) are widely available.

6.1.1 Ecological characteristics

ODP and GWP

The ozone depletion potential (ODP) and global warming potential (GWP) are the primary factors used for comparison of the ecological characteristics of refrigerants. Hydrocarbons have negligible ODP and GWP values compared to synthetic refrigerants (see Table 6.2).

Hydrocarbons that escape into the atmosphere have no potential for depleting the ozone layer (ODP = 0). Furthermore, hydrocarbons have only low global warming potential (e.g., $GWP_{R290} < 1$).

Table 6.2 ODP and GWP (100 years) of some hydrocarbons and reference refrigerants according to RTOC 2018 [6-1]

Refrigerant	ODP	GWP (20 years)	GWP (100 years)
R170	0	5,2	1,4
R290	0	<1	<1
R600a	0	<1	<1
R1270	0	<1	<1
R134a	0	3810	1360
R404A	0	6600	4200
R410A	0	4400	2100

TEWI /LCCP

The total equivalent warming impact (TEWI) – sometimes referred to as life cycle climate performance (LCCP) and life cycle warming impact (LCWI), etc. – indicates the amount of "direct" (refrigerant-related) and "indirect" (energy generation related) CO_2 equivalent emissions to the atmosphere throughout the life time of a refrigeration system.

Direct emissions correspond to refrigerant leakage during system operation, venting, etc. Indirect emissions arise during system operation due to the consumption of electrical energy which is generated by fossil fuels, as well as those during production of system components and materials, etc.

Generally (although it depends upon the energy mix in the selected country), indirect CO_2 emissions tend to dominate TEWI values for most types of refrigeration systems; see Figure 6.1. In the future, where medium or high GWP refrigerants are used, it is likely that the energy-related contribution of TEWI will be less pronounced due to systems being more efficient and with energy being provided mostly or fully by renewable energy sources.

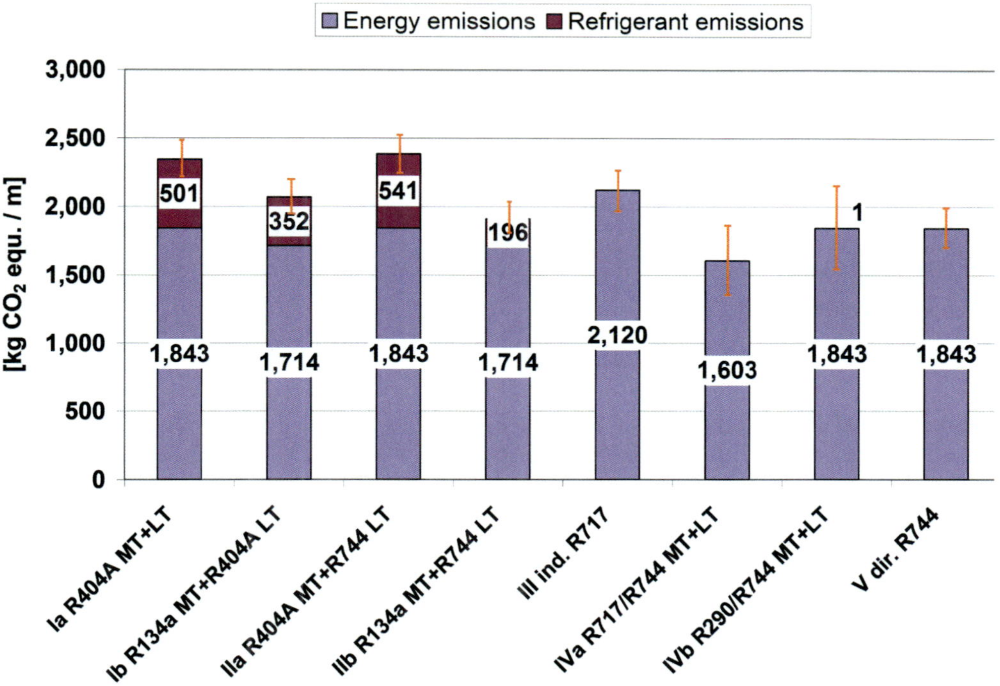

Figure 6.1 Annual greenhouse gas emissions per metre of refrigerated shelf in supermarkets for various refrigeration systems and refrigerants (assumed annual leak rate of 6,15 %) [6-5]

POCP

The photochemical ozone creation potential (POCP) indicates how much ground-level ozone is formed by substances emitted into the atmosphere. Ground-level ozone is primarily responsible for the phenomenon known as "summer smog". Near the ground, ozone is harmful and poses risks to plants, animals and humans. The chemical reactions that lead to formation of ozone are facilitated by solar radiation, which leads to an increased occurrence of the phenomenon in summer.

Formation of ground-level ozone involves various compounds originating from anthropogenic emission. When these precursors react chemically, atmospheric ozone is formed. Precursors include nitrogen oxides, carbon monoxide and hydrocarbons. The reacting hydrocarbons are classed as volatile organic compounds (VOC) and are often referred to as non-methane hydrocarbons (NMHC) [6-8].

However, in the potential quantities used as refrigerants, the VOC contribution of hydrocarbons is minimal compared to other sources (see Table 6.3).

Also, bearing in mind that the POCP of the leading hydrocarbon refrigerant, R290, is much lower than the majority of NMHC VOCs.

Hydrocarbons are strongly involved in formation of ground-level ozone when they are emitted into the atmosphere, under certain environmental conditions.

Table 6.3 VOC emission by refrigerants in 2017 [6-17]

VOC emission	Percentage	amount in		
		10^3 t	10^3 $tons_{US}$	10^3 $tons_{IMP}$
NMHCs global total	100 %	8300	9146.6	8169.7
Refrigerants total	0,72 %	60	66.1	59.1
HC refrigerants	0,3 %	25	27.6	24.6

Water hazard class

Hydrocarbon refrigerants are not hazardous to water.

6.2 Thermodynamic aspects

Thermophysical properties determine a fluid's applicability as a refrigerant. Individual hydrocarbons are suited across a wide range of technical applications, see Figure 6.2. This range can be extended further by the use of refrigerant blends.

This chapter describes how the properties of hydrocarbons affect the thermodynamic performance of the refrigeration cycle and influence energy efficiency of real processes in comparison to fluorocarbons.

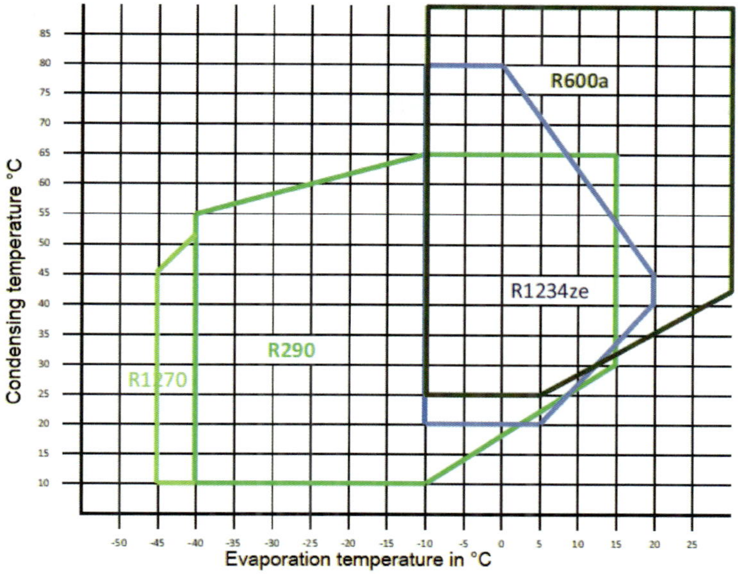

Figure 6.2 Application limit diagrams of some hydrocarbons and reference refrigerant

For comparison with fluorocarbons, R134a is used as a reference refrigerant for intermediate evaporation temperatures, and R404A for higher evaporation temperatures.

Data concerning thermodynamic properties was generated using REFPROP 10.0 [6-9] and CoolProp 6.4.1 [6-10].

A comparison of the refrigeration cycle is carried out using an example process with the following boundary conditions:

saturated condensation temperature	40 °C (104 °F)
subcooling of liquid refrigerant	0 K
evaporation temperature	varies
superheating in evaporator	10 K

Furthermore, this example process is assumed to be ideal, with isentropic compression and without pressure losses, as shown in Figure 6.3.

6.2.1 Evaporation latent heat

During evaporation, latent heat is transferred, enthalpy increases and phase transitions from liquid to vapour. The specific enthalpy difference during evaporation of a refrigerant is greater at lower temperatures. This can be seen in the pressure-enthalpy diagram in the upward-tapering two-phase region in Figure 6.3. If more evaporation latent heat is available, more specific heat can be transferred during phase transition.

In Table 6.4 and Table 6.5, evaporation latent heat for various temperature and thus pressure levels are shown for selected refrigerants.

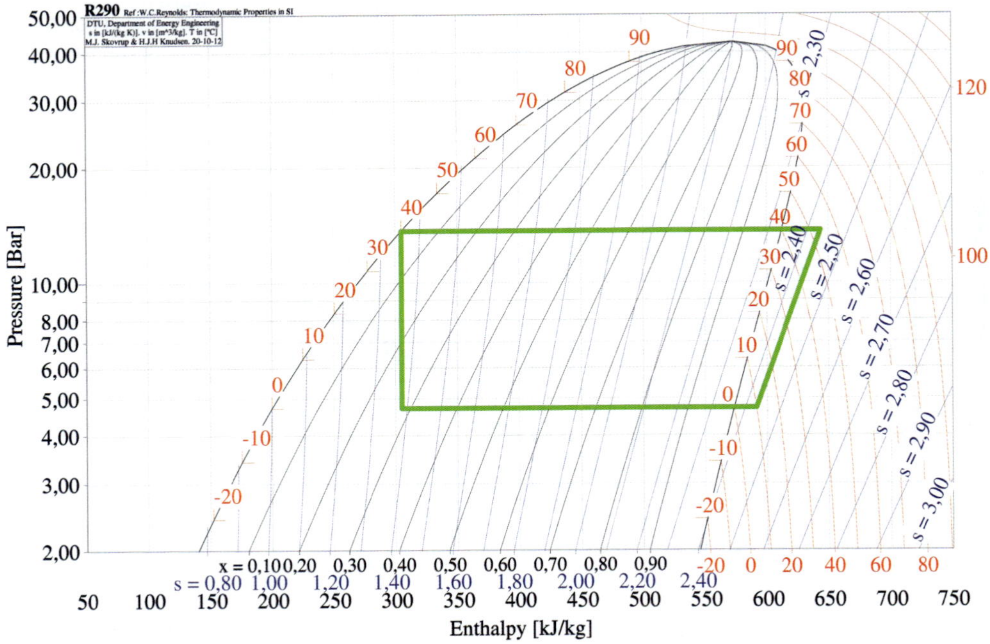

Figure 6.3 Pressure-enthalpy diagram showing example R290 process (green) with temperature (red), entropy (blue), and two-phase vapour quality (black), calculation and diagram with [6-11]

Table 6.4 Evaporation latent heat of some hydrocarbons and reference refrigerants in kJ/kg

Refrigerant	R170	R290	R600a	R1270	R134a	R404A	R410A
–20 °C	364	401	372	406	213	183	244
0 °C	303	375	354	378	199	167	221
20 °C	206	344	334	344	182	147	194

Table 6.5 Evaporation latent heat of some hydrocarbons and reference refrigerants in Btu/lb

Refrigerant	R170	R290	R600a	R1270	R134a	R404A	R410A
–4 °F	156.5	172.4	159.9	174.5	91.6	78.7	104.9
32 °F	130.3	161.2	152.2	162.5	85.6	71.8	95.0
68 °F	88.6	147.9	143.6	147.9	78.2	63.2	83.4

6.2.2 Evaporation temperature and pressure

Common evaporating temperature range of applications are listed below:

–40 to –15 °C / –40 to –5 °F	low temperature (freezing)
–15 to 0 °C / –5 to 32 °F	medium temperature (chilled)
0 to 15 °C / 32 to 59 °F	high temperature (air conditioning)

The evaporation temperature of any refrigerant rises with increasing saturation pressure: Operation below atmospheric pressure, i.e. in low pressure down to vacuum applications, are technically manageable (e.g., hermetic circuits in plug-in units). Hydrocarbon refrigeration circuits are operated below the critical point. For R600a, by comparison, saturated vapour pressure falls below atmospheric for evaporation temperatures below –10 °C (14 °F) whilst critical pressure of 36 bar (522.13 psi) is reached at a temperature of 135 °C (275 °F). Operation with pressures above that of atmospheric pressure prevents ingress of air (oxygen and moisture) into the refrigeration circuit in the event of leaks. Normal Boiling Point (NBP) temperatures of hydrocarbons are shown in Table 6.6.

Hydrocarbons are available for applications in a wide range of pressure levels.

The saturated vapour pressures of R290, R600a and R1270 are shown with corresponding temperatures in Figure 6.5 as an overview and in comparison to R134a, R404A and R410A.

Table 6.6 Normal boiling point temperatures of some hydrocarbons at 1,001 bar/ 14.518 psi (absolute)

Refrigerant	**R170**	**R1270**	**R290**	**R600a**
Boiling point in °C / °F	-89 / –128.2	-48 / –54.4	-42 / –43.6	-12 / 10.4

For R1270, the saturated vapour pressure lies above atmospheric pressure above –47 °C (–52.6 °F), and reaches critical pressure of 45,6 bar at 91 °C (942.7 psi at 195.8 °F). Saturated vapour pressure of R290 runs below that of R1270. At around –30 °C (86 °F) and below, the vapour pressure of R290 is below that of R1270. Saturated vapour pressure of R290 is above atmospheric pressure from –42 °C (–43.6 °F) and reaches critical pressure of 42,5 bar at 97 °C (616.4 psi at 206.6 °F). Therefore, taking into account compressor application limits, the compressor envelope, R290 is well suited for applications with relatively low evaporation temperature as well as high condensation temperatures.

The saturated vapour pressure curve of R134a runs approximately centrally between R290 and R600a. Across most of the temperature range, R1270 has a saturated vapour pressure that is just below that of R404A, although the curves converge below 0 °C (32 °F). R410A has the highest saturated vapour pressure over the whole temperature range.

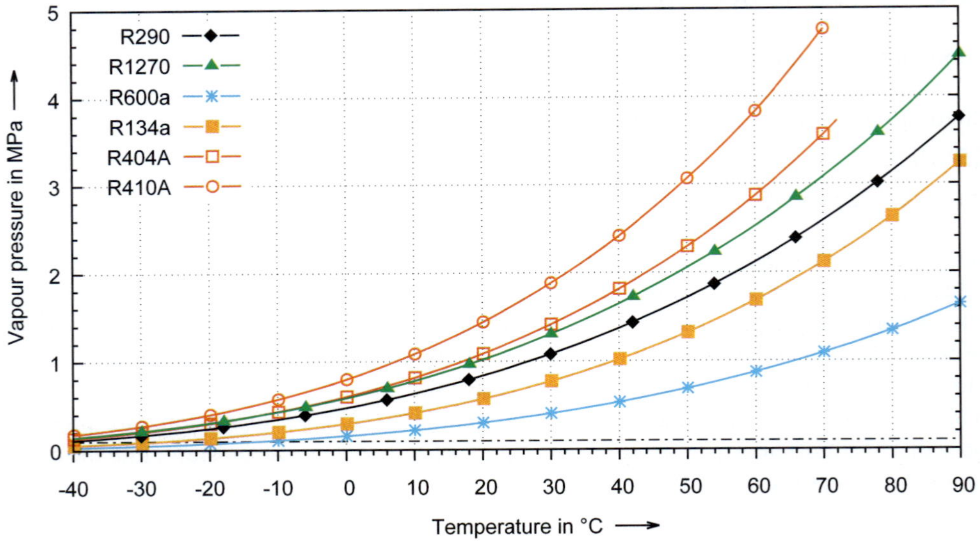

Figure 6.4 Saturated vapour pressure of some hydrocarbons and reference refrigerants

6.2.3 Isentropic expansion exponent κ

For a given substance, the isentropic expansion exponent κ is calculated as the quotient of the specific heat capacity at constant pressure c_p and the specific heat capacity at constant volume c_v. The isentropic exponent κ determines the curvature of the isentropes in the pressure-enthalpy diagram. When ideal compression is assumed, volume compression work is considered as adiabatic and without friction, hence such process runs isentropically.

The higher the value of κ, the flatter the isentropes are in the pressure-enthalpy diagram. Thus, a relatively large portion of compressor work leads to an increase in temperature when gas is being compressed. This leads to higher discharge temperatures.

The smaller the values of κ, the steeper the isentropic curves are, and a larger portion of compressor work yields an increase in pressure, while temperature increase is less pronounced.

Hydrocarbon refrigerants have, as shown in Table 6.7, isentropic exponents typically larger than the fluorocarbon reference refrigerants, with exception of R600a. When comparing isentropic discharge temperature and pressure of hydrocarbons after compression with reference hydrofluorocarbons, those of hydrocarbons are higher at the same evaporation and condensation temperatures.

Table 6.7 Isentropic exponents κ of some hydrocarbon and reference refrigerants

Refrigerant	R170	R290	R1270	R600a	R134a	R404A	R410A
κ	1,20	1,12	1,14	1,08	1,10	1,10	1,14

For vapour compression processes (without intercooling), isentropic curves in a pressure-enthalpy diagram represent the theoretical maximum of energy efficiency of compression.

In real processes, further factors influence the compression efficiency, such as friction from moving parts, which produces heat and further temperature rise, heat transfer from the discharge to suction-side and leakage between high and low pressure sides. Therefore, the actual "polytropic" curve, representing the state change of a refrigerant during real compression, proceeds to the right of the ideal isentropic curve in the pressure-enthalpy diagram.

6.2.4 Coefficient of Performance / Energy Efficiency Ratio

The Coefficient of Performance (COP), also referred to as Energy Efficiency Ratio (EER), is defined in EN 14825. This metric serves to quantify the energy efficiency of a refrigeration system; it represents the ratio of thermal capacity to electric power to drive (and operate) the system. EER can be expressed for the cycle only (cycle thermal capacity divided by compressor power) or the system (where the power denominator also includes power for fans, pumps, controls, etc.). Cycle EER depends on the pressure difference between evaporation and condensation of the refrigerant, subcooling and superheating at the heat exchangers, piping and components pressure losses as well as losses during the compression process. For part-load performance (as covered by EN 14825) it additionally considers the effect of cycling losses and different temperature lifts.

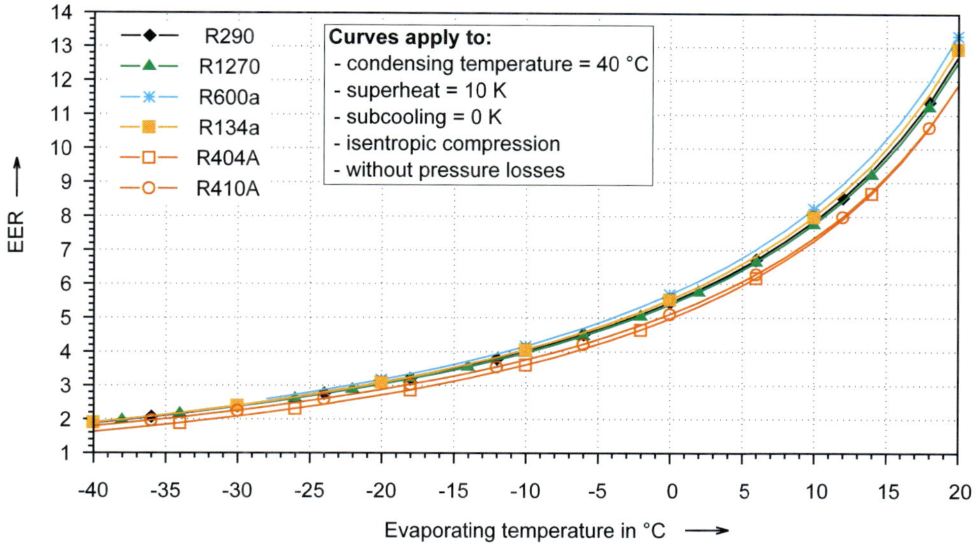

Figure 6.5 Ratio of refrigeration capacity to required electric power of ideal example process with isentropic compression (here: EER) of some hydrocarbons and reference refrigerants

As a simplistic comparison, ratios of refrigeration capacity to required compressor power for an ideal cycle are shown in Figure 6.5, and of hydrocarbons relative to those of the fluorocarbon baseline refrigerants in Figure 6.6.

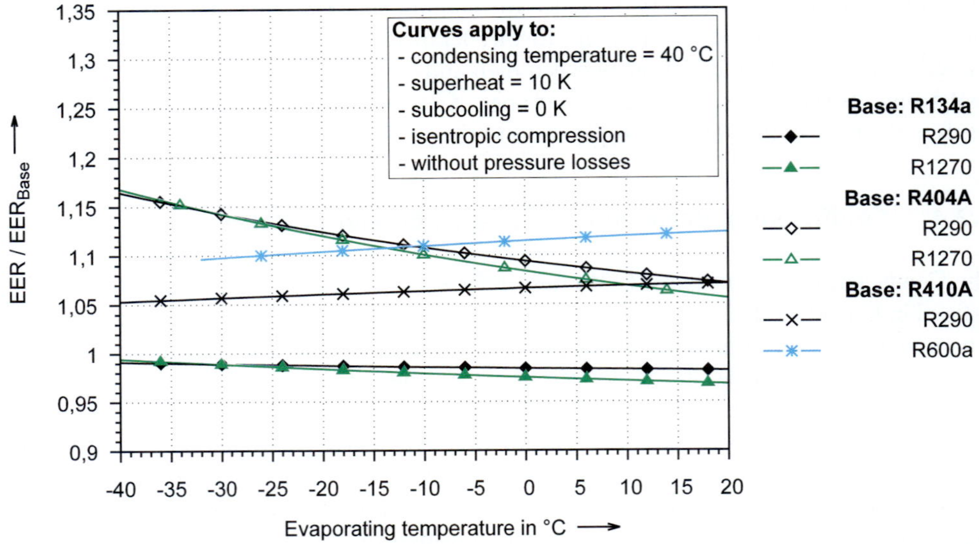

Figure 6.6 Ratio of refrigeration capacity to required electric power (here: EER) of some hydrocarbons and reference refrigerants to the Base of reference refrigerants

The ratio of an ideal cycle with R600a is 2 % to 6 % higher, depending on the evaporation temperature, when compared to R134a. EER of R290 and R1270 are 1 % to 3 % lower than that of R134a, whereas the EER for R290 and R1270 are between 5 % and 17 % higher than R404A. Compared to R410A, EERs for R290 and R600a are between 5 % and 12 % higher.

This simplistic comparison of EERs is indicative of the strong potential for energy efficient operation of HC systems.

6.2.5 Compression or pressure ratio

Figure 6.7 shows the trend of compression ratios (condensation pressure relative to evaporation pressure) for some hydrocarbons and reference refrigerants. This is based on the ideal cycle as described at the beginning of Chapter 6.2, whereby the condensation temperature is constant, with the trend of compression ratio versus evaporation temperature shown in the graph.

Hydrocarbons are compressed at lower pressure ratios than R134a. A lower compression ratio indicates less work by the compressor – under the same condensing temperature. Furthermore, low compression ratios allow use of an intermediate heat exchanger (liquid-suction HX) to increase the evaporating capacity by means of subcooling liquid refrigerant (see Chapter 6.3).

Compared to compression ratios of R404A, R1270 and R290 are slightly lower. Pressure ratios of R600a are higher than those of R290, but a little lower than pressure ratios of R134a which are the highest.

Even though R410A yields the highest absolute pressure, the compression ratio is just above those of R290 and R600a.

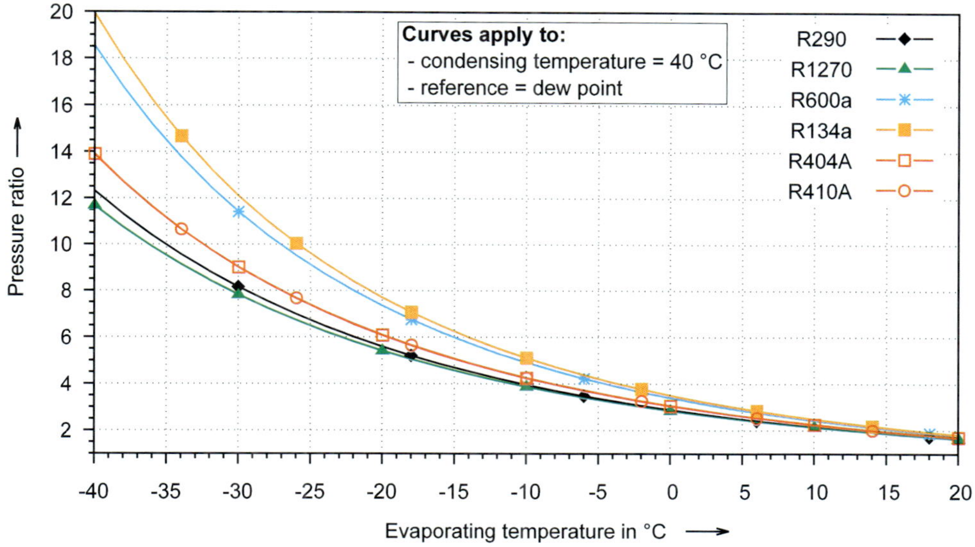

Figure 6.7 Compression ratio of some hydrocarbons and reference refrigerants

Figure 6.8 shows curves representing the absolute pressure differential between suction and discharge pressure, or condensation and evaporation pressure, of the ideal cycle that neglects pressure loss.

Values of pressure differencial depending on evaporation temperature, as shown in Figure 6.8, are the lowest when using R600a, and higher when using R290. For R1270, pressure differencial is highest compared to R600a and R290.

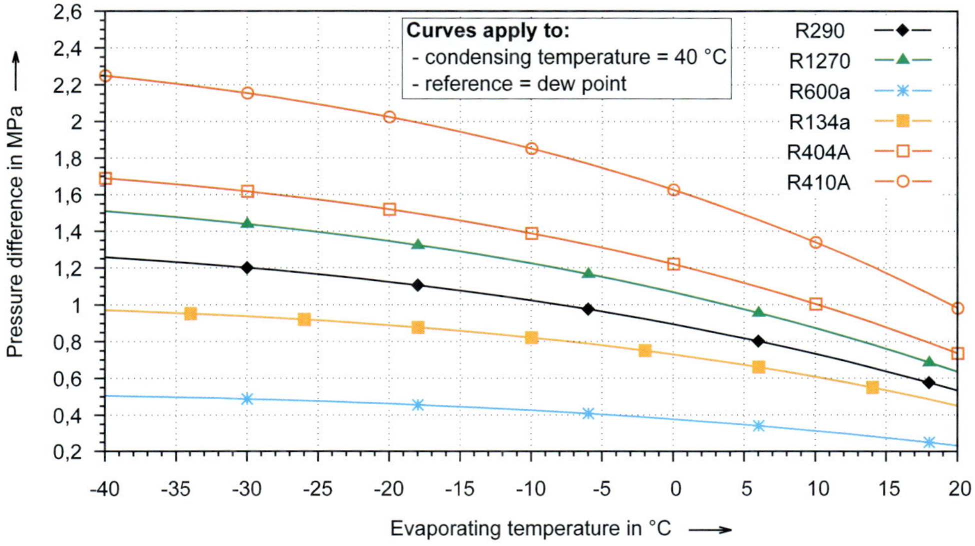

Figure 6.8 Pressure differencial of some hydrocarbons and reference refrigerants

For reference, R410A and R404A exhibit the highest pressure difference, whereas those of R134a run between those of R600a and R290.

6.2.6 Volumetric cooling capacity

The volumetric cooling capacity relates to the volume of the refrigerant suction gas at the compressor inlet to the achievable cooling capacity at the evaporator. Thus, the volumetric cooling capacity influences the required compressor displacement for a given cooling capacity.

Figure 6.9 shows curves of volumetric cooling capacity of the ideal cycle. Figure 6.10 shows curves of hydrocarbons relative to those of the fluorocarbon baseline refrigerants.

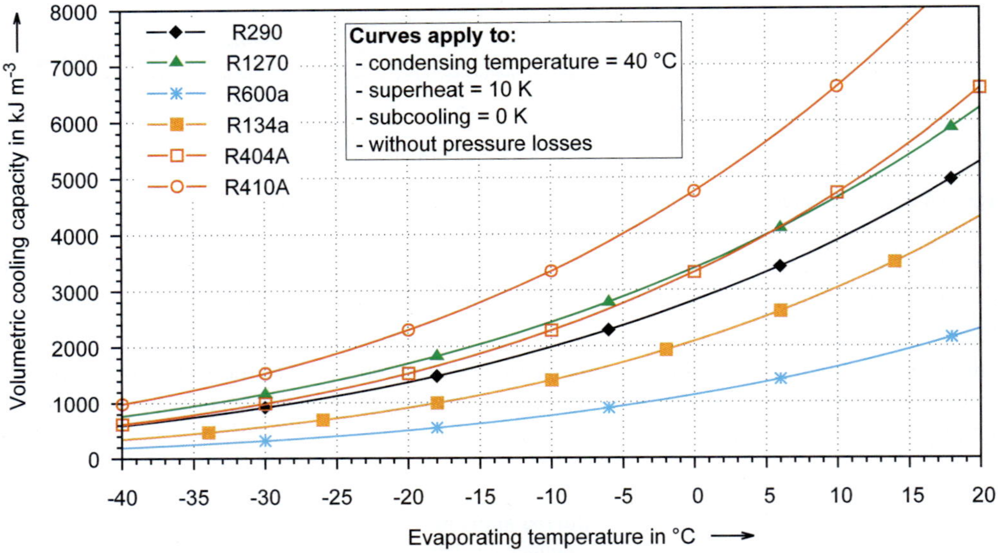

Figure 6.9 Volumetric cooling capacity of some hydrocarbons and reference refrigerants

The volumetric cooling capacity of R600a is lower than other refrigerants. For R290, the volumetric cooling capacity is comparatively high, above R134a and in the range of low-temperature application similar to R404A. R1270 has the highest volumetric cooling capacity of the HCs shown, even higher than for R404A for temperatures below 5 °C. For reference, R410A has the highest volumetric cooling capacity over the entire range of evaporation temperatures.

6.2.7 Discharge temperature

The isentropic exponent κ of a refrigerant and losses during the compression process determine the compression efficiency and the discharge temperature.

If the discharge temperature is too high, decomposition and coking of lubricating oil occurs and oil sludge may form. Lubrication properties deteriorate and the amount of oil in the refrigerant circuit decreases and in the worst case the compressor can be damaged by lubricant starvation.

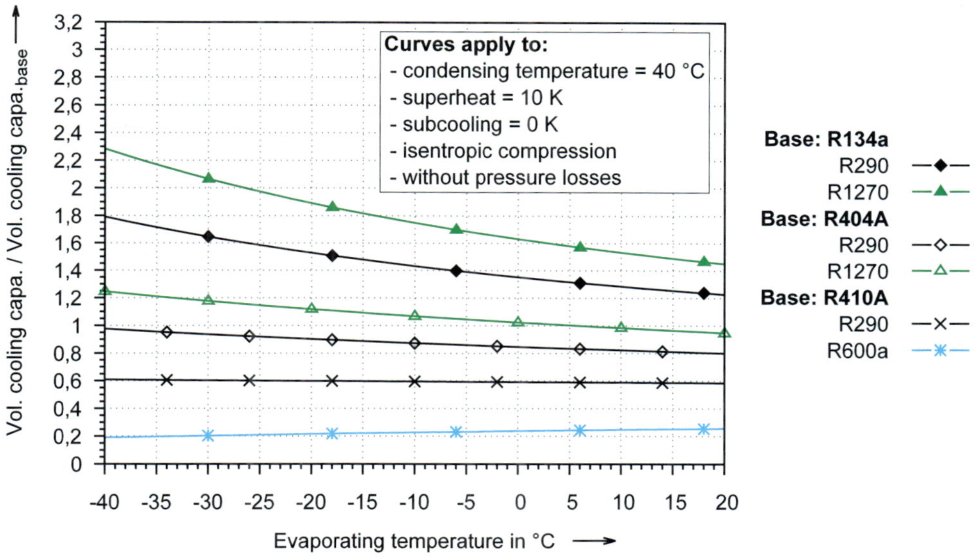

Figure 6.10 Ratios of the volumetric cooling capacity of some hydrocarbons to reference refrigerants

Too high discharge temperatures can cause damage to materials. A low discharge temperature therefore has a positive effect on compressor materials, components and stability of the oil.

Figure 6.11 shows curves of discharge temperature of the ideal cycle for some hydrocarbons and reference refrigerants. R600a has comparatively low discharge temperatures and those for R290 are slightly higher. For hydrocarbon refrigerants, discharge temperatures are lower than those of the reference refrigerants. Discharge temperatures of R1270 are slightly higher than of other HCs, but considerably lower than of R410A.

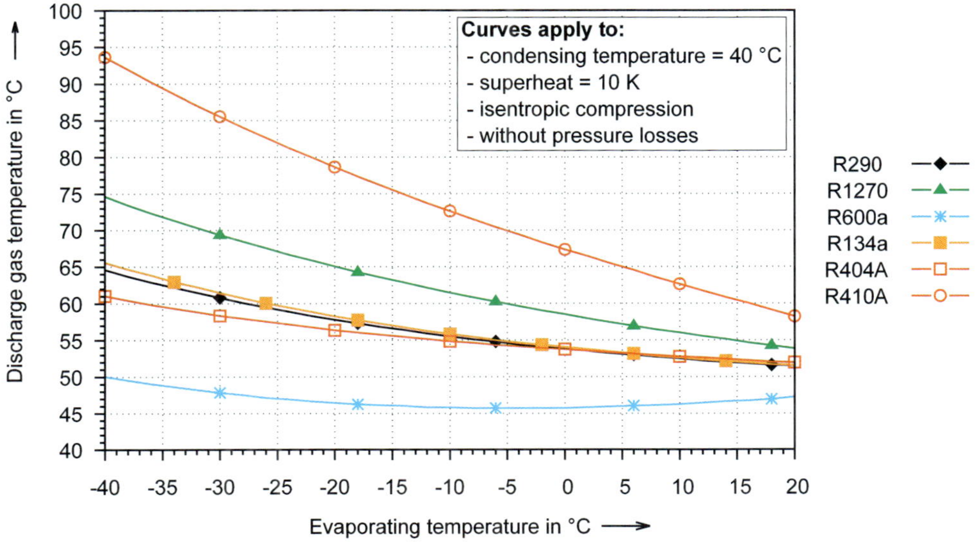

Figure 6.11 Discharge temperatures of some hydrocarbons and reference refrigerants

6.2.8 Density

Hydrocarbons have lower densities than fluorocarbons. For this reason, despite higher specific latent heat, hydrocarbons do not show higher volumetric refrigeration capacity.

Due to the low density, refrigerant charges for hydrocarbons are correspondingly small.

In Table 6.8 and Table 6.9, the density of hydrocarbons in saturated vapour and liquid phase are shown.

Table 6.8 Density of some hydrocarbons and reference refrigerants in kg/m³ at saturation at various temperatures

Refrigerant	R170	R290	R600a	R1270	R134a	R404A	R410A
ρ_l (T = –20 °C)	442	555	603	574	1358	1222	1245
ρ_g (T = –20 °C)	25,8	5,5	2,1	6,6	6,8	15,4	15,4
ρ_l (T = 0 °C)	400	529	581	546	1295	1150	1170
ρ_g (T = 0 °C)	46,1	10,3	4,3	12,3	14,4	30,5	30,6
ρ_l (T = 20 °C)	339	500	557	515	1225	1067	1083
ρ_g (T = 20 °C)	86,2	18,1	7,9	21,4	27,8	56,3	56,8

l = liquid phase, g = gaseous phase

Table 6.9 Density of some hydrocarbons and reference refrigerants in lb/ft³ at saturation at various temperatures

Refrigerant	R170	R290	R600a	R1270	R134a	R404A	R410A
ρ_l (T = –4 °F)	27.6	34.6	37.6	35.9	84.8	76.3	77.7
ρ_g (T = –4 °F)	1.6	0.3	0.1	0.4	0.4	1.0	1.0
ρ_l (T = 32 °F)	25.0	33.0	36.2	34.1	80.8	71.8	73.0
ρ_g (T = 32 °F)	2.9	0.6	0.3	0.8	0.9	1.9	1.9
ρ_l (T = 68 °F)	21.2	31.2	34.8	32.1	76.5	66.6	67.6
ρ_g (T = 68 °F)	5.4	1.1	0.5	1.3	1.7	3.5	4.1

l = liquid phase, g = gaseous phase

As a rule-of-thumb, alternative refrigerant charge (m_2) can be approximated from the original refrigerant charge (m_1) using the ratio of liquid densities (ρ_2/ρ_1)at condensing temperature, i.e.:

$$m_2 \approx m_1 \cdot \frac{\rho_2}{\rho_1}$$

With the values from Table 6.8 or Table 6.9, hydrocarbon refrigerant charge is around 40 to 50 % of the fluorocarbon charge (assuming the similar system geometry).

6.2.9 Viscosity

Higher viscosity leads to increased pressure losses in the refrigerant flow. When a real process is shown in the pressure-enthalpy diagram, negatively sloping curves occur during de-superheating, condensation, subcooling, evaporation, and superheating. The more viscous the refrigerant, the greater the detriment to efficiency compared to an ideal process.

Hydrocarbons generally have comparatively low viscosity, which ultimately has a positive effect on the achievable COPs.

Table 6.10 and Table 6.11 show dynamic viscosity of some hydrocarbons in saturated vapour and liquid phase at 20 °C (68 °F) and of reference refrigerants.

Table 6.10 Dynamic viscosities of hydrocarbons in 10^{-6} Pa s at 20 °C at saturation

Refrigerant	R170	R290	R600a	R1270	R134a	R404A	R410A
η_l	42	102	159	101	207	137	128
η_g	12	8	7	9	11	12	13

Table 6.11 Dynamic viscosities of hydrocarbons in 10^{-6} lb/ft/s at 68 °F at saturation

Refrigerant	R170	R290	R600a	R1270	R134a	R404A	R410A
η_l	0.9	2.1	3.3	2.1	4.3	2.9	86.0
η_g	0.3	0.2	0.1	0.2	0.2	0.3	8.7

It should be noted that where refrigeration machine oil is present, the mixture of refrigeration machine oil and refrigerant influences the material properties. In particular, the viscosity of the oil-refrigerant mixture determines the thickness of the fluid film adhering to walls. This fluid film forms a thermal boundary layer in which heat is transmitted via ("slow", ineffective) heat conduction and not via ("fast", effective) convective material transport. If the viscosity of the refrigerant-oil mixture is too low, the lubrication capacity is impaired by insufficient lubricating film thickness. If the viscosity is too high, friction is not reduced and heat transfer is impaired. Therefore, a compromise must always be found between good heat transfer properties (low viscosity) and good lubrication (high viscosity).

Since the viscosity of hydrocarbons is low and the miscibility with and solubility in mineral oils are good, either highly viscous oils or special, less soluble oils are used, e.g., polyol ester (POE), polyalkylene glycol (PAG) or polyalphaolefin (PAO) oils. More details are described in Chapter 9 "Lubricants".

6.2.10 Material compatibility

Hydrocarbon refrigerants are chemically relatively stable chemicals and non-corrosive to common materials. They are compatible with most system materials, although certain elastomers may need to be avoided. For example, neoprene / chloroprene (CR) may not be sufficiently resistant to R1270 and R600a, and none of the discussed hydrocarbon refrigerants are compatible with ethylene propylene diene monomer (EPDM). Hydrocarbons absorb little water and do not form acids with water, unlike fluorocarbons.

6.3 Refrigeration system components

System components for hydrocarbons are broadly equivalent to those for fluorocarbons. All necessary components are currently available from usual suppliers. Components that have an electrtical input (e.g., solenoid valves, electronic expansion valves, crankcase heaters, oil level switches, pressure switches, etc.) may be assessed and approved according to the relevant standards (IEC 60079-series) for ex-type equipment.

6.3.1 Compressor

For hydrocarbons, reciprocating and rotary and scroll compressors are used whenever small and medium refrigeration capacities are required. Compared to screw compressors, they convey lower displacement volumes up to 300 m^3/h (10600 ft^3/h). Semi-hermetic reciprocating compressors with 2, 4, 6 or 8 pistons, have a comparatively smaller housing volume and a small oil quantity. Lubricants include polyalkylene glycole (PAG) and polyol ester (POE) oils.

Screw compressors are generally used when larger refrigeration capacities are required. They use a semi-hermetic design and, for hydrocarbons, convey displacement volumes from 100 m^3/h to over 1000 m^3/h (3530 ft^3/h to 35300 ft^3/h). Hydrocarbon refrigerants are generally highly soluble in mineral and alkylbenzene (AB) refrigeration oils, which is why the oil in the compressor may contain a considerable amount of refrigerant.

The available scroll compressors for hydrocarbons range up to 115 m^3/h (4061 ft^3/h), although capacity ranges are increasing.

The required motor power for the compressor differs at the same displaced volume depending on the design operating point of the refrigerant circuit. Reciprocating compressors require up to around 100 kW (341413 Btu/h), and screw compressors 300 kW (1024240 Btu/h), respectively.

Maximum allowable discharge temperature depends on the refrigerant oil. In reciprocating compressor systems, superheating may be limited, cylinder heads may be fan-cooled, or both measures may be adopted. In screw compressor systems with economisers, compression may also be cooled at intermediate-pressure level via flash gas.

6.3.2 Piping and fittings

Piping and fittings for hydrocarbons do not differ from those used with fluorocarbons, although (as discussed in Chapter 6.4.1), it is preferable to avoid inherently leaky fittings such as flares. Detachable connections may be used only where no other joining method is possible. Pipe sizes are similar to common fluorocarbons with a comparable vapour pressure, although due to lower viscosity, diameters can be incrementally smaller. Sizing of oil return lines depend upon the specific conditions.

6.3.3 Heat exchangers

Due to advantageous thermodynamic and transport properties of hydrocarbons, comparatively compact heat exchangers (HX) may achieve the required thermal performance. Highly efficient HX with small liquid content are often preferred to assist with system charge reduction. To further lower refrigerant charge, dry expansion HX may by preferred over flooded operation.

Use of a liquid suction (LSHX) or intermediate or internal heat exchanger (IHX) compressor increases suction gas temperature and therefore discharge temperature. Refrigerants with high isentropic exponents (e.g., ammonia) are not well suited for high suction temperatures, as discharge temperature increases comparatively rapidly and compressor volumetric efficiency decreases. Hydrocarbons are more suitable. LSHX can improve the cycle efficiency of some systems with hydrocarbons and can help reduce refrigerant charges by 5 % to 15 % [6-12]. Due to their high oil solubility, hydrocarbon refrigerants require high superheating values (e.g., 15 K to 20 K) at the compressor inlet, which may be efficiently produced in the LSHX. Additional subcooling of the liquid refrigerant in the LSHX after the condenser and ahead of the throttle device results in increased evaporation latent heat; also, liquid receivers may be dispensed with, this further helps reducing system charge. However, selection of a LSHX should avoid high suction pressure drops, which can inadvertently offset the expected benefits to cycle efficiency.

6.3.4 Throttle device

Throttle devices are selected to suit application, refrigerant and operating conditions. Smaller plug-in systems that are mostly air-cooled, direct-expansion use capillary tubes and thermostatic expansion valves (TEV). In medium and larger systems evaporators, electronic expansion valves (EEV) are often used, e.g., for dry expansion where dynamic capacity control is expedient.

6.3.5 Receivers

If it is necessary to reduce and limit the refrigerant charge of a hydrocarbon refrigeration system, the machine is designed without a refrigerant receiver. If there is no refrigerant reservoir it is necessary to adjust the refrigerant amount more precisely to achieve the optimum charge.

6.3.6 Multi-stage systems

Hydrocarbons are suitable for reaching very low temperatures in single- and multi-stage systems and can be used both for lower-temperature and for high-temperature stages.

R290 and R600a are characterised by relatively low discharge temperatures. Comparatively large temperature differentials can therefore be realised even in systems with single-stage refrigeration circuits.

Very low-temperature systems use R170 in the low stage of a cascade system, with higher boiling hydrocarbons in the high stage.

6.4 Safety of refrigeration systems with hydrocarbons

A safety concept considers safety, health and environmental hazards during the life cycle of a hydrocarbon refrigeration system, e.g. production, transport, operation, service/ maintenance, disposal/ recycling, etc. Regarding flammability, there are regulations and responsibilities for manufacturers, planners, users, operators, etc. and specific requirements apply to smaller and larger systems for system engineering, design, production as well as safety devices and measures. There are training courses where expert knowledge of safe use of flammable refrigerants may be acquired by technical personnel.

A thorough and well-regulated process of hazard identification, risk analysis and mitigation yields quick identification of weaknesses and reaction before a serious damage incident occurs.

The manufacturer is responsible for the operational safety comprising inherently safe construction, suitable safety devices and other protective measures. Relevant regulative requirements concerning product safety must be complied with during production and manufacturing of a refrigeration system. Once a refrigeration unit is assembled, its conformity to product standards, i.e. technical, safety, etc., must be declared; rules and regulations are established at international / national levels, and local requirements may apply.

For operators of larger systems, there are further obligations with regard to occupational, health and environmental safety on site, as well as system maintenance and documentation.

Limitation of the extent of damage

Flammable gases form explosive atmospheres with air. If a hazardous area is of non-negligible extent, where concentration throughout the space exceeds the LFL it is appropriate to include construction features to relieve an explosion, in the unlikely event of ignition, to minimise the possibility of structural damage. This is usually achieved by the use of a frangible part of the ceiling or the wall, providing relief to the outside, corresponding to at least 20 % of one wall area. (Although it should be evaluated on a case-by-case basis.)

Risk assessment of refrigeration systems

Risk assessment requires an iterative approach and includes:

- identification of hazards and unintended use of the refrigeration system during all life cycle phases
- analysis of risks arising from the identified hazards
- implementation and documentation of constructive measures to eliminate hazards and of safeguards to reduce or minimise the risks of remaining hazards

Risk assessment of larger systems must also consider that dangerous situations arise when persons are on site who lack knowledge of the intended use and operation of refrigeration systems.

Depending on a hazard situation and probability of its occurrence, accumulation of several hazards must be considered as well.

Regardless of size, all hydrocarbon systems should be subject to risk assessment

Of course, risk assessment procedure is required for all systems, regardless of the refrigerant type (flammable or non-flammable, etc.). Numerous different techniques are available for risk assessment – the most appropriate for the situation should be selected.

Safety class A3

Hydrocarbon refrigerants belong to the safety class A3. The designation describes the:

- toxicity class A (low toxicity or no significant adverse health effects on humans at concentrations below 400 ppm)
- flammability class 3 (higher flammability or LFL of the refrigerant-air mixture at less than 3,5 % v/v)

The banned chlorofluorocarbons belong to safety class A1. Hydrochlorofluorocarbons, fluorocarbons, and hydrofluoroolefins (unsaturated fluorocarbons) belong to safety classes A1, A2, A2L and A3. Hydrocarbons are lower toxicity but higher flammability, so it is flammability that dictates their application in terms of charge limits.

Table 6.12 Selected safety data of some hydrocarbons according to EN 378–1, Table E.1

Refrigerant no	Chemical name	Safety classification	ATEL/ODL in kg/m^3	LFL in kg/m^3	Normal boiling point in °C	Normal vapour density in kg/m^3	Auto-ignition temperature in °C
170	ethane	A3	0,0086	0,038	−89	1,23	515
290	propane	A3	0,09	0,038	−42	1,80	470
1270	propene	A3	0,0017	0,046	−48	1,72	455
600a	isobutane	A3	0,059	0,043	−12	2,38	460

Table 6.13 Selected safety data of some hydrocarbons according to EN 378–1, Table E.1

Refrigerant no	Chemical name	Safety classification	ATEL/ODL in lb/ft^3	LFL in lb/ft^3	Normal boiling point in °F	Normal vapour density in lb/ft^3	Auto-ignition temperature in °F
170	ethane	A3	0.00054	0.00237	192.2	0.07679	959
290	propane	A3	0.00562	0.00237	−43.6	0.11237	878
1270	propene	A3	0.00011	0.00287	−54.4	0.10738	851
600a	isobutane	A3	0.00368	0.00268	10.4	0.14858	860

6.4.7 System tightness

Refrigeration systems should be as tight as possible, irrespective of the refrigerant used; for safety, environmental and cost reasons. For hydrocarbons, the emphasis is for minimisation of flammability risk. Components, joints and fittings should as far as possible be brazed and comply with ISO 14903. Other aspects associated with the construction of the system, such as vibration (from compressor and other moving machinery), external corrosion, fretting, external impacts, etc. should also be considered. Minimising leakage is addressed in other safety standards such as EN 378, ISO 5149 and EN 1127-1.

6.4.8 Charge minimisation

When using flammable refrigerants, an effective measure to reduce flammability risk is to minimise the refrigerant charge. It is desirable to minimise the system charge as much as possible, some options to this end are presented.

In order to yield the lowest possible refrigerant charge, systems are optimised for least charge per kW capacity. Refrigerant circuits are charged precisely to suit given operating points so excess charge is avoided. Dry expansion systems yield lower charges; flooded systems usually demand much greater charge. Components with small internal volumes, e.g., air-cooled condenser with microchannels, are used to reduce charge. Some applications like heat recovery employing a desuperheater require a receiver to account for all operating points.

To further minimise refrigerant charge per circuit, it is possible to increase the number of circuits at a given capacity. For example, it may be possible to design a unit with multiple refrigerant circuits with smaller capacity, each requiring a smaller charge, instead of just one circuit with large capacity and large charge.

6.4.9 Determination of limited refrigerant charge

Determination of charge limits is handled differently across the various safety standards. EN/IEC 60335-2-40 and EN/IEC 60335-2-89 apply charge limits according to the design of the equipment. Charge limits in EN 378 differ according to the design of the equipment, but also the type of occupancy within which the equipment is located; these are described by the installation area "access category" and system "location classification".

Installation site access category of the refrigeration system depends on the potential movement and competence of the persons allowed in the location of refrigerant-carrying equipment. Table 6.14 summarises those access categories.

Table 6.14 Categories of access areas of installation sites of refrigeration systems in accordance with EN 378–1:

Access category	Explanation
a) General access area	The following persons may be present: • sleeping people • people with limited mobility • unknown number of people • people without knowledge of the hazard Examples: hospital, prison, theatre, train station, hotel, residential apartment
b) Monitored access area	The following persons may be present: • limited number of people • at least some people are familiar with the safety arrangements of the facility Examples: office or business premises, manufacturing and workplaces
c) Access area for authorised persons only	Restricted access where the following persons may be present: • all are familiar with the safety precautions of the facility • materials or goods are manufactured, processed or stored there Examples: refineries, cold stores, dairies

The location classification refers to the arrangement of the refrigerant-carrying parts as shown in Table 6.15.

Table 6.15 Location classification in accordance with EN 378

Location classification	Explanation
Class I: Mechanical devices in occupied areas	Class II does not apply, and refrigeration equipment is located in the occupied area.
Class II: Compressors in the machine room or outdoors	Class III does not apply, and refrigerant-containing parts of the refrigeration system are arranged as follows: • all compressors and pressure vessels within the machine room (according to EN 378–3) or outdoors • Pipes and valves possibly in public areas
Class III: Machinery room or outdoors	All refrigerant-carrying parts arranged in the machinery room (according to EN 378–3) or outdoors.
Class IV: Ventilated housing	All refrigerant-containing parts inside ventilated housing (according to EN 378–2 and –3).

In general, the maximum allowable charge of a refrigeration system using any refrigerant is the lower of the two limits determined by toxicity and flammability. For flammable refrigerants, there are then two limits:

- The allowable charge limit, which is usually based on the room size · LFL · a factor smaller than 1, thereby ensuring the entire room volume cannot become flammable.
- An upper charge limit, or "cap" limit (m_1, m_2, m_3), which constrains the charge below a prescribed quantity, is assumed to reflect a threshold of releasable energy (from combustion). This upper charge limit applies regardless of the room size.

Depending on the location of a system's refrigerant-carrying parts and compressor, on its application and on the access category of the location, the maximum allowable charge is determined.

Note that the charge limits are based on individual system (circuit) charges, rather than the total charge. In other words, maximum allowable amount of refrigerant would be based on the largest system (circuit) charge of a multi-circuit unit and not the summation of all the circuits.

Table 6.16 Maximum allowable refrigerant charge for human comfort cooling refrigeration systems with flammability class 3 refrigerants according to EN 378–1, Table C.2

<table>
<tr><th rowspan="2">Category of access area</th><th colspan="4">Classification of location</th></tr>
<tr><th>I</th><th>II</th><th>III</th><th>IV</th></tr>
<tr><td>a</td><td colspan="2" rowspan="3">m_1 or m_{max} and maximum m_2 or 1,5 kg (3.3 lb)</td><td>5 kg (11 lb)</td><td rowspan="3">m_3</td></tr>
<tr><td>b</td><td>10 kg (22 lb)</td></tr>
<tr><td>c</td><td>no limit on the refrigerant charge</td></tr>
</table>

Direct systems in occupied areas

Sealed refrigeration systems may be installed without requirement for minimum room size or access restrictions if charge is less than 150 g (5.29 oz) (or cap factor $m_1 = 4\ m^3 \cdot LFL$ / $m_1 = 141.25\ ft^3 \cdot LFL$). It must be noted that cap factor m_1, on which the value limit of 150 g (5.29 oz) for sealed refrigeration systems is based, is not considered as state of the art anymore but rather as obstructive and detrimental to development of smaller systems with higher charge. After decades of experience with smaller, direct, sealed hydrocarbon refrigeration systems for applications such as household, commercial, human comfort, etc., research has demonstrated that higher charge limits up to 300 to 500 g (10.58 to 17.63 oz) can be safely applied and a revision of capping factor m_1 is necessary.

Of direct systems with refrigerant-carrying parts in occupied spaces, the location classification is either I or II. It is, similarly as discussed above, considered obstructive that current standard EN 378-1 differentiates between direct refrigeration systems providing cooling for human comfort and other direct systems when determining the maximum allowable charge.

In case of human comfort application, the allowable charge limit must not exceed the greater of 1,5 kg (3.3 lb) or cap factor $m_2 = 26\ m^3 \cdot LFL$ ($m_2 = 918.18\ ft^3 \cdot LFL$) and is also restricted below:

$$m_{max} = 2{,}5 \cdot LFL^{1{,}25} \cdot h_o \cdot \sqrt{A}$$

Where:

m_{max}	maximum allowable refrigerant charge in a room
A	room surface area
LFL	lower flammable limit, according to Table 6.12 and Table 6.13
h_o	a height representing the location of the unit: 0,6 m (1.96 ft) when installed on the floor; 1,8 m (5.9 ft) for wall mounting; 1,0 m (3.28 ft) for window mounting; 2,2 m (7.2 ft) for ceiling mounting)

whereas for all other applications of direct systems with location classification I and II, the allowable charge is determined from 20 % of the LFL and room size:

$$m_{max} = 0{,}2 \cdot LFL \cdot V_{room}$$

where V_{room} is the room volume. Depending on the access category and location of the installation site, maximum allowable charge of a direct system must also not exceed a total amount of:

- access category a below / above ground: 1 kg / 1,5 kg (220 lb / 3.30 lb)
- access category b below / above ground: 1 kg / 2,5 kg (220 lb / 5.51 lb)
- access category c below ground: 1 kg (2.20 lb)
- access category c above ground with location classification I / II: 10 kg / 20 kg (22.04 lb / 44.09 lb)

As mentioned above, both formulas for m_{max} will be revised in forthcoming versions of safety standards. They are expected to be superseded with others that more accurately represent the factors involved.

Conversely, if the system charge quantity is known, then the formula for allowable charge limit m_{max} is transposed and the minimum room size for that system should be determined.

Indirect systems in machinery rooms

For location classification III of systems in machinery rooms or in the open air according to EN 378-3, there is no restriction of refrigeration charge as long as the room may be considered as an unoccupied area with access restricted to authorized personnel. If the charge exceeds that for an equivalent occupied area (as in equation for direct systems, see above) then additional measures, such as emergency ventilation, is necessary.

If a machinery room is occupied for any period of time (except for service, maintenance and similar activities) it must be treated as an occupied area. For machinery rooms with general access (category a) the refrigerant charge is limited to below 5 kg (11.02 lb), and below 10 kg (22.04 lb) if access is categorised as monitored (category b).

Indirect systems in ventilated enclosures

The maximum allowable charge of all systems with ventilated enclosures that fall under location classification IV is limited by cap factor m_3 = 130 m^3 (4590.9 ft^3) · LFL (e.g., R290 with m_3 = 4,94 kg / 9.89 lb). There is no differentiation between access categories for such systems.

6.4.10 Mechanical ventilation

Use of mechanical ventilation is a fundamental element of many safety features of HC installations.

Extract ventilation can be extremely useful in removing and dispersing potentially flammable mixtures that arise from refrigerant leaks.

For machinery rooms and ventilated enclosures, extract ventilation must be provided, either continuously or upon demand of a leak or gas detection system.

A background ventilation rate of four air changes per hour is required for any machinery room if workers are present (and this applies regardless of the refrigerant type). Similarly, an "emergency" ventilation rate is required of $0{,}0014 \cdot m^{2/3}$, where m is the system refrigerant charge. Air must be exhausted to a safe area.

However, within the hazardous area standards, EN/IEC 60079-10-1, defines the size of hazardous zones according to the amount ("degree of dilution") and "availability" of ventilation.

IEC 60070-10-1 addressed the relationship between potentially hazardous and ventilated areas differently from the RACHP (refrigeration, air conditioning, and heat pump) standards. Since the ex-standards tend to take precedence in most countries, it is pertinent to accommodate the approaches detailed therein.

Recommended ventilation rate is

$$\dot{V}_{min} = \frac{\dot{m}_{leak}}{\rho_g} - \frac{f \cdot \dot{m}_{leak}}{\rho_g \cdot \phi \cdot LFL_v}$$

where $\dot{V}_{min}$ is the minimum airflow rate, $\dot{m}_{leak}$ is the assumed leak mass flow rate, ρ_g is the gas density, ϕ is a dimensionless multiplier for the lower flammability limit (LFL), dictating the ventilation outlet concentration (i.e. < 1 to ensure the mixture is non-flammable), LFL_v is LFL as volume fraction and f is a factor used to account for the internal mixing efficiency that may range from 1 (e.g., an empty room) up to 5 (e.g., a highly congested room).

IEC 60079-10-1 provides a detailed discussion on dilution effectiveness. However, for RACHP applications and corresponding charge amounts, it has been found that a mean room air speed of at least 0,3 m/s (0,98 ft/s) will provide a high degree of dilution. Low dilution occurs around < 0,05 m/s (0.16 ft/s) (i.e. "still" conditions).

"Good" availability is where ventilation is either continuously present or rapidly upon demand of leak detection. For natural ventilation it may be outside in a non-confined area. "Poor" availability is when airflow is intermittent or outside in a sheltered area.

According to EN 378-3, if a machinery room has more than 80 % of one wall "open", it's deemed as in the open air.

6.4.11 Ventilated enclosures

Construction of units with ventilated enclosures housing the compressor and other refrigerant-carrying parts is a good way to avoid formation of a flammable atmosphere. The extraction of air creates a negative pressure inside the enclosure (> 20 Pa / 0.002 psi) and any leaked refrigerant can be diluted and dispersed; thus formation of a flammable atmosphere is safely avoided during operation. Following minimum ventilation flow Q_{min} is required

$$Q_{min} = 15 \cdot s \cdot \left(\frac{m_c}{\varrho}\right) \geq 2\ m^3/h$$

Where 15 in 1/h is a conversion constant that defines a "catastrophic" leak rate of the whole refrigerant charge in four minutes, s = 4 is a safety factor, m_c is the refrigerant charge mass and ρ is the density of the refrigerant at atmospheric pressure at 25 °C (77 °F).

An interpretation of the leak rate has been approved for IEC 60335-2-40 regarding smaller systems with enhanced tightness, shown in Table 6.17. Therein, calculation of the assumed leak mass flow rate considers the choked flow mass flux of the specific refrigerant from the circuit at above atmospheric pressure to the ambient environment.

Table 6.17 Refrigerant leak rate for smaller systems with enhanced tightness

Refrigerant	Refrigerant leak rate	
R290	0,00112 kg/s	0.002469174 lb/s
R600a	0,00048 kg/s	0.001058218 lb/s
R1270	0,00135 kg/s	0.002976237 lb/s

If fans are ex-type, they can be directly placed in the extract airflow. The extract air ducts are sealed at seams and joints, free from ignition sources and lead from the enclosure directly to open air. If extract air is exhausted to a room, maximum charge is restricted and room volume must be as specified above for direct systems. Manufacturers are required to specify the exhaust ventilation duct dimensions, the maximum length and number of bends.

Enclosure ventilation can be continuous with sufficient air flow to dilute leaking flammable gas below LFL and an air flow switch to ensure permanent fan operation at all times. Permanent fan operation may not be desired in noise-sensitive environments. If there is gas detection inside the enclosure, it must be ventilated when hydrocarbons are detected (> 25 % LFL). The unit itself is switched off if a certain threshold is exceeded (> 50 % LFL) in order to deactivate possible ignition sources, while ventilation keeps running, and alarm is raised to warn persons to leave the vicinity and further safety measures may be triggered.

The function of all components to ventilate enclosures must be ensured at all times. If the enclosure ventilation or associated components fail, it is treated as if a flammable atmosphere is present. Faults must be repaired immediately, whilst sufficient ventilation of housing and the installation site must be ensured and only persons who are equipped with mobile gas warning devices can service the unit.

6.4.12 Machinery room ventilation

Requirements for machinery rooms are specified in EN 378-3. Specifically mentioned for systems with flammable refrigerants is that machinery rooms shall be assessed and classified according to the requirements of IEC 60079-10-1 for the hazardous zone. Such an assessment considers LFL as well as type of release, and it may conclude, e.g., in case of a unit with sufficiently ventilated and appropriate enclosure to avoid formation of an explosive atmosphere, that the hazardous area is of negligible extent. Yet, if a hazardous area has been identified and gas detection is required in the machinery room, the emergency mechanical ventilation has to be activated by the gas sensor(s).

For normal operation and when occupied, machinery rooms need mechanical ventilation according to national regulations with a minimum of 4 air changes per hour. For emergency mechanical ventilation, the required airflow $\dot{V}$ increases and is calculated as follows

$$\dot{V} = 0{,}014 \cdot m^{2/3}$$

where m is the mass of the mass of the refrigerant charge, in the refrigeration system with the largest charge, and 0,014 is a conversion factor.

Note that the calculation of ventilation rate prescribed in IEC 60079-10-1 differs somewhat from that in EN 378. The ventilation rate is

$$\dot{V} = \frac{\dot{m}_{leak}}{\rho_g}\left(\frac{f}{\phi \cdot LFL_v} - 1\right)$$

where $\dot{V}$ is the minimum airflow rate, $\dot{m}_{leak}$ is the assumed leak rate, ρ_g is the density of refrigerant vapour, ϕ is a dimensionless multiplier for the lower flammability limit (LFL) dictating the ventilation outlet concentration (i.e. < 1 but preferably < 0,5 to ensure the mixture is non-flammable), LFL_v is LFL as volume fraction (0,021 for R290) and f is a factor used to account for the internal mixing efficiency that may range from 1 (e.g. an empty room with "piston" type ventilation) up to 5 (e.g., a highly congested room with air inlet locations that lead to excessive recirculation).

Note that unlike for EN 378-3 formula, the absolute quantity of refrigerant is not considered, as ventilation is deemed to be a steady flow process. Also, the EN 378 formula does not consider the quantity of the airflow within the space.

It is recommended that the higher result for the ventilation volume flow rate of the two calculations is used.

If a unit with ventilated enclosure is installed in a machinery room, care must be taken that enough additional supply air is available to this purpose.

6.4.13 Gas/leak detection

Mobile personal and stationary gas warning devices provide means for leak detection by determining if HCs are present. Gas detection is important during service and maintenance for personal protection and to locate existing leaks. During normal operation gas detection may be required depending on a system's design, its location and charge. Gas warning devices and systems are monitored continuously, and there are thresholds for pre-alarm (e.g., above 25 % LFL) and gas-alarm (e.g. above 50 % LFL) if a flammable atmosphere is forming.

Gas detection sensors must be regularly calibrated, serviced and their function tested according to regulation as well as local and the manufacturer's requirements. Service and calibration is due multiple times per year and intervals may need to be shorter if ambient atmosphere is contaminated.

There are several principles that are used for measuring the presence of HCs in air. Chemical principles are employed for testing strips that are non-reusable and replaced after HCs have been detected. They don't provide precise measurement and are thus not implemented as critical safety devices but are rather employed to prove the occurrence of leakage. For more precise measurements of the amount of HCs in the ambient atmosphere, e.g., to determine % LFL, infrared transmitters, pellistors and molecular property spectrometers are used.

Gas sensors must be placed where hydrocarbon refrigerants will likely accumulate, e.g., close to ground surfaces beneath refrigerant circuits.

6.4.14 Safety devices

Safety devices of hydrocarbon systems generally correspond to those for other systems.

Pressure limiting devices are pressure switches with safety function that respond to a predetermined pressure by stopping the operation of the compressor to protect against excessive pressure. Pressure limiters or cut outs are manually reset (PS) or reset manually by using a tool (PZ) which is considered more safe. They may be mounted on the high-pressure side (PSH/ PZH) or low-pressure side (PSL/ PZL) of a refrigerant circuit.

Pressure relieve devices, e.g. safety valves, rupture members, etc. relieve excessive pressure automatically. Refrigerant is released to housing/ ambient air in case of smaller systems and low charge, or in case of larger systems, a pipe is connected to the pressure relieve valve that directly leads to a safe blow-off area (outside). If a pipe is connected to a pressure relieve valve, the refrigerant flow through the pipe must not be obstructed and the maximum pressure drop must not exceed a certain limit, e.g., according EN 13136.

If a unit is not in operation, also in case of emergency, valves can be used to shut off components with large inner volumes holding refrigerant. Thus, if leakage occurs refrigerant is retained and only partly released.

6.4.15 Ignition sources

Any area, where a hazardous flammable atmosphere might occur that is not negligible, should be free of (effective) ignition sources to avoid catastrophic incidents. Sometimes it is not feasible to eliminate all ignition sources in such an area, e.g., electric components situated in a machine room if there is the possibility that ventilation could fail at the same time of a catastrophic leak.

A gas detection system provides continuous monitoring of hydrocarbon levels in the air and ignition sources may be switched off safely before LFL is reached, therefore making them ineffective. Also, especially during service/ maintenance or when repairing a component of a refrigerant circuit, there can be a non-negligible risk of a hazardous flammable atmosphere forming.

Possible ignition sources are discussed in EN 1127-1 and include for example:

- Hot surfaces
- Flames and hot gases (including hot particles)
- Mechanically generated sparks
- Electrical apparatus
- Stray electric currents, cathodic corrosion protection
- Static electricity
- Lightning

Electric and mechanical components should be ex-type (IEC 60079 series) if they are located inside a potentially flammable zone and cannot be switched off before LFL is reached, e.g., > 50 % LFL.

Table 6.18 Minimum ignition energy, auto-ignition temperature, minimum experimental safe gap and flashpoint of some hydrocarbons

Refrigerant	Minimum ignition energy	Auto-ignition temperature	Minimum safe gap	Flashpoint	Most flammable concentration
	mJ	°C	mm	°C	Vol. %
R170	0,25	515	0,91	−135	6,5
R290	0,24	470	0,92	−104	5,2
R600a	> 0,18	460	> 0,9	−83	n.a.
R1270	> 0,18	485	0,91	−108	n.a.
	mJ	°F	in	°F	Vol. %
R170	0.25	959	0.0358	−211	6.5
R290	0.24	878	0.0362	−155.2	5.2
R600a	> 0.18	860	> 0.0354	−117.4	n.a.
R1270	> 0.18	905	0.0358	−162.4	n.a.

Protection against fire and explosion hazards is discussed in EN 378-2 (see Figure 6.12). As stated above, ex-type equipment is safe for use when located in potentially flammable zones where leaked refrigerant could flow or stagnate; but it is cost-intensive. An example of intrinsic safety is electrical equipment of which the maximum possible energy (spark/arc) within its circuit will not ignite the most flammable concentration, e.g. ex-type air flow sensors. If ventilation is facilitated to dilute and disperse leaked refrigerant, airflow can be permanent or initiated prior to energising components/apparatus.

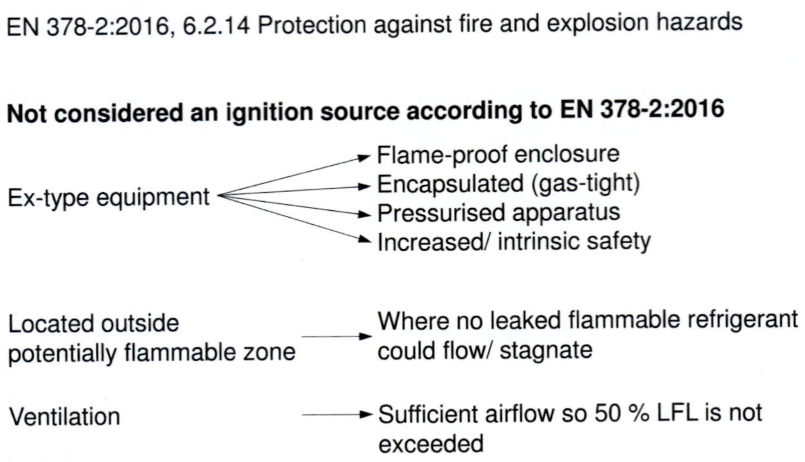

Figure 6.12 Ineffective ignition sources according to EN 378-2

6.5 Safe handling of flammable refrigerants

Specific safety measures need to be implemented when handling flammable refrigerants.

As with handling any refrigerant, a site risk assessment must be carried out by technicians before executing their work. When involving flammable refrigerants that risk assessment should additionally consider the applicable hazards.

Where appropriate, specific safety measures are taken so that any resulting risks may be reduced.

In particular, the simultaneous occurrence of three factors must be prevented:

(1) presence of refrigerant,

(2) formation of hazardous explosive atmosphere with air oxygen and

(3) presence of ignition sources.

Odorants which are usually added to fuel gases as a hazard warning are seldom used with hydrocarbon refrigerants, due to detrimental effects on refrigeration systems. Further, they have been found to degrade over relatively short periods of time so the odour may no longer be present. Generally, whenever a gaseous ("oily") odour is perceived from hydrocarbons, it is highly likely that the concentration is well above the LFL. An exception is R1270, which has a strong olefin smell and can be detected by most noses before it reaches LFL.

Tools and equipment

Tools and equipment used with flammable refrigerants must be suitable. In general they must not produce arcs or sparks or very high temperatures, i.e., act as potential sources of ignition. Today, most producers offer refrigeration service tools and equipment that are suitable for use with flammable refrigerants (when relevant).

As with all refrigerants, but more importantly with flammables, ventilation (units) should be used to ensure adequate airflow of indoor working areas.

Oxygen-free dry nitrogen (OFDN) is to be used for inerting and flushing refrigerant circuits before they are worked on.

Additionally, CO_2 or powder fire extinguishers should always be present.

Site preparation

When handling flammable refrigerants, the working environment and its surroundings must be checked for the presence of any sources of ignition, which must then be removed.

For all activities, adequate ventilation rate must be ensured so that any significant release will be quickly diluted and exhausted from the area.

The working area must be suitably demarcated and unauthorised entry into the area prevented.

Venting or recovery

It should be noted that flammable refrigerant is to be vented into a safe area, outdoors, without ignition sources. The blow-off site must be safely shut off and provided with warnings. Refrigerant vapour must not drift into buildings or similar after venting, nor must it be allowed to sink into ground sinks or shafts and accumulate there.

After venting, the blow-off line ought to be flushed with nitrogen.

If refrigerant is recovered, the recovery machine must be suitable for flammable refrigerants. Special refrigerant cylinders suitable for flammable refrigerant and bearing appropriate marking shall be used for recovery.

6.6 Smaller systems

Hydrocarbons are widely used in smaller capacity systems, with charge amounts up to about 1,5 kg (3.30 lb).

Direct systems, i.e. those that have direct contact with the occupied space, can be used with hydrocarbons, although the size of systems (individual refrigerant circuits) tend to be limited in size, compared to indirect systems that have refrigerant-containing parts outside or within a ventilated enclosure or unit housing.

The primary reason for such limitations is that a rapid release of the charge can potentially fill a closed space with a flammable mixture. For instance, an instantaneous release of 1 kg (3.3 lb) of R290 homogeneously mixed would fill a 26 m^3 (918.2 ft^3) space with a flammable concentration. Any potential source of ignition within that space could ignite the mixture with severe consequences. Therefore, indoor direct systems tend to be limited in capacity and charge amount and accompanied with minimum space sizes (room floor area or room volume) to ensure non-negligible flammable concentrations do not arise. Significant work has been carried out to identify what quantities present low risk, in relation to indoor space sizes.

Historically, hydrocarbons had been used in smaller appliances prior to the development of chlorofluorocarbons. Subsequently, in the form of fuel gas (liquefied petroleum gas or LPG), they have occasionally been used (informally) as “drop-ins” in a variety of systems. However, the modern advent of hydrocarbons began with smaller systems in the early 1990s with their introduction to domestic refrigerators. Today these are by far the most common type of hydrocarbon systems.

Modern use of hydrocarbons was initiated in 1993 with the Greenpeace "Greenfreeze" campaign [6-18] and the conversion of a German refrigerator producer (DKK Scharfenstein/Foron) from CFC-12 to an R290/R600 blend (to match the pressure-temperature characteristics of CFC-12). The market for hydrocarbon refrigerators rapidly expanded and today almost all of the one billion domestic refrigerators use R600a [6-1]. Most appliances have a refrigerant charge of between 20 and 100 g (0.70 and 3.52 oz), although the applicable safety standard, IEC 60335-2-24, permits up to 150 g (5.29 oz) without the need for a minimum room area.[1] There have been very few reported accidents involving ignition of leaked hydrocarbon refrigerants in domestic refrigerators; historically a few per tens of millions. Today, analysis of national fire statistics[2] does not indicate any such accidents for many years. Indeed, fire accidents as a result of electrical faults are thousands of times more frequent than those caused by ignition of leaked refrigerant.

Given the vast number of hydrocarbon appliances, this outstanding safety record demonstrates that hydrocarbon refrigerants can be safely used and applied, if suitable design and technician practices are developed and implemented.

Room air conditioners

"Room air conditioners" include window units, small split systems, portable AC and dehumidifiers and similar. HC charge amounts range from 100 to 1000 g (3.5 to 35.3 oz). "Specific charges", i.e. refrigerant mass per unit of thermal capacity, are typically in the range of 50 to 100 g/kW (0.116 to 0.232 lb/Btu_{IT}/s) for charge-optimised systems across all these types of ACs.

Portable air conditioners and dehumidifiers

Currently there is an extensive market for R290 portable air conditioners and dehumidifiers; in Europe – partly augmented by the F-gas regulations – the majority of such units employ hydrocarbon refrigerants. Whilst this applies to products found in Europe, it also includes products sold throughout Asia and to a lesser extent, in other parts of the world. Such R290 products were first introduced by an Italian manufacturer in the late 1990's and subsequently gained widespread acceptance. Today the vast majority are manufactured in Asia.

Charge quantities range from 100 to about 300 g (3.5 to 10.6 oz).

Window air conditioners

In principle, window air conditioners are well suited to be used with hydrocarbons [6-19]. They offer the advantages described below for split air conditioners, but notably because they are factory-sealed the probability of leakage is further reduced. Whilst they are integral units, the condenser and compressor are separated to the outdoor section, so leaks are more likely to be directly to the outside. Because of the relatively large indoor enclosure housing the evaporator, leaks tend to "pre-mix" well before flowing into the room, meaning that concentrations (for a given room size) are significantly lower than with other types of equipment.

Manufacturers in some Asian countries are developing R290 window ACs and charge amounts tend to be less than 300 to 400 g (10.6 to 14.1 oz).

[1] Originally, the 150 g (5.3 oz) charge limit for hydrocarbon refrigerators was set to enable all current (at the time) domestic refrigerator and fridge/freezer appliances and their anticipated required charge amounts.
[2] E.g., for the UK, https://www.gov.uk/government/collections/fire-statistics-monitor

Split-type air conditioners and heat pumps

Despite the case for using hydrocarbons in split-type air conditioners being compelling, it is surprising to find that there are relatively few products currently available. Indoor parts of these systems are factory-sealed, they are usually installed at a high level (so refrigerant leaks will not stratify), high-pressure parts including the compressor are outside and modern designs have no sparking electrical components, i.e. no potential sources of ignition.

Over the past 20 years various smaller companies have dabbled with hydrocarbon split air conditioners, but widespread introduction to the market appears to have been challenging. This has been largely due to two reasons. Firstly, there has been limited availability and otherwise, higher cost of suitable compressors. Secondly – and more significantly – many of the requirements of applicable safety standards have hitherto been highly obstructive.

Whilst in the late 1990s and 2000's, several split air conditioner manufacturers had products on the market in various countries, the introduction of a revised edition of the product standard, IEC 60335-2-40, effectively put a brake on such activities. Flawed and questionable assumptions were used to develop requirements such as those concerning refrigerant charge limits, effectively inhibiting the wider use of hydrocarbon refrigerants.

Despite this, in 2010 an Indian manufacturer, Godrej, introduced R290 split air conditioners and by 2014, more than 100000 had been installed [6-20], again (as with domestic refrigerators) exhibiting an exceptional safety record. Today there are almost one million units on the market.

Various Chinese manufacturers are working towards implementation of hydrocarbons into room air conditioners, although despite the policy of the Chinese government to phase in R290 [6-21], the development of the sector remains embryonic. One company, Midea, has on the other hand vigorously pursued this objective and has launched R290 splits on the European market.

Examples of some Godrej, Midea and Chunlan R290 split air conditioner products are given in Table 6.19.

Table 6.19 Selected split AC products

Manufacturer	Model	Nominal capacity	Efficiency*	Charge
Godrej	GIC 12TGC5-WUA	3450 W	4,57	0,305 kg
Godrej	GIC 18TGC3-WUA	5100 W	3,74	0,370 kg
Midea	MSAEBU-12HRFN7-QRD6GW	3500 W	n.a.	0,380 kg
Chunlan	CS-09/AZ3BPDAWB-2	2638 W	3,89	n.a.
Chunlan	CS-12/AZ3BPDAWB-2	3517 W	3,66	n.a.

* Note that determination of "efficiency" (COP) is different between regions so one manufacturer's product cannot necessarily be compared against another.

Although the current requirements of IEC 60335-2-40 are obstructive to the widespread uptake of R290 in air conditioners, studies have found that the flammability risk of a split air conditioner (400 g / 14.10 oz) of R290) is around 100 times lower than that of a R600a (100 g / 3.5 oz) domestic refrigerator [6-22]. The reasons for this are several-fold:

a) Leakage is more likely from compressor and high-pressure parts of the system, which for the air conditioner are outside and for the indoor parts, there is a fan which (when operating) helps to rapidly dilute leaked refrigerant to below LFL;

b) A leak that occurs indoors from the air conditioner does so at high level thus enabling it to disperse easily, whereas a leak from the refrigerator will usually be close to the floor and to some extent confined, leading to a higher density refrigerant layer on the room floor which persists for much longer;

c) In the event of ignition of a leak from the evaporator, the overpressure from the indoor unit will be negligible, but potentially severe for ignition within the refrigerator due to confinement of the box.

Detailed examination of the factors that affect flammability risk in hydrocarbon products and supporting empirical evidence, strongly suggests that the widespread application of R290 split air conditioners would not result in a negative safety impact to the sector.

Heat pump tumble dryers

Within Europe, the "ecodesign" directive is effectively nudging the consumer from tumble dryers which utilise resistance heating for those employing heat pump systems. Earlier models used R134a as the refrigerant, but today an increasing number of models employ R290. Charge quantities tend to be small, around 100 to 150 g (3.5 to 5.3 oz).

Commercial refrigeration

There has been a gradual increase in the number of commercial refrigeration appliances – typically plug-in display cabinets, but also catering storage cabinets and small coldstore systems – using hydrocarbons. In the 1990's several companies in a few countries began to offer hydrocarbon products as part of their ranges. Since then the proportion has grown. Initially this was augmented by the "Refrigerants, Naturally!" initiative, which saw global brands (such as Coca-Cola, PepsiCo, Red Bull, SABMiller, Unilever, etc.) making statements and implementing purchasing policies for only R744 and hydrocarbon plug-in appliances. Many supermarket and other retail chains began to follow suite. More recently, with the publication of the revised commercial refrigeration appliance product safety standard (IEC 60335-2-89), a wider range of plug-in units are being offered with hydrocarbons. Estimates from compressor manufacturers indicate almost half of light commercial refrigeration appliances are today being produced using hydrocarbons.

A major driver for this trend appears to be the demands of supermarket chains, who historically have suffered with major CO_2-eq emissions from centralised chlorofluorocarbon, hydrochlorofluorocarbon and fluorocarbon systems. Not only can extended use of air-cooled or liquid-cooled integral hydrocarbon (and R744) systems reduce refrigerant-related emissions to near zero, but notable reduction in energy use have also been reported.

Commercial hydrocarbon appliances include:

- refrigerated display cases
- retail freezer cases
- bottle/drinks coolers
- ice cream freezers
- sorbet/ice cream/slush dispensers
- professional kitchen storage cabinets
- ice makers

- vending machines
- draught beer coolers

Most units have hydrocarbon charge sizes of up to 150 g (5.29 oz) on account of the (previous) safety standards, although many larger appliances comprise several circuits (thus the total charge of the unit exceeds this amount). Equipment constructed according to the horizontal standards, EN 378 and ISO 5149, use up to 1,5 kg (3.30 lb), although with the advent of the new IEC 60335-2-89 it is likely that most units will settle for charges of around 500 g (17.63 oz). As with other product groups, manufacturers are implementing continuous efforts to reduce refrigerant charge. Specific charges generally range from about 80 g/kW (0,186 lb/Btu_{IT}/s) for chiller cabinets, up to about 200 g/kW (0,465 lb/Btu_{IT}/s) for low-temperature (freezer) systems.

Heat pumps

With the increased focus on adoption of heat pumps for domestic heating and hot water in Europe as a means to displace fossil-fuel boilers, there is increasing recognition on the importance of employing environmentally benign refrigerants.

Of the European heat pump manufacturers, nearly all offer at least some products using R290. In 2021, there were about 170 heat pumps with R290 on the European market. This number is expected to increase significantly in subsequent years. Globally, a smaller proportion of heat pump manufacturers offer models with hydrocarbons, although the number seems to be increasing.

The majority include air-source or water/ground-source models that have the refrigeration system located outside and supply hot water to the building. Charge amounts are generally less than about 2 kg (4.40 lb) but some systems reach 5 kg (11.02 lb) (for commercial products). Additionally there are a large number of exhaust-air-source heat pumps that draw heat from vented building air. These have a charge size of between 200 g and 1 kg (7.05 oz and 2.20 lb). Regardless of the type, specific charge(s) are usually around 50 to 80 g/kW (0.116 to 0,186 lb/Btu_{IT}/s), although some highly charge-optimised models use as little as 20 g/kW (0.046 lb/Btu_{IT}/s).

A particular concept using a ventilated enclosure or housing has been employed by some manufacturers, where the entire refrigeration system of a water-to-water heat pump is located indoors. The housing is vented to the outside, either continuously or in response to leak detection. With the growth of the heat pump market and increased uptake of R290 as an environmentally preferred refrigerant, alternative safety concepts are being developed and implemented with the intention of providing lower cost units with minimal risk.

Coldrooms (monoblock units)

Whilst the use of hydrocarbons in coldrooms is not uncommon, smaller capacity applications using hydrocarbons in "monoblock" units is becoming more prevalent. There are several manufacturers offering monoblock systems for coldrooms, both covering medium- and low-temperature applications. Some manufacturers offer monoblock units that use more than 150 g up to about 2 kg (5.3 oz up to about 4.4 lb) of hydrocarbon charge.

For lower temperature applications, there may be a conflict between the allowable charge permitted by safety standards and the charge required to achieve the desired cooling capacity of the application. Approaches for working around this include ensuring a high internal airspeed within the coldroom to guarantee homogeneous dispersion of any leaked refrigerant and further, employing a "limited releasable charge" concept, where safety shut-off valves are fitted to prevent a proportion of refrigerant from being released into the coldroom.

6.7 Larger systems

Use of hydrocarbons in larger refrigeration systems has become more prevalent in Europe during the past two decades, as components have become readily available. F-Gas Regulation and the phase down of halogenated hydrocarbons has further increased demand for systems with natural refrigerants including hydrocarbons. Globally, there has been less demand and there are not many manufacturers producing larger hydrocarbon units.

Whether larger refrigeration systems are factory assembled, or parts of it that are assembled on site, they are more carefully designed according to the particular application operating at conditions defined by source and sink media

Hydrocarbons are used in larger indirect refrigeration systems so protective measures for flammability safety can be applied more easily. All refrigerant-carrying parts of indirect systems are located in the open air, in a machinery room and often additionally in a ventilated enclosure. A system with its location in a ventilated enclosure, that adheres to the respective restriction concerning the maximum allowable charge, may be located within occupied areas due to their design. Larger direct systems are not (yet) entering the market, which typically require comparatively high liquid refrigerant charges to be distributed to evaporators that are spaced out, and often, refrigerant-carrying parts are located in occupied areas.

Outer dimensions of a refrigeration unit are limited as it is assembled in a factory and transported by road to the installation site. How many refrigeration circuits fit inside one unit depends on the size of the enclosure as well as components and on what kind of additional equipment is required. Modular system design comprising multiple units allows scaling up of the thermal capacity.

Such systems range in thermal capacities from around 10 to 20 kW up to about 400 kW (2.8 to 5.7 TR up to about 114 TR) per circuit, using reciprocating compressors, at evaporation temperatures of +10 °C. (50 °F) With screw compressors and economisers, capacity of a single refrigerant circuit can be as large as 1000 kW (284.3 TR).

Larger systems often employ one or more particular functions, for example:

- indirect systems requiring additional hydronic equipment
- reversible systems, providing a heating function
- systems with additional heat recovery
- systems with free cooling

For each of these, there are special considerations when applying hydrocarbon refrigerants. Indirect refrigeration utilises hydronic systems which transport a heat transfer fluid between the refrigerant circuit and the location where cooling (or heating) is needed, e.g., a cold room. Additional hydronic equipment such as piping, buffer tanks, expansion vessels, pumps, valves, sensors etc. may be included inside the equipment housing. It is useful to design the housing with multiple compartments, with one comprising all refrigerant carrying parts which may be easily ventilated and is separate from the compartment containing the hydronic components. As refrigerant can enter the hydronic system in case of leakage, e.g., in case of freezing transfer fluid in an evaporator, the purger of the secondary system should vent to an unoccupied area without ignition sources.

As for any refrigerant, if hydrocarbon heat pumps use air as heat source, and if they can function reversely as chillers with air as heat sink, it is challenging to dimension the ambient air heat exchanger for the dual purpose. An air heat exchanger which is designed for heat pump function (heat source) can be relatively small when machine function switches to cooling (heat sink). Therefore energy efficiency decreases as more fan power is required to account for the increased air flow in cooling mode. The task is challenging, for example, to achieve higher energy efficiency in cooling mode of heat pumps, the air heat exchanger could be increased by size which may impede operation in heating mode, e.g., decreased pressure drop may lead to maldistribution of refrigerant. In Europe, the Ecodesign Directive 2009/125/EC and Energy Labelling regulation set up a general framework and regulation was revised in 2019 with more ambitious requirements concerning energy efficiency of chillers. Currently, the process of updating designs of such reversible heat pumps for larger hydrocarbon systems in order to fulfil the new requirements and developing technical solutions is ongoing.

Systems employing heat recovery have an additional (water-cooled) heat exchanger installed in parallel to the (air-cooled) condenser which results in a higher refrigerant charge because of additional internal volumes and furthermore to account for additional inner volume of the HX and system operating points during heat recovery. If charge needs to be minimised receivers should be avoided and systems with heat recovery may not be feasible.

Transport

Usually, hydrocarbon systems are pre-filled after assembly at the factory. For larger hydrocarbon systems, it is important to pump down the refrigerant from the low- to the high-pressure side, liquefy it and keep it shut-off during transport.

Installation

Air-cooled systems use ambient air as a heat sink and are installed outdoors. Hydrocarbon refrigerant vapour is denser than air and therefore, if released, will sink to the ground (in absence of wind or high airspeeds). Outdoors, natural air movement is sufficient to dilute and disperse leaked refrigerant, even in the event of pressure relief valves being activated, which involves rapid releases and may occur in the event of malfunction of a unit during very brief periods of time. Installation sites are safely located outdoors in places away from pits, manholes, doorways, and fresh air inlets for ventilation systems etc. to avoid accumulation of refrigerant.

Furthermore, areas in close vicinity of hydrocarbon systems should be free from ignition sources. If drainpans with electric parts, e.g. shut-off valves, glycole sensors, etc., are situated underneath units where hydrocarbons may sink to when released, all electric components should be ex-type.

Water-cooled systems use a liquid heat transfer medium such as water or brine as a heat sink. They may be installed outdoors or indoors in machinery rooms with sufficient ventilation.

6.7.16 Application examples

Air conditioning of a truck dynamometer – Example 1

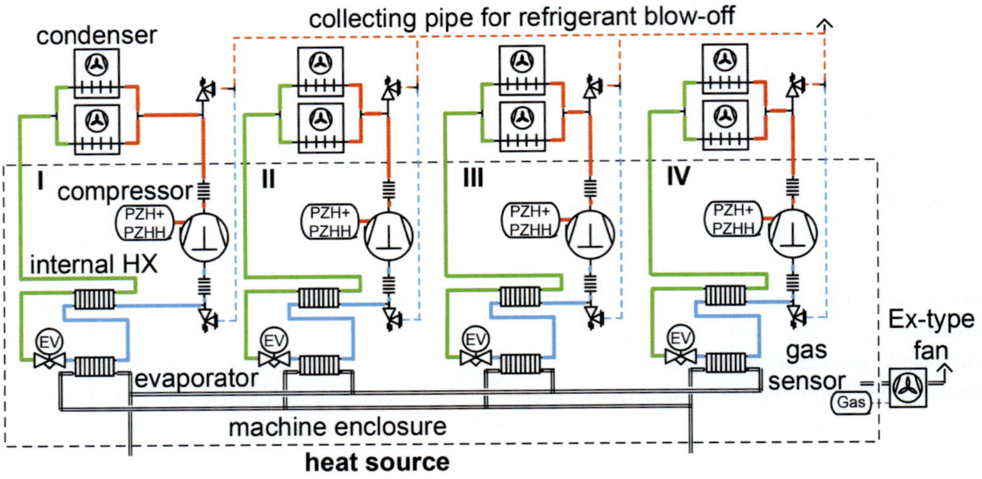

Figure 6.13 Process flow diagram of a refrigeration unit with four separate circuits for use in the automotive industry

Figure 6.13 shows one of three air-cooled units of a refrigeration system for the automotive industry that is used for air conditioning of a newly built dynamometer for trucks, where driving with wind speed up to 90 km/h (56 mph) is simulated in real-time. The entire system comprises three units with identical layout, which are operated with R290; each unit functions as depicted with following technical data:

Technical data	
Chiller 1 / 2	
Type	air-cooled
Refrigerant	R290
Capacity control	step-wise (4 steps)
Compressor	reciprocating 8 cylinder
Number of circuits	4
Refrigerant charge per circuit	8,1 kg to 8,8 kg
Cooling capacity	max. 672 kW
Cooling capacity per kg R290	20 kW/kg
Temperatures carrier medium (evaporator)	15/9°C to –12/–15°C
Max. power consumption	323 kW

Technical data	
Chiller 3	
Refrigerant charge per circuit	8,3 kg to 8,5 kg
Total cooling capacity	608 kW
Cooling capacity per kg R290	18 kW/kg
Specific charge	0,055 kg/kW
Temperatures carrier medium (evaporator)	11/5°C
Max. power consumption	323 kW

Two units (Chiller 1 / 2) are used exclusively to supply the cooling coils of wind turbines, which poses a challenge in the system design because of different temperature levels at the evaporators and rapid load changes whilst in operation. The third unit (Chiller 3) cools electrical components and the AC-system of the building. R290 was chosen to account for the relatively wide evaporative temperature range.

There are two parallel connected air-cooled microchannel heat exchangers per circuit giving a total of eight condensers. Energy efficient EC fans, with a diameter of 910 mm (35.8 in), are used as condenser fans. Airflow is 221000 m^3/h (7804550 ft^3/h) at ambient air temperature of 32 °C (89.6 °F).

An IHX is implemented in each refrigerant circuit for subcooling and superheating.

Four evaporators are connected in parallel on the secondary side. The system is designed with the same pipe lengths (sum of inlet and outlet piping) for all the evaporators to ensure balanced pressure drop and thus flow rates.

There are pressure relief valves on the high- and low-pressure side of the system, which discharge through a common header pipe.

On the compressor, a pressure limiter and pressure cut out are provided to protect against high pressure. The compressor housing ventilation is carried out by means of an Ex type fan, which is operated if a gas sensor inside the housing detects an R290 concentration of > 20 % LFL. Whilst the housing fan keeps running, the unit is switched off if concentration reaches > 40 % LFL and a visual and acoustic alarm is raised.

Figure 6.14
Refrigeration system installation for air conditioning of a truck dynamometer.
Source: Secon GmbH

Industrial refrigeration with heat-recovery for food processing – Example 2

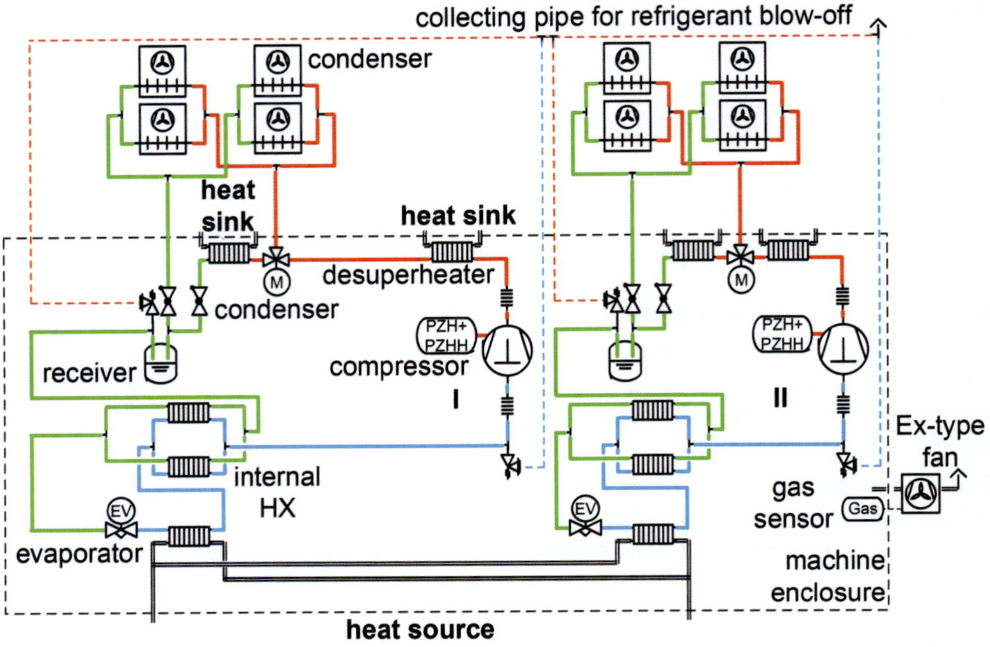

Figure 6.15 Process flow diagram of a refrigeration system for industrial food processing with heat recovery for heating well water

Figure 6.15 shows one of three air-/liquid-cooled units of a refrigeration system for a food processing plant. Air conditioning and heating systems of a building are supplied as well as condensers of subcritical R744 refrigeration units, that supply blast chillers and freezers for food processing, are cooled in a cascade cycle. In two of the three units, as depicted, brazed plate heat exchangers (BPHE) are incorporated as desuperheaters for heat recovery to supply warm water to heat the building. Ground water from the nearby river Rhine is used to cool plate heat exchangers that are connected as condensers in parallel to air-cooled micro-channel-heat exchangers. Water-cooling of the BPHE condensers enables low condensation temperatures, year-round, below 25 °C (77 °F).

Technical data	
Chiller 1 / 2	
Type	air- and water-cooled
Refrigerant	R290
Compressor	reciprocating 8 cylinder
Capacity control	stepless (2 x inverter)
Number of circuits	2
Refrigerant charge per circuit	14 kg

Technical data	
Total cooling capacity air-cooled water-cooled	 305 kW 360 kW
Cooling capacity per kg R290	11 kW/kg to 13 kW/kg
Specific charge	0,077 to 0,091 kg/kW
Temperatures carrier medium (evaporator)	-2/-6 °C
Max. power consumption	166 kW
Temperatures carrier medium (desuperheater)	45/65 °C
Temperatures carrier medium (parallel BPHE condenser)	15/25 °C
Chiller 3	
Refrigerant charge per circuit	11 kg
Total cooling capacity air-cooled water-cooled	 304 kW 360 kW
Cooling capacity per kg R290	14 kW/kg to 16 kW/kg
Specific charge	0,063 to 0,071 kg/kW
Temperatures carrier medium (evaporator)	12/6 °C
Max. power consumption	174 kW
Temperatures carrier medium (parallel BPHE condenser)	15/25 °C

Compressor capacity is continuously controlled via frequency inverters from 25 Hz to 70 Hz. The discharge gas flows through a plate heat exchanger desuperheater. Downstream of the desuperheater, the cooled vapour flows via a three-way valve either through four parallel air-cooled microchannel condensers or alternatively, through another water-cooled plate heat exchanger, if ambient temperature is warm during summertime. The 910 mm (35.8 in) diameter EC fans of the air-cooled heat exchanger coils convey a volume flow of 160000 m^3/h (5650352 ft^3/h) at an air temperature of 35 °C (95 °F).

After condensation, the refrigerant flows into a receiver via check valves and is routed via two parallel IHX. The system has two evaporators – one per refrigerant circuit – that are connected in parallel on the secondary side as described in Example 1.

The safety equipment of the refrigeration units also corresponds to the previous Example 1.

Figure 6.16
Hydraulic connection lines of the refrigeration system for industrial food processing with heat recovery. Source: Secon GmbH

Figure 6.17
Two units of a refrigeration system for industrial food processing with heat recovery. Source: Secon GmbH

Retrofit application of multiple compact refrigeration units with supervisory control – Example 3

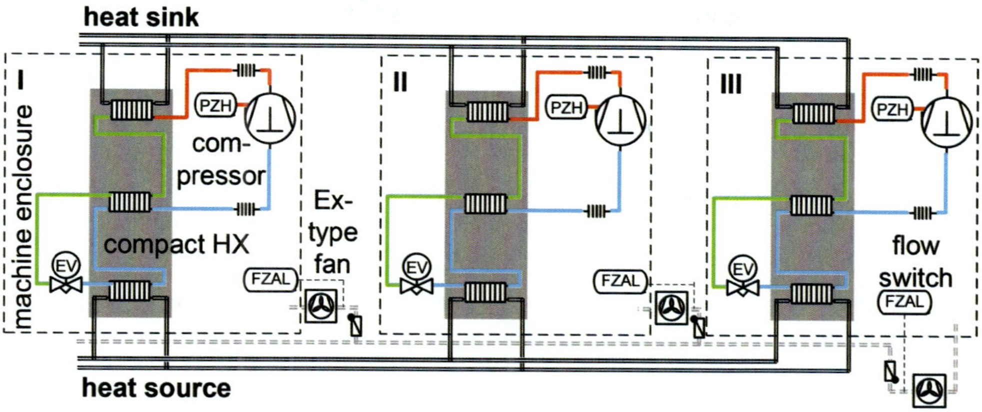

Figure 6.18 Process flow diagram of a refrigeration system comprising three compact units with supervisory control

Figure 6.18 shows three refrigeration units that are installed as a retrofit-application. The three identical units are run via supervisory control. The system is used for air conditioning as well as for machine cooling in laboratories of a university with following technical data:

Technical data	
Chiller 1 / 2 / 3	
Type	water-cooled
Refrigerant	R290
Compressor	reciprocating 4 cylinders
Capacity control	step-wise (2 steps)
Number of circuits	1
Refrigerant charge per circuit	2 kg
Cooling capacity	52 kW
Cooling capacity per kg R290	26 kW/kg
Specific charge	0,038 kg/kW
Temperatures carrier medium (evaporator)	14/8°C
Max. power consumption	24 kW
Temperatures carrier medium (condenser)	30/35 °C

The individual units operate as needed, depending on how much cooling capacity is required. The condenser-side heat transfer medium is provided by an adiabatic cooling module at 30/35 °C (86/95 °F) during summer and 14,5/18 °C (58.1/64.4 °F) during winter. A separate module facilitates free cooling which is also operated by supervisory control at ambient temperatures below 12,5 °C (54.5 °F). The heat exchangers of a unit, i.e. condenser, IHX and evaporator, are integrated in one compact highly efficient plate heat exchanger. The compact refrigeration system has outer dimensions (w · l · h) of 590 mm · 1180 mm · 1621 mm (23.2 in · 46.5 in · 63.8 in).

On the secondary sides, the three condensers are connected in series, as well as the three evaporators. The system is designed with identical pipe lengths at the heat exchangers to balance the pressure drop and obtain equal volume flow rates.

For safety equipment, a pressure limiter is provided in the refrigerant circuit of each refrigeration unit to protect against excessive pressure. Ex-type fans continuously ventilate the enclosures with an airflow switch for monitoring.

Figure 6.19 Retrofit application with three compact refrigeration units with supervisory control for air conditioning and machine cooling in laboratories of a university. Source: Secon GmbH

6.8 Outlook

An observation that can be drawn from other sections within this chapter, is that safety rules broadly dictate the extent of use of hydrocarbons in RACHP systems. Current safety standards inhibit their use and application across many subsectors. Whilst in Europe, regional regulations (directives) effectively allow their use in almost every sub-sector, European and International safety standards impose obstructive requirements which act as a major hindrance to their wider use in many applications. Similarly, national safety standards – often based on the ISO 5149 / EN 378 / Ashrae 15 group – similarly restrict wider application of hydrocarbons. Conversely, in recognising the importance of adopting flexible RACHP safety standards for the benefit of environmental protection, certain countries have developed improved standards; examples include Colombia (NTC 6228) and Indonesia (SNI 6500). Similarly, due to the obstructions presented by the established international and regional safety standards, extensive work has been carried out globally to try to lift these restrictions by developing improved requirements. Table 6.20 lists the key safety standards and gives a brief summary of their current status (2021) with regards to improved requirements for hydrocarbons.

Table 6.20 Revision status concerning hydrocarbon refrigerants of various safety standards

Standard	Status (September 2021)
IEC 60335-2-40	Final Draft International Standard (FDIS) being prepared to increase allowable charge limit (more g of hydrocarbons per m^2 or room area) and to establish more realistic assumed leak rates.
IEC 60335-2-89	New edition published in 2019 to increase upper charge limit from 150 g to about 500 g (5.3 oz to about 17.6 oz) of hydrocarbons.
EN 378-1, -2, -3*	New editions being prepared to increase allowable charge limits (more g of hydrocarbons per or room area), establish more realistic assumed leak rates and to eliminate below ground level prohibition.
ISO 5149-1, -2, -3, -4*	No notable changes currently proposed.

* Note: EN 378-4 is being deleted and republished as ISO 5149-4, since they are almost identical.

There are several aspects that have until recently not been suitably handled within RACHP safety standards and these are summarised as follows:

- differences between systems for air conditioning, heat pumps and refrigeration
- assumed leak rates
- improved tightness systems
- use of integral airflow to disperse leaks
- determining minimum room sizes based on releasable charge
- design of equipment enclosures or unit housing to more effectively disperse leaks
- "black box" assessment

These aspects are summarised in the following sub-sections and it is expected that these will be addressed in new and forthcoming revisions of the safety standards.

6.8.1 Differences between purpose of systems

Currently there are differences in safety requirements between systems intended for air conditioning, heat pumps and refrigeration. EN 378:2016 and ISO 5149:2014 utilise the phrase "human comfort", to distinguish air conditioning from refrigeration and non-comfort air conditioning applications. This reflects a deep misunderstanding and has led to erroneous interpretations of the risk associated with flammable refrigerants in different types of systems. Moreover, it has resulted in a deterrent to the wider application of HCs in air conditioning applications.

Further explanation of this misunderstanding is given. Take a large integral refrigeration system; condensing unit located at floor level and a refrigerated chamber just above, housing the evaporator. According to EN 378-1 or ISO 5149-1, the maximum charge amount in an access category 'a' location would be 1,5 kg (3.30 lb). The corresponding minimum room area would be about 75 m^2 (807 ft^2) (assuming a room height of 2,5 m / 3.28 ft). Now, take a split AC as an example for a "human comfort" system. The maximum charge would be 1 kg (2.20 lb) and if the indoor unit was a wall type, the minimum room area would be 175 m^2 or 1579 m^2 (1883.7 ft^2 or 16996 ft^2) in the case of a floor type unit.

Such a large difference in maximum charge and minimum room area is counter to common sense; most leaks occur from condensing units, which for the AC is outside. If a leak occurs at a lower level it (as with the non-comfort refrigeration system) disperses less effectively thus necessitating a larger floor area to compensate. Yet the system positioned at lowest height, with the leakiest part indoors is permitted to have the larger charge amount and located in the smaller room. This is counter to sound reasoning.

It is hoped that the next revisions of EN 378 and ISO 5149 will have such contradictions rescinded.

6.8.2 Assumed leak rates

The assumption of leak size is critical to determination of allowable charge limits: broadly, a smaller (lower mass flow) leak means that the refrigerant mixes more readily with the air and more refrigerant would be required to form a flammable layer on the room floor. Whereas a larger (higher mass flow) leak forms a flammable layer with less refrigerant. In older versions of the standards (e.g., EN 378), there was no consideration given to sizes of refrigerant leaks; in fact the assumption of leak mass flow rate was to some extent "hidden" within the 20 % of LFL used to define the Practical Limit.

When various safety standards were first being revised in response to anticipated increased use of flammable refrigerants, the so-called "four-minute leak" fallacy was introduced. This was an assumption that the largest likely leak hole in systems would be such that the entire charge is released in four minutes. For a system with 1 kg (2.2 lb) charge (regardless of refrigerant), the leak mass flow would be 250 g (8.8 oz) per minute. The consequence of this assumption is that only a small amount of refrigerant charge would form a flammable layer across the room floor and consequently the allowable charge limits are congruently stringent (such as those in IEC 60335-2-40:2002,2013 and EN 378:2008 for "human comfort").

However, studies such as those in the EU Life Front project [6-23] where physical sizes of leak holes from a variety of equipment were measured, they were found to be much smaller than previously assumed (i.e. for development of allowable charge limits). Most leaks were smaller than 10 g (0.35 oz) per minute with the largest (without human interference) at about 100 g (3.5 oz) per minute. Further, there was no distinction between leak hole sizes from small (say, 100 g / 3.5 oz) and large systems (say, 10 kg / 22 lb). This meant that alternative ways of determining allowable refrigerant charge should be developed.

A further aspect is the variation of leak rate amongst different refrigerants. Previously, RACHP safety standards effectively assumed the same leak mass flow regardless of refrigerant; in IEC 60335-2-40:2018, 1 kg (0.35 oz) of R1270 would leak at 250 g/min (8.8 oz/min), the same as 1 kg (2.2 lb) of R600a, and EN 378-2:2016 uses the same for allowable charge of human comfort systems and 60 g/min or 180 g/min (2.1 or 6.3 oz/min) for determination of ignition sources, irrespective of refrigerant.

However, these assumptions are inconsistent with conventional thermodynamic concepts. Until there is a minimal amount of refrigerant left in the system (and the pressure is about 2 bar / 29 psi), refrigerant will leak under choked flow conditions. From conservation of energy, it is known that even if internal pressure is higher, the velocity of a leak cannot exceed sonic velocity. Given the wide variation in refrigerant specific heats, densities and thus sonic velocities, mass flow through a leak hole (of a given size) will vary accordingly. Figure 6.20 shows the variation of leak mass flow at choked conditions for various flammable refrigerants, which exceeds a factor of 10.

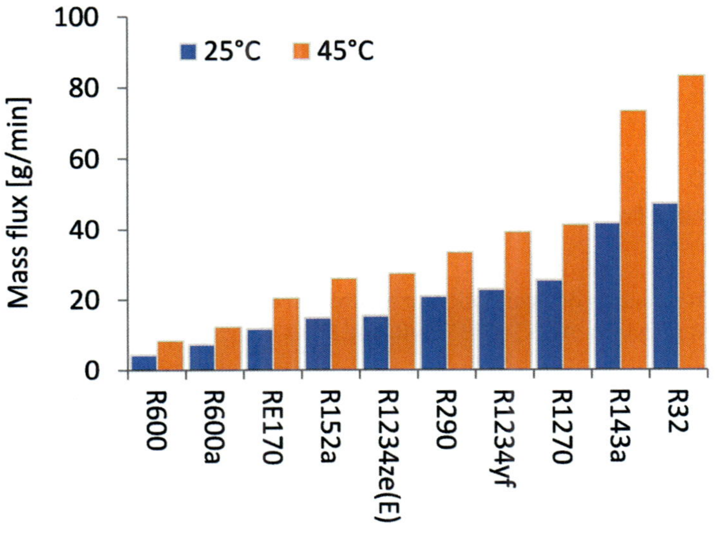

Figure 6.20
Mass flow rates of vapour leaks through a 0,5 mm² (0,000775 in²) hole for different flammable refrigerants

An example of how the choice of assumed leak mass flow rate affects important safety parameters is shown in Figure 6.21, using 1 kg (2.20 lb) of R1270 and leaking from a 1 m (39.37 ft) high unit. The left y-axis shows the size of the room needed to avoid a flammable floor layer (assuming certain unit housing characteristics such as internal volume, area of openings and degree of internal congestion, that dictate the outflow or exiting concentration). The right y-axis shows the extent of the "zone", as described in the hazardous area classification standard, IEC 60079-10-1. For both parameters, a lower assumed flow rate corresponds to a smaller minimum room area or extent of zone. For instance, assuming the four-minute leak fallacy – 250 g/min (8.8 oz/min) – the minimum room area would be about 60 m^2 (645.8 ft^2), whereas a more realistic 65 g/min as reported in the EU Life Front project, the room area would be 35 m^2. Corresponding extent of zones would be 1,5 and 0,9 m (4.9 and 2.95 ft). In general, basing the assumed leak rate on more realistic values leads to more flexible and cost-effective installations. Differences can further be recognised when considering the release mass flow of the various refrigerants (see Figure 6.21).

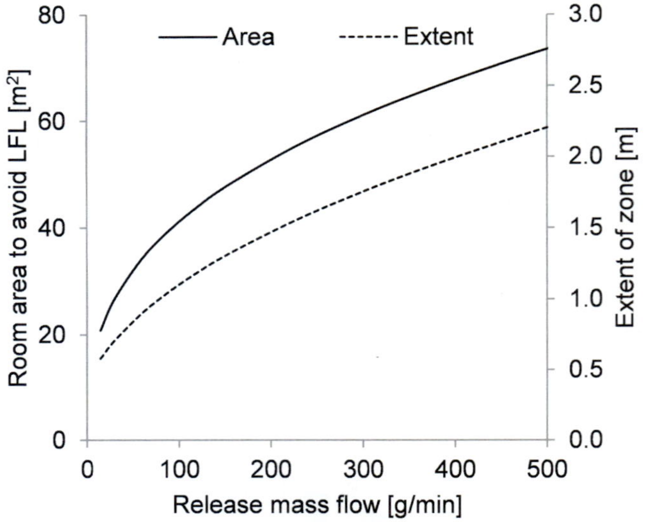

Figure 6.21
Example of how assumed release mass flow rate affects the minimum room area [6-24] needed to avoid a layer at or above the LFL and the extent of the hazardous zone [6-25] according to IEC 60079-10-1.

Revised IEC 60335-2-89: 2019 enables determination of assumed leak mass flow rate according to refrigerant type. Proposed revisions of IEC 60335-2-40 and EN 378 similarly include for this.

6.8.3 Improved tightness systems

Generic safety standards emanating from the process and fuel gas sectors, such as EN 1127-1, identified differences between "basic" installations and those using "technically durably tight" or latterly "enhanced tightness" systems. For such systems, improved piping design, installation methods, appropriate types of fittings and so on are recognised as measures that will usually ensure that leaks occur less frequently and if they do occur are relatively small and thus unlikely to lead to flammable mixtures large enough to pose any concern. (IEC 60079-10-1 identifies these as "zones of negligible extent"; leading to a mixture volume of less than 0,1 m^3 (3.53 ft^3).) Current versions of IEC 60335-2-40 and EN 378 recognise such systems for so-called A2L refrigerants, but neglect to acknowledge them for HCs. It is planned for forthcoming revisions of these standards to do so; the advantage of such systems is the assumption of smaller leak mass flow rates and therefore less restrictive minimum room areas for a given charge amount (see Figure 6.21).

As an example, the formula in the proposed IEC 60335-2-40 is:

$$m_{max} = F \cdot LFL \cdot h_o \cdot A$$

where LFL is the lower flammability limit, h_o is the unit height, A is the area and F is a "concentration factor" used to limit the quantity of refrigerant within the stratified layer assumed to form below the unit so that LFL is avoided at floor level. Presently, F = 0,35, representing the lowest achieved value across a wide number of experiments, although it is known to vary depending upon unit housing geometry, mass flow rate and other factors. This formula ("new") is compared against the formula in IEC 60335-2-40: 2018, ISO 5149-1:2014 and EN 378-1:2016

[6-14] Colbourne, D., Proklima: *Guidelines for the safe use of hydrocarbon refrigerants*. GIZ, BMZ, Bonn, 1st Edition, Reprint, 2012

[6-15] Maclaine-cross, I., Leonardi, E.: *Performance and Safety of LPG Refrigerants*. Proceedings of the 'Fuel for Change' Conference of the Australian Liquefied Petroleum Gas Association Ltd, ISBN 064624884 7. Surfers' Paradise, Australia, pp 149–168, 28th February to 2nd March, 1995

[6-16] Wolfer, M.; Seitz, E.; Seiler, H.: Ammoniak und Kohlenwasserstoffe als Kältemittel: Risikoanalyse, Produkthaftpflicht und Strafrecht. Schweizer Bundesamt für Energie, Bern, 1999

[6-17] Altenstedt, J., Pleijel, K.: *POCP for individual VOC under European conditions*, IVL report B-1305, IVL Swedish Environmental Research Institute, Göteborg, 1998, https://www.ivl.se/download/18.343dc99d14e8bb0f58b7368/1445515409320/B1305.pdf, accessed 24.10.2021

[6-18] https://www.greenpeace.org/international/story/15323/how-greenpeace-changed-an-industry-25-years-of-greenfreeze-to-cool-the-planet/ accessed 24.10.2021

[6-19] https://hydrocarbons21.com/articles/10107/strong_environmental_technical_arguments_seen_for_r290_window_acs accessed 24.10.2021

[6-20] D. Rajadhyaksha, B.J. Wadia Abhijit A. Acharekar, D. Colbourne: *The first 100 000 HC-290 split air conditioners in India*, International Journal of Refrigeration, Vol. 60, 2015, p. 289-296, https://www.sciencedirect.com/science/article/abs/pii/S014070071500239X

[6-21] *New environmentally-friendly low-carbon air conditioner products made its debut at Appliance World Expo*, Foreign Economic Cooperation Office, Ministry of Environmental Protection of China, Beijing, 2014, http://en.fecomee.org.cn/News_1011/201404/t20140414_563583.html

[6-22] D. Colbourne, Suen, K.O.: *Comparative evaluation of risk of a split air conditioner and refrigerator using hydrocarbon refrigerants*, International Journal of Refrigeration, Vol. 59, 2015, p. 295-303, https://www.sciencedirect.com/science/article/abs/pii/S0140700715001978

[6-23] Colbourne, D., Pitarch Mocholi, M., Munzinger, P., Oppelt, D., Paetzold, B., Vince, I.: 2021. *Leak hole sizes from refrigeration, air conditioning and heat pump systems*. International Journal of Refrigeration, https://doi.org/10.1016/j.ijrefrig.2021.07.003

[6-24] Colbourne, D., Suen, K.O.: 2021. *Hydrocarbon refrigerant charge lim-its for quiescent rooms*. Int. J. Refrig., vol. 125, pp. 75–83

[6-25] Santon, R., Ivings, M., Webber, D., Kelsey, A.: 2012. *New methods for hazardous area classification for explosive gas atmospheres*. Proc. Hazards XXIII, The Institution of Chemical Engineers, UK

[6-26] Colbourne, D., Suen, K.O.: 2018. *Minimum airflow rates to dilute R290 con-centrations arising from leaks in room air conditioners*. Proc. 13th IIR Gustav Lorentzen Conference on Natural Refrigerants (GL2018). Valencia, Spain

[6-27] Colbourne, D., Hühren, R., Rajadhyaksha, D.D., et al.: 2013. *Refrigerant leakage control for R-290 in split type refrigeration and air conditioning systems*. Compressors 2013: 8th International Conference on Compressors and Coolants, Castá Papiernicka, Slovak (Republic)

[6-28] Colbourne D., Suen, K.O.: 2021. *Hydrocarbon refrigerant charge limits for quiescent rooms*. Int. J. Refrig. Vol. 125, pp 75–83

EN/ IEC 60079-0:2017 Explosive atmospheres – Part 0: Equipment – General requirements

EN/ IEC 60079-1-10:2020 Explosive atmospheres – Part 10-1: Classification of areas – Explosive gas atmospheres

EN 1027-1:2011 Explosive atmospheres – Explosion prevention and protection – Part 1: Basic concepts and methodology

EN 378-1:2016+A1:2020 Refrigerating systems and heat pumps – Safety and environmental requirements – Part 1: Basic requirements, definitions, classifications and selection

EN 378-2:2016-02 Refrigerating systems and heat pumps – Safety and environmental requirements – Part 2: Design, construction, testing, marking and documentation

EN 378-3: 2016+A1:202 Refrigerating systems and heat pumps – Safety and environmental requirements – Part 3: Installation site and personal protection

EN 378-4: 2016+A1:2019 Refrigerating systems and heat pumps – Safety and environmental requirements – Part 4: Operation, maintenance, repair and recovery

EN 1127-1:2019 Explosive atmospheres – Explosion prevention and protection – Part 1: Basic concepts and methodology

EN ISO 12100:2010 Safety of machinery – General principles for design – Risk assessment and risk reduction

EN 13313:2010 Refrigerating systems and heat pumps – Competence of personnel

EN 14825:2018 Air conditioners, liquid chilling packages and heat pumps, with electrically driven compressors, for space heating and cooling – Testing and rating at part load conditions and calculation of seasonal performance

prEN 14825:2020 Air conditioners, liquid chilling packages and heat pumps, with electrically driven compressors, for space heating and cooling, commercial and process cooling – Testing and rating at part load conditions and calculation of seasonal performance

ISO 51149-1:2014 Refrigerating systems and heat pumps – Safety and environmental requirements – Part 1: Definitions, classification and selection criteria

ISO 14903:2017-07 Refrigerating systems and heat pumps – Qualification of tightness of components and joints

7 Water as a refrigerant

Florian Hanslik, Jürgen Süß, Michael Kauffeld

7.1 Introduction

Water was actually the first refrigerant [7-1]. The first attempts to create a vacuum with a hand pump to vaporise a substance and cool it down were made in 1810 by the Scotsman John Leslie in Edinburgh. Leslie used water as a working fluid. Nevertheless, the breakthrough in compression refrigeration machines came by switching to more volatile substances such as ethyl ether, ammonia or carbon dioxide. For this reason, water had initially disappeared as a refrigerant for many decades, quite in contrast to heat engines with their clockwise cycles: From the beginning, steam engines were dominated by water [7-1].

What are the reasons for this, and why is it worthwhile to focus on water as a working fluid for counterclockwise cycles, i.e. cooling processes?

The environmental damage caused by refrigeration systems should be mentioned as one argument. Water has no ozone depletion potential (ODP), it has no greenhouse effect (GWP) and it enables – at least in theory – high coefficients of performance. Refrigeration systems with water as the refrigerant therefore only have a minor influence on the global environment. Since water is also non-toxic and non-flammable, refrigeration systems with water as the refrigerant do not affect the local environment. This is probably one of the reasons for the popularity of steam jet refrigeration systems with water as a working substance in the chemical industry. In addition, steam or waste heat is often available as a waste product to drive the steam jet compressor, first patented by Maurice Leblanc at Westinghouse in 1905 [7-1].

So why are there so few vapour compression refrigeration systems with water as refrigerant? Let us first look at the properties of water compared to common refrigerants. Water is particularly challenging when it comes to the system pressure, pressure ratio and volumetric heat of vaporisation. The very low saturation pressure means that the refrigeration system operates in a vacuum - both in the evaporator and in the condenser. The high-pressure ratio would actually call for displacement compressors, ideally designed as reciprocating piston machines, whereas the low volumetric refrigeration capacity and the resulting high volume flows ask for turbo machines to be able to compress the large volume flow. In all other aspects, water has clear advantages over all other refrigerants.

7.2 Properties and area of application

7.2.1 Thermodynamic properties and comparison to common refrigerants

As an introduction to the topic, we consider the thermodynamic properties and the direct impact on the environment in comparison to common refrigerants as the basis for the further evaluation of water (R718) as a refrigerant. The curves of the following diagrams were created with the program REFPROP [7-2].

Figure 7.1 shows the comparison of the specific enthalpy of vaporisation of selected substances. The majority of the substances shown are close together at a very low level compared to water. Three natural refrigerants (propane, ammonia and water) deserve special mention, with water having almost six times the evaporation enthalpy of propane (R290) and twice that of ammonia (R717).

The comparison of the vapour pressures over the evaporation temperature of the selected refrigerants is shown in Figure 7.2. There are two extreme characteristics. While CO_2 (R744) is a high-pressure refrigerant, which preferably is used at low evaporation temperatures, water is a low-pressure refrigerant, which should be used for higher evaporation temperatures.

The property relevant for the dimensioning of the compressor, the volumetric refrigeration capacity, is shown in Figure 7.3. The significantly lower values of water compared to the usual refrigerants require significantly higher volume flows for the same evaporation temperature and cooling capacities in order to transport the required mass flow of refrigerant. The low values are directly related to the low density of water vapour at the low vapour pressures found in refrigeration systems with water as refrigerant.

Figure 7.4 shows the pressure ratio required for a temperature rise of 30 K between the evaporator and the condenser. Close to the freezing of water, the pressure ratio of water is more than twice that of common refrigerants. However, a clear decrease can be seen with increasing evaporation temperature.

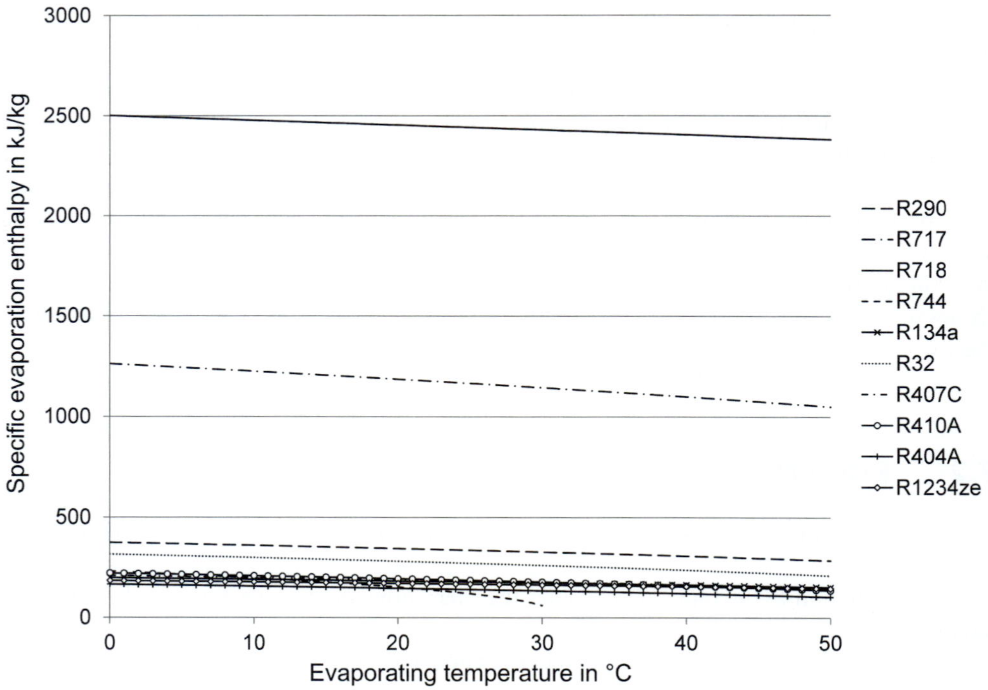

Figure 7.1 Specific enthalpy of vaporisation of selected refrigerants

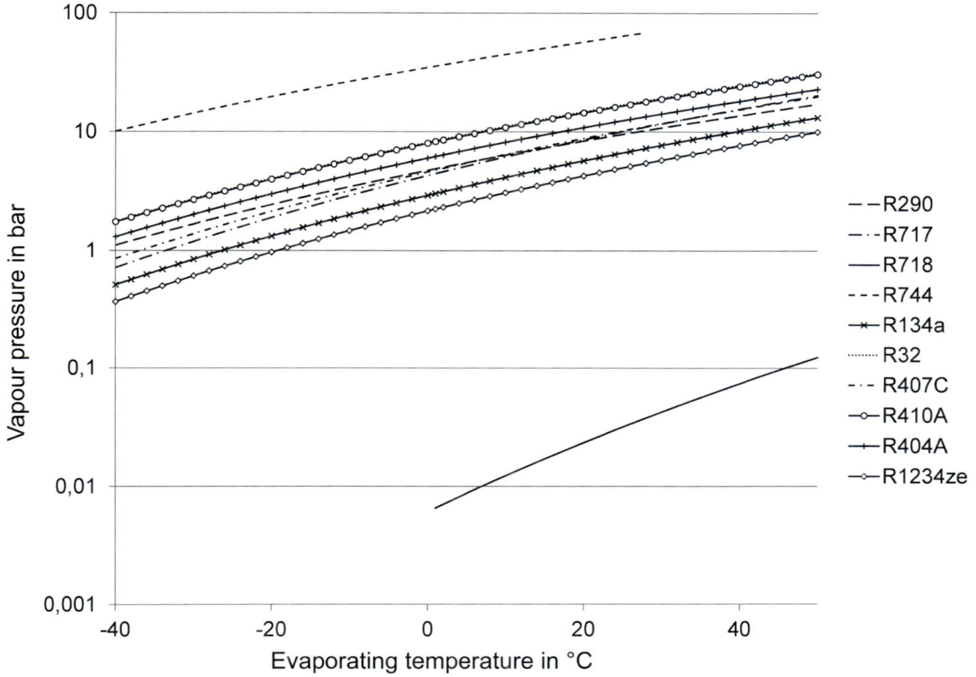

Figure 7.2 Vapour pressure curves of selected refrigerants

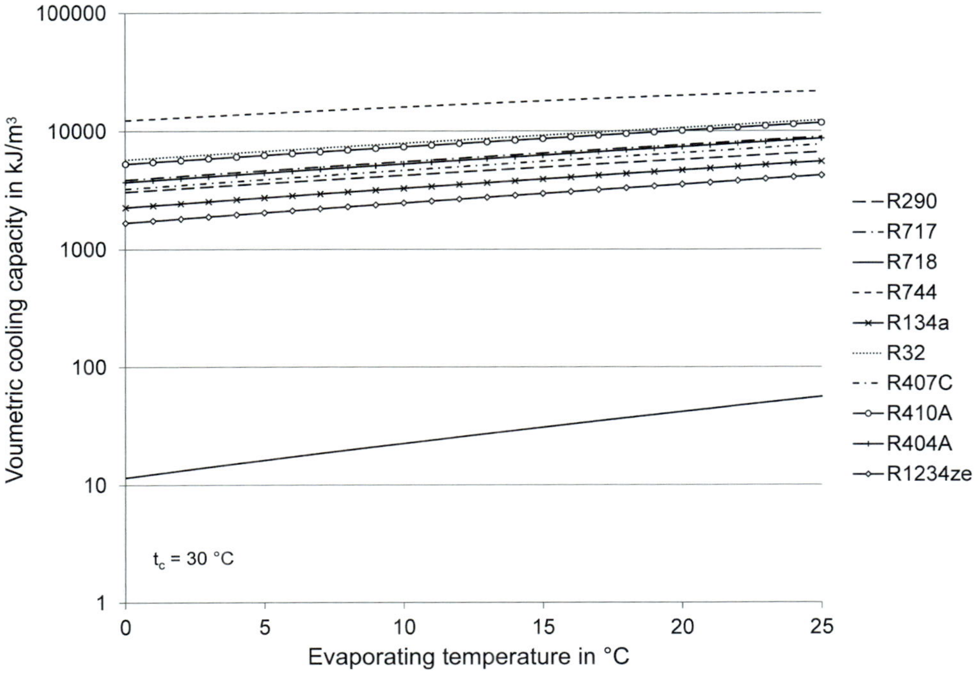

Figure 7.3 Volumetric cooling capacity of selected refrigerants

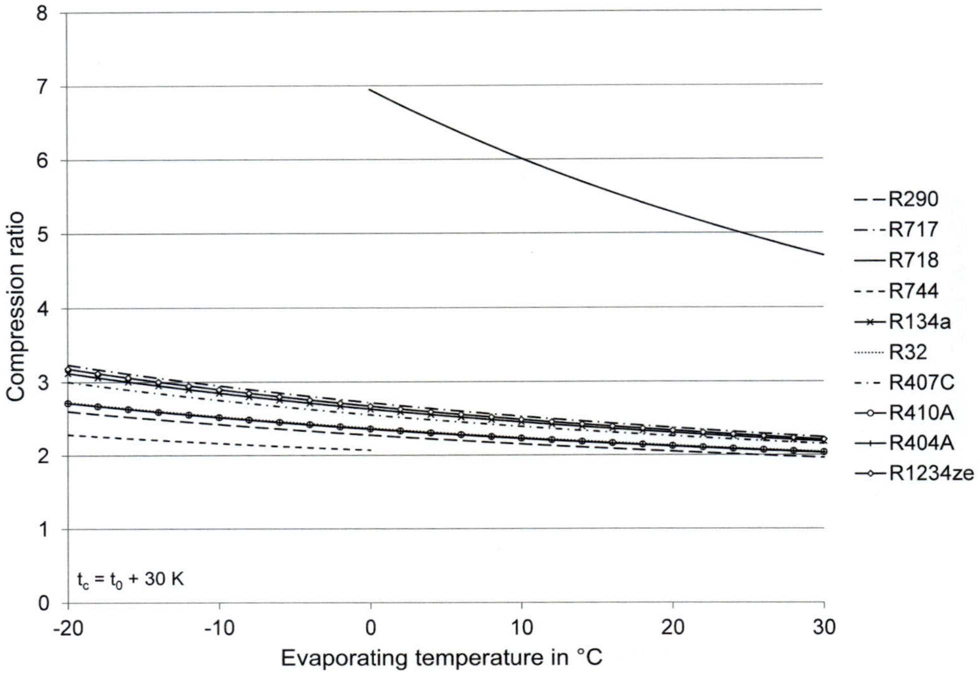

Figure 7.4 Required pressure ratio of selected refrigerants

Table 7.1 Material properties of selected refrigerants

	GWP_{100} (IPCC IV)	Flammable	Toxic	Safety class	ODP
R404A	3922	no	no	A1	0
R134a	1430	no	no	A1	0
R407C	1774	no	no	A1	0
R410A	2088	no	no	A1	0
R1234ze(E)	7	yes	no	A2L	0
R32	675	yes	no	A2L	0
R744	1	no	no	A1	0
R717	0	yes	yes	B2L	0
R290	3	yes	no	A3	0

In addition to the thermodynamic properties, the effects on the environment and the properties with regard to flammability and toxicity are also important aspects for a refrigerant. Table 7.1 shows the most important properties of common refrigerants. Water is the only refrigerant that is neither toxic nor flammable and has both a GWP and an ODP value of 0.

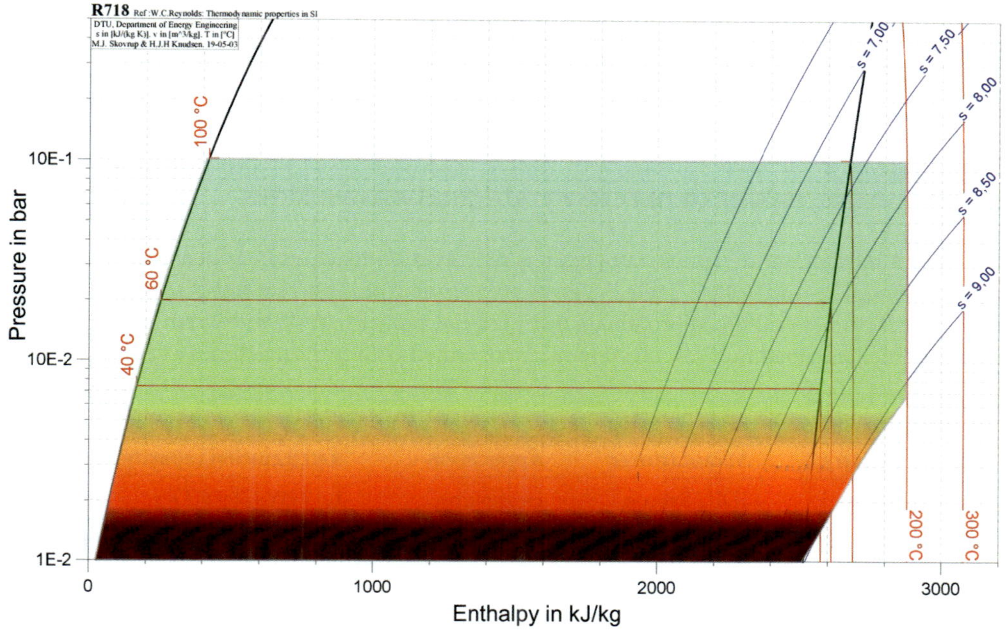

Figure 7.5 Pressure-enthalpy diagram of water including the range of a useful application as a refrigerant

In addition to some challenges, the use of water as a refrigerant also has many advantages. It has no negative effects on the environment and, under certain conditions, also enables a significant gain in efficiency compared to common refrigerants.

Furthermore, a useful range of water as a refrigerant can be derived based on the properties mentioned. This is shown in Figure 7.5. The colour gradient from dark red (conditionally good area of application) to light green (very good area of application) shows the field of application. The limitation to a hot gas temperature of 200 °C (392 °F) depends on the material compatibility of the compressor components.

7.2.2 Use as a refrigerant

Water is the refrigerant most widely used in absorption refrigeration systems. The process developed to market maturity by the Carré brothers in the middle of the 19th century works in almost all installed systems with a combination of water and lithium bromide, where water is the refrigerant and lithium bromide acts as absorbent. Also in another variant of thermally driven systems, the adsorption systems, water is used as the working substance in most systems due to its large enthalpy of vaporisation.

In the chemical industry, applications with steam-jet refrigeration systems using water as the refrigerant are being used.

Another application of water as a refrigerant are high-temperature heat pumps [7-14] [7-15] [7-16]. However, absolute pressures above 1 bar (14.5 psi) occur in this application, and the benefit of the system is the heat generation and not cooling.

Vapour compression refrigeration systems with water as the refrigerant have so far not been of great relevance due to the restrictions mentioned but are becoming increasingly attractive due to the positive environmental balance. The most important projects in this area are presented below.

7.2.3 Use in vapour compression refrigeration machines

Water as a refrigerant in vapour compression refrigeration machines has been used in various projects since the mid-1980s. Table 7.2 gives an overview of most relevant projects of the various manufacturers and the approximate refrigeration capacity. With the exception of the cooling machine described in [7-3], all systems are operated with turbomachines as compressors.

Table 7.2 Vapour compression refrigeration systems with R718 as refrigerant

Year	Country	Manufacturer	Application	Refrigeration capacity	Source
1990	Denmark	IDE / Sabroe	Heat pump	1,3 MW (369.6 TR)	[7-4]
1995	Denmark	IDE / Sabroe / TI	Plastic molding	2 MW (568.7 TR)	[7-4]
1995	South Africa	IDE / Integral	Mining	6 x 3 MW (853 TR)	[7-4]
1999	Germany	ILK / GHS Essen	Chilled water	1 MW (284.3 TR)	[7-5]
1999	Germany	ILK / VW	Chilled water	1 MW (284.3 TR)	[7-6]
2000	Germany	ILK / Daimler	Chilled water	1 MW (284.3 TR)	[7-5]
2002	Japan	DE / Sanken	Vacuum ice	350 kW (99520 TR)	[7-4]
2007	Japan	IDE / Nissan	Vacuum ice	1,5 MW (426.5 TR)	[7-4]
2008	RSA	IDE	Mining	3 x 3 MW (853 TR)	[7-4]
2008	Switzerland	IDE	Artificial snow	1,4 MW (398 TR)	[7-4]
2008	Austria	IDE	Artificial snow	1,4 MW (398 TR)	[7-4]
2011	USA	MSU	Chilled water	–	[7-7]
2012	Denmark	Rotrex / Topsøe	Heat pump	350 kW (99.5 TR)	[7-8]
2013	Japan	Kawasaki HI	Chilled water	350 kW (99.5 TR)	[7-9]
2012	Germany	ILK / Engie	Chilled water	800 kW (227.5 TR)	[7-10]
2014	Japan	Sasakura	Chilled water	17,5 kW (5 TR)	[7-3]
2015	Denmark	TI / JCI	Chilled water	800 kW (227.5 TR)	[7-11]
2015	Germany	ILK	Vacuum ice	50 kW (14.2 TR)	[7-10]
2015	Germany	Efficient Energy	Chilled water	35 kW (10 TR)	[7-12]
2019	Japan	Panasonic	Cold water	100 kW (28.4 TR)	[7-13]

7.3 Functionality of compression refrigeration systems with the refrigerant water

A vapour compression refrigeration system with water as refrigerant essentially consists of the same functional components – evaporator, compressor, condenser and expansion device – as systems for conventional refrigerants, even if these have special features.

Turbo machines with oil-free bearings are mostly used as compressors with water as refrigerant. Examples include water-bearing systems, life-lubricated hybrid ball bearings or magnetic bearing systems. The evaporator or condenser is designed with an internal or external heat exchanger. Due to the properties of water described in Chapter 7.2.1, both variants are operated under vacuum if it is assumed that both the evaporation and the condensation temperature are below 100 °C (212 °F). With an assumed pressure ratio of 4, the pressure difference between the condenser and evaporator is in the absolute range from 50 to 70 mbar (0.7 to 1.0 psi), which corresponds to a temperature lift of approximately 24 K. On the one hand, this makes it possible to dispense with a controlled expansion valve and use a simple overflow with the corresponding water column from 500 to 700 mm (19.7 to 27.6 inch) instead. On the other hand, special attention should be paid to keep the flow losses of the water vapour flow as low as possible. A pressure loss of, for example, 5 to 10 mbar (0.07 to 0.14 psi) on the suction side of the compressor results in a temperature drop that is reduced by up to 3 K.

Because of the operation in a vacuum and the mostly non-hermetic design of the systems, in addition to the components mentioned, an external vacuum pump is required in order to keep the system free of non-condensable gases. On the one hand, the pressure in the vacuum system is increased by air entering from the outside; on the other hand, inert gases enter the system with the air, which have a negative influence on evaporation or condensation. This means that the compressor consumes more energy.

In systems with external heat exchangers (Figure 7.6), the complete system between the chilled water and the cooling water heat exchangers is evacuated. The pressure in the respective areas depends on the prevailing temperature. The heat transfer between the refrigerant and the chilled water as well as the cooling water typically takes place in plate heat exchangers. The phase change of the refrigerant takes place inside a vessel that contains the evaporator and condenser area, the compressor and the expansion device. The evaporation takes place at the interface between the liquid and vapour phase in the lower region of the vessel, due to the lowered pressure. For condensation, the gaseous refrigerant passes through a packed bed together with liquid refrigerant. The gaseous refrigerant condenses out at the interface with the liquid refrigerant. The packed bed serves to increase the surface area of the liquid refrigerant. As an alternative to the plate heat exchangers, the system described can also be implemented with a water-air heat exchanger on the cold-water side as well as on the cooling water side. In cold air applications, this can save transmission loss as well as an additional secondary circuit. The integration of an air coil on the cooling water side enables operation as an air-cooled machine for outdoor installation, provided that the system is guaranteed to be frost-free.

In systems with internal heat exchangers (Figure 7.7), the heat exchangers are located inside the evacuated container, which includes the same areas as the system described above. The heat transfer between chilled water and the refrigerant takes place in a classic evaporator with a finned tube design. Likewise, the refrigerant is condensed on a finned tube heat exchanger located in the vacuum container, through which the cooling water flows internally.

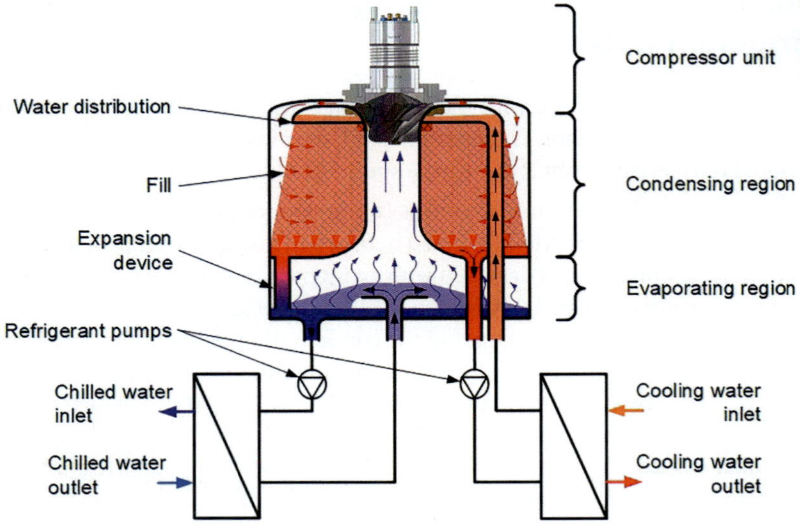

Figure 7.6
Single-stage construction with external heat exchangers

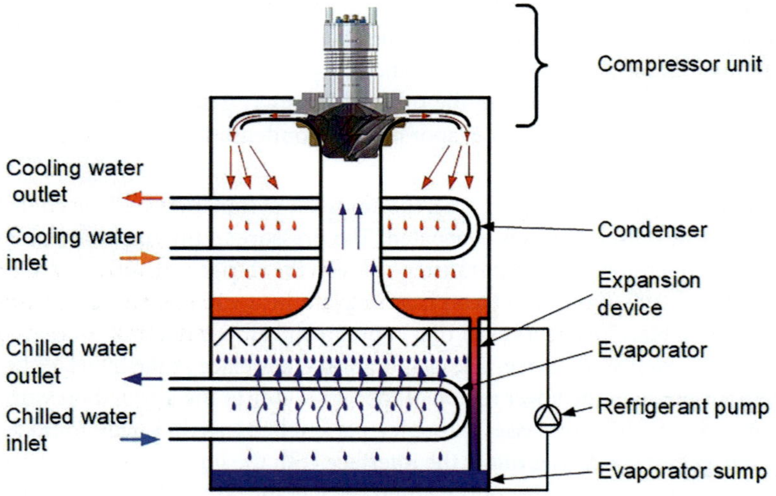

Figure 7.7
Single-stage construction with internal heat exchangers

7.4 Application in the low power range

7.4.1 Compact water chillers

The majority of the projects shown in Chapter 7.2.3 are in the cooling capacity range over 500 kW (2.8 TR) and at chilled-water temperatures below 10 °C (50 °F). Exceptions are Kawasaki and Sasakura. In addition to the cooling capacity, systems from Efficient Energy GmbH also differ from the other manufacturers in the temperature range of the chilled water generated since they specialise in chilled-water temperatures above 16 °C (60.8 °F). This made it possible to offer a compact system that closes the gap between classic chilled-water generation and cooling tower applications. It can therefore be assumed that the relevance of R718 in the field of vapour compression refrigeration systems will continue to increase and will establish itself as a serious refrigerant in the field of vapour compression refrigeration machines.

The chiller essentially consists of two vessels, two plate heat exchangers, two hydraulic shunts and the connecting pipes (Figure 7.8). The two vessels in the system work according to the principle described in Chapter 7.3 with external heat exchangers, whereby the inlet of the evaporator of the second vessel is connected to the outlet of the condenser of the first vessel and is therefore connected as a cascade to the first stage. The two plate heat exchangers represent the delimitation of the vacuum system and the interface to the secondary side. The two hydraulic shunts integrated in the hydraulic connection lines can be used to switch between different operating

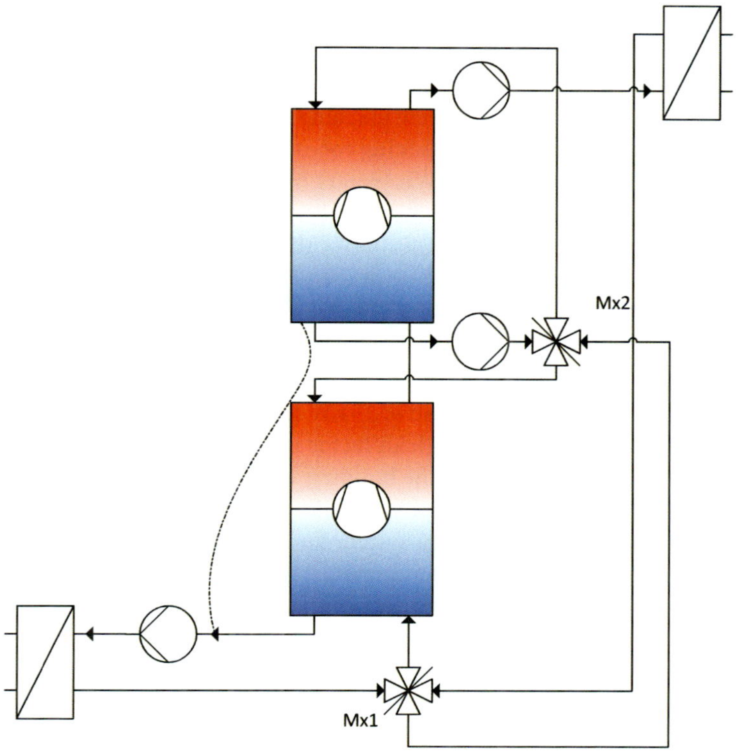

Figure 7.8
Schematic structure of the chiller from the manufacturer Efficient Energy

states. This happens automatically depending on the cooling water temperature in relation to the desired chilled-water temperature and ranges from pure free cooling operation to single-stage and two-stage operation at high cooling water temperatures.

7.4.2 Practical examples

Cooling server rooms [7-18]

The task was to cool a highly efficient data centre with a maximum cold air temperature of 25 °C. The special thing about this application was the slowly growing demand for cooling power due to the gradual expansion of the servers. The challenge was the initially low cooling capacity of around 8 to 10 kW (2.2 to 2.8 TR), which corresponds to approximately 10 to 15 % of the installed cooling capacity.

As a solution, the server racks were set up in a warm aisle enclosure in the server room. The cooling air supply is provided to the servers by side coolers.

After commissioning in summer 2016, it quickly became clear that operation was highly efficient even at this low partial load. This is due to the infinitely variable turbo compressors in connection with the integrated free cooling.

In spite of the initially low cooling capacity required, this installation was able to achieve a coefficient of performance of more than 18 based on measurement data over a period of 18 months.

Cooling of building [7-19]

In this application, a R718 chiller was combined with a photovoltaic system for the energy supply.

Water as a refrigerant has the advantage that no separate machine room is required for installation. This enabled the chiller to be installed in an adjoining room to the staircase. To compensate for fluctuations in electricity generation, a 28 kWh (95600 Btu) electricity storage system was installed, which can cover a complete daily cycle in summer. The combination of electrical and thermal storage (chilled water hydraulics) in connection with the part-load efficient turbo compressor covers the cooling during the day, including pre- and post-cooling of the building structure. At favourable outside temperatures, partial or complete free cooling is carried out at night, which significantly extends the useful life of the battery storage.

By using ceiling cassettes, the optimum chilled-water inlet temperature of 16 °C (60.8 °F) could be set for this system. In summer, there is a cooling load of 25 kW (7.1 TR). The gradual increase in the utilisation of the office space initially led to a load factor of the refrigeration system of only 40 to 60 %. Using the pre- and post-cooling of the building structure results in an annual performance factor (averaged EER) for the refrigeration system of over 12. This is unique in combination with a vapour compression refrigeration system for cooling office space with the possibility of unrestricted indoor installation of the refrigeration system.

Process cooling [7-20]

As an example of process cooling, the cooling of an extruder for the production of a continuous product made of plastic, such as a pipe, is presented. Because of the processing of various product recipes and the resulting changes in the heat load and temperature control, the extruder is not connected to the central chilled-water supply.

The required temperature range of the chilled water generated is between 16 °C (60.8 °F) and 20 °C (68 °F), and the cooling capacity is around 30 kW (8.5 TR). Further criteria were the installation as close as possible to the extruder, an integrated free cooling system to reduce energy costs and the use of an environmentally friendly and sustainable refrigerant.

The chiller has been in continuous operation since commissioning in July 2017. The operator on site confirms the easy handling when changing the product and thus changing flow temperatures between 16 °C (60.8 °F) and 18 °C (64.4 °F), depending on the dimension. After one year of operation, an annual refrigeration coefficient of performance of over 16 was determined.

7.5 Generation of temperatures below freezing

Due to the freezing point of water at 0 °C (32 °F), this refrigerant cannot be used for cryogenic temperatures without the addition of other substances. This is only possible if water is combined in a cascade as a high-temperature stage with other natural refrigerants in the low-temperature stage. For the use of such cascade systems even in the lower capacity range, the refrigerants CO_2 (R744) and propane (R290) are used. The possible setups are described below.

7.5.1 Description of the systems

A distinction is made between partial and full cascades as well as direct and indirect designs. Figure 7.9 shows an indirect full cascade with a secondary circuit between the condenser of the propane stage and the evaporator of the water stage on the left. In the middle, one can see a direct full cascade, which liquefies the CO_2-stage directly in the evaporator of the water stage. A partial cascade is shown on the right. In this configuration, the supercritical CO_2 gas is subcooled further after the gas cooler using the water stage.

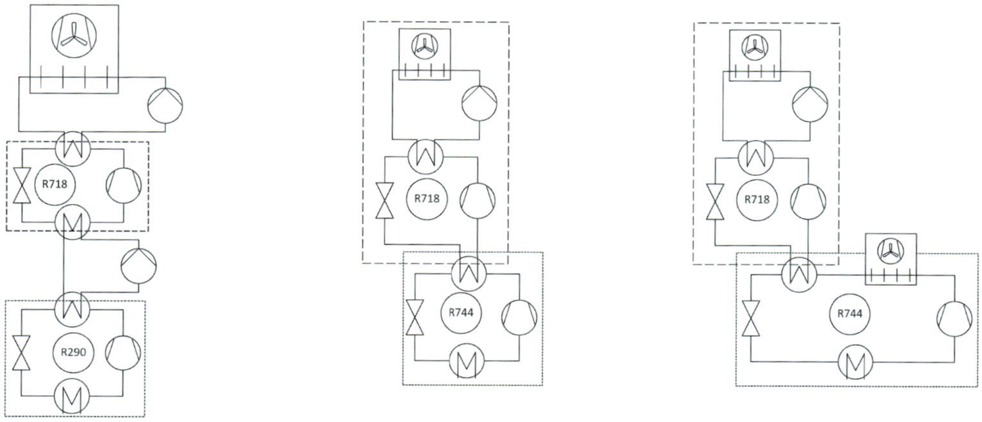

Figure 7.9 Possible cascades with water in the high temperature and propane or CO_2 in the low-temperature level

The systems shown here are considered in more detail in the next section with regard to the areas of application.

7.5.2 Possible applications

For the two full cascade systems, low-temperature (LT) and medium-temperature (MT) temperatures of –30 and –0 °C (–2 and 14 °F) and cooling capacities of 10 and 30 kW (2.8 and 8.5 TR) were assumed for the efficiency comparison.

Indirect full cascade

The indirect full cascade is suitable for use in supermarkets. The condensers of the MT and LT refrigeration cabinets – with integrated propane circuits with a maximum refrigerant charge quantity of 150 g (5.3 oz.) – are connected to a secondary circuit. In addition to the re-cooler, which is installed as standard in such systems, a high-pressure stage with water as refrigerant is also integrated. The high temperature stage keeps the secondary circuit at a certain constant temperature, regardless of the ambient temperature.

If the standard system is compared with the cascade, one can see that an increase in efficiency in pure cooling mode can only be achieved for use at comparatively high ambient temperatures such as in southern Europe and around the Equator.

However, if the waste heat is extracted from the system for heating operation from an ambient temperature of 15 °C (59 °F), the efficiency advantage at high condensing temperatures can also be exploited in cooler regions. In this way, significant energy savings are possible with the cascade layout over the year.

Direct full cascade

The direct full cascade is just as suitable for supermarkets as the indirect one. The differences lie in the refrigerant used, in this case CO_2, and in the absence of a secondary circuit. With this layout, the CO_2 is liquefied directly in the evaporator of the water stage, and the CO_2-booster system with direct evaporators can be operated subcritically all year round, regardless of the outside temperature.

Compared to transcritically operated booster systems, the cascade layout can also save up to 10 % energy over a year.

Partial cascade

In the partial cascade, the gaseous CO_2 is subcooled after the gas cooler using the water stage, thereby increasing the efficiency. According to Hanslik et al. [7-17], this subcooling results in an efficiency increase of more than 35 % compared to a pure transcritical process. In addition to large hypermarkets, the area of application is also in processes in which temperatures below –20 °C (–4 °F) are required for cooling capacities greater than 100 kW (28.4 TR).

7.6 Conclusion

Due to the currently valid EU F-Gas Regulation [7-21], the EU Ecodesign Directive [7-22] and the rapidly increasing environmental awareness of users, the trend in refrigeration with vapour compression refrigeration machines is moving towards natural refrigerants. Water (R718) occupies a special position in this group because it is neither flammable nor toxic and has an ODP and GWP value of zero. In addition to the future-proof and inexpensive availability of water, neither national nor international restrictions can be expected in any way, as is the case with so-called safety refrigerants today.

With these prerequisites and when used in the correct temperature and cooling capacity range, water, as the oldest known refrigerant, is a future-proof solution to guarantee the cooling supply safely, efficiently and in an environmentally friendly manner.

7.7 Literature

[7-1] Kauffeld, M., Süß, J.: *Kaltdampfkompressionskältemaschinen mit Wasser als Kältemittel – Vor- und Nachteile dieser Technologie (Cold vapor compression chillers with water as the refrigerant – advantages and disadvantages of this technology)*, in: KI Kälte-, Luft- und Klimatechnik (2016), H. 11, pp. 44–48

[7-2] Lemmon, E. W., Huber, M. L., McLinden, M. O.: *NIST Reference Fluid Thermodynamic and Transport Properties Database (REFPROP)*, Version 9.1 (2013)

[7-3] Sasakura, T.: *Report 2014 Sasakura Engineering Co., Ltd.* Sasakura Engineering Co., Ltd. 7–32 Takejima 4-chome, Nishiyodogawa-ku, Osaka 555-0011, Japan

[7-4] Ophir, A.: *Mechanical Heat Pumps Using Water as Refrigerant for Ice Production and Air Conditioning*. In: IDEA 99th Annual Conference & Trade Show (2008), Orlando, Florida

[7-5] Albring, P.: *Kältetechnik mit Wasser als Kältemittel (Refrigeration technology with water as a refrigerant)*, Netzwerk Kältetechnik (2009)

[7-6] Albring, P., Honke, M.: *Ice-Making and Ice Storage, with Water as Refrigerant*, in: Proceedings of the 23rd IIR International Congress of Refrigeration (2011)

[7-7] Li, Q., Piechna, J., Müller, N.: *Thermodynamic potential of using a counter rotating novel axial impeller to compress water vapor as refrigerant*, in: International Journal of Refrigeration 34 (2011), H. 5, pp. 1286-1295

[7-8] Borup, J., Jensen, J.: *Udvikling af Rotrex turbokompressor til vanddamp compression (Development of Rotrex turbocharger for water vapor compression)*, No. 344-009, PSO Program 2012

[7-9] Murayama, S.: *Report 2014 Kawasaki Heavy Industries, Ltd.* Kawasaki Heavy Industries, Ltd. Tokyo Office, 14-5 Kaigan 1-chrome, Minato-ku, Tokyo 105-8315, Japan

[7-10] Honke, M., Safarik, M., Herzog, R.: *R718 turbo chillers and vacuum ice generation - Two applications of a new generation of high speed, high capacity R718 centrifugal compressors*, in: Proceedings of the 24th IIR International Congress of Refrigeration (2015)

[7-11] Madsbøll, H.: *Competitive chiller concept with water as refrigerant*, Danske Køledage 2013

[7-12] Süß, J.: *Eine kompakte Kälteanlage mit Wasser als Kältemittel (A compact refrigeration system with water as a refrigerant)*, in: Deutsche Kälte- und Klimatagung 2014, Düsseldorf

[7-13] Shoyama, T. et al: *2019 IOP Conf.* Ser.: Mater. Sci. Eng. 604 011010

[7-14] Chamoun, M., Rulliere, R., Haberschill, P., Peureux, J.L: *Experimental Investigation of a New High Temperature Heat Pump Using Water as Refrigerant for Industrial Heat Recovery*, International Refrigeration and Air Conditioning Conference (2012), Purdue

[7-15] Bantle, M., Tolstorebrov, I., Hafner, A.: *Energierückgewinnung mittels Brüden-Kompression in Trocknungssystemen mit überhitztem Dampf (Energy recovery using vapor compression in Drying systems with superheated steam)*, Deutsche Kälte- und Klimatagung 2014, Düsseldorf

[7-16] Ponweiser, K., Schlögl, T.: *Wärmepumpen mit Wasser als Arbeitsmedium – Perspektiven und Grenzen (Heat Pumps with Water as Working Medium – Perspectives and Limits)*, VDI-Forum 2015, University of Linz

[7-17] Hanslik, F., Süß, J., Koehler, J.: *Efficiency Enhancement by Subcooling the Carbon Dioxide Process with Water as Refrigerant*, International Refrigeration and Air Conditioning Conference (2018), Purdue

[7-18] Efficient Energy: *RZ-Kühlung, die mit der IT wachsen kann und im Teillastbetrieb effizient ist (Data centre cooling that can grow with IT and is efficient in part-load operation)*, https://efficient-energy.de/referenzen/sparkasse-calw/, accessed on: 05.01.2021

[7-19] Efficient Energy: *100 % autarke, solarbetriebene Gebäudekühlung mit Wasser als Kältemittel (100 % self-sufficient, solar-powered building cooling with water as the refrigerant)*, https://efficient-energy.de/referenz/combitherm/, accessed on: 05.01.2021

[7-20] Efficient Energy: *Extruderkühlung Gardena (extruder cooling Gardena)*, https://efficient-energy.de/references/gardena/, accessed on: 05.01.2021

[7-21] European Parliament, Council of the European Union: *Regulation (EU) No. 517/2014 of the European Parliament and of the Council of 16 April 2014 on fluorinated greenhouse gases and repealing* Regulation (EC) No 842/2006 Text with EEA relevance

[7-22] European Parliament, Council of the European Union: *Directive 2009 / 125 / EG of the European Parliament and of the Council of October 21, 2009 establishing a framework for the setting of ecodesign requirements for energy-related products*

8 Refrigeration systems for temperatures below –50 °C (–58 °F)

Michael Kauffeld

Since January 1st, 2020, the ban on the installation of new stationary refrigeration systems, which contain HFCs with a global warming potential (GWP) of 2500 or more, has been enforced. The EU F-Gas Regulation (No. 517/2014) permits an exemption for devices intended for applications to cool down products below –50 °C (–58 °F). This exemption had to be stated in the EU F-Gas Regulation because the only non-combustible refrigerants for such low temperatures were so far R14 and R23. With their 100-year time horizon global warming potentials of 7390 and 14800, respectively, these refrigerants are no longer environmentally acceptable. The high global warming potential also means considerable price increases for the respective refrigerants in the course of the European and a little later worldwide HFC phase-down. A significantly reduced availability of high GWP refrigerants such as R23 at a later point in time led therefore to price increases.

8.1 Hydrocarbons and mixtures with CO_2

Conversely, there are alternatives for refrigeration applications below –50 °C (–58 °F). These are presently the two flammable refrigerants ethane (R170) and methane (R50) or mixtures of these or with other (natural) refrigerants. If the refrigerant is to be rendered non-flammable, a certain amount of an inert gas, e.g. CO_2, can be added. However, high CO_2 content (up to 50 %) is required. This raises the triple point of the mixture and thus limits possible uses for reaching temperatures below –50 °C (–58 °F). Göpfert and Hesse (2015) suggest a mixture of 50 mol-% CO_2 /Ethane, which has a triple point, calculated from material data, of approximately –83 °C (–117.4 °F) [8-1].

8.2 Nitrous oxide and mixtures with CO_2

Presently under research and in development is the use of a nitrous oxide / carbon dioxide mixture (N_2O/CO_2) [8-2]. The CO_2 addition is necessary here to suppress the possible exothermic decomposition of the N_2O. At least 15 % CO_2 is required for this. This increases the triple point, which for pure N_2O is –90,8 °C (–131.44 °F). For a 30 % CO_2 share, triple point increases to approximately –78 °C (–108.4 °F) [8-3]. Hydrocarbon-based lubricants should not be used here [8-4]. The use of so-called perfluoropolyethers, i.e. lubricants in which all hydrogen atoms have been replaced by fluorine atoms, would be possible. These lubricants are used for oxygen compressors, but are very expensive and have a density that is about twice as high as conventional lubricants. In addition, they can cause premature wear and tear on aluminium friction surfaces. At the University of Applied Sciences Karlsruhe, an appropriate compressor was operated for 1000 hours without any signs of wear [8-5].

8.3 Air

Another option for generating low temperatures is the use of air or nitrogen in a gas process. The American doctor John Gorrie had already used a corresponding cold air-cooling system in Florida in the 1840s [8-6]. In a gas refrigeration machine, the expansion does not take place in a throttle, but in a work-performing machine. Although the Joule-Thomson effect would also cool the gas flowing through a throttle to below the inversion temperature of that respective gas, this cooling would be comparatively slight with the isenthalpic change of state present in a throttle process. It is only when work is carried out in an expansion machine that the gas cools down appreciably. Gas cooling machines contain at least one compressor, one or two heat exchangers and one work-producing machine. This machine could provide work input into the compressor. Figure 8.1 shows a corresponding system that has an internal heat exchanger. With the working medium being air, the cold heat exchanger can be omitted if the working pressure on the low-pressure side corresponds to 1 bar (14.5 psi). The air is extracted directly from the space to be cooled and is then blown back into it after having passed through the compressor, the heat exchanger(s) and the expansion machine.

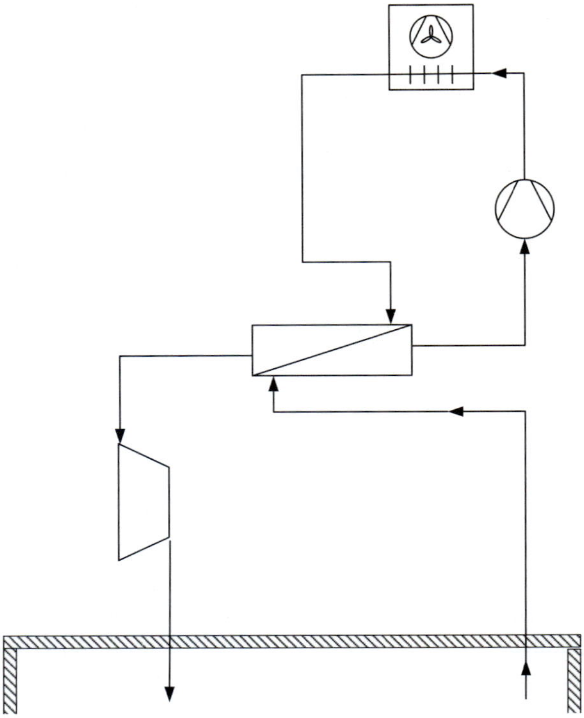

Figure 8.1 Open overpressure process flow diagram for cold air refrigeration machine

Such cold air refrigeration systems have been used for cooling aircraft cabins for decades. At the relatively high temperatures required for air conditioning, the energy efficiency of a cold air cycle is inferior to a vapour compression refrigeration machine. However, due to the relatively low operating pressures – typically below 2 bar (29 psi) absolute on the high-pressure side – the cold air air conditioner is relatively lighter in weight, compared to the vapour compression system. This characteristic is more important for airplanes than aspects of energy efficiency of the air conditioning system. There is also a small batch of air conditioning systems for carriages in Deutsche Bahn Intercity Express (ICE) trains, which work with air as a refrigerant. Here, too, the lower system weight was a decisive design factor.

However, energy comparisons at temperatures below –50 °C (–58 °F) look quite different. Here, the energy efficiency of vapour compression machines, including those in cascade mode, decreases faster with decreasing temperature than the efficiency of cold air chillers. At a minimum of –100 °C (–148 °F), cold air refrigeration machines are more energy efficient than vapour

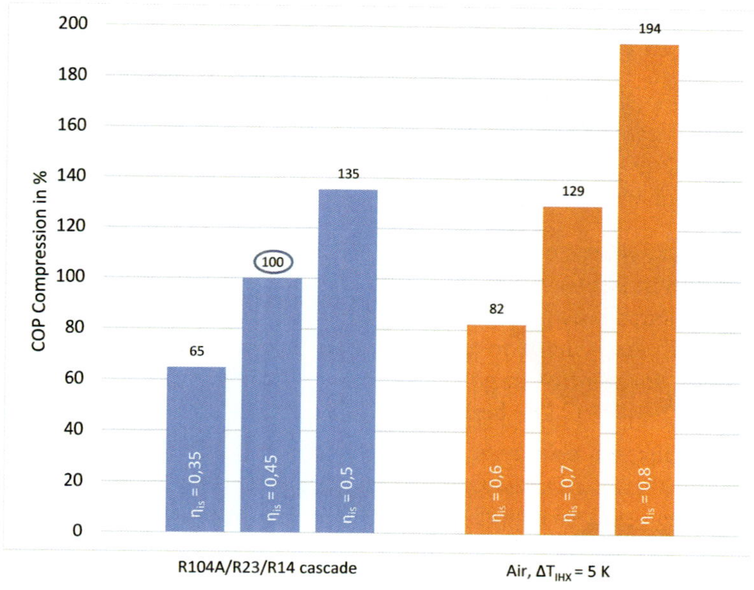

Figure 8.2 Comparison of a cold air refrigeration machine (right side) with a three-stage cascade cooling system (left side) for a cooling chamber temperature of −100 °C (−148 °F). Evaporation temperature of −120 °C (−184 °F) in the lowest stage of the cascade system; chamber inlet temperature of cold air machine at −110 °C (−166 °F). η_{is} refers to the isentropic efficiency of the compressors and expansion machine.

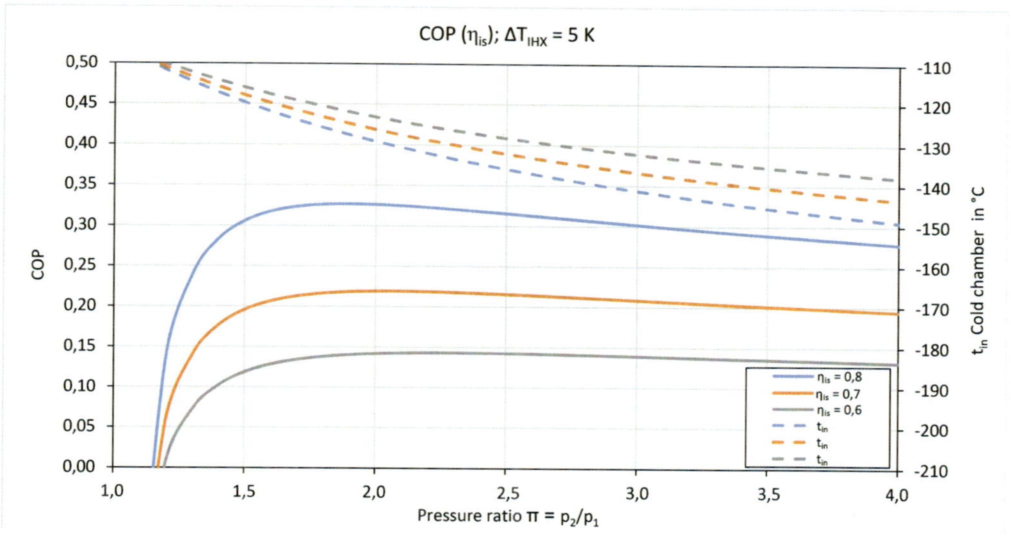

Figure 8.3 Influence of isentropic efficiency on cooling capacity of a cold air refrigeration machine. Temperature difference at internal heat exchanger assumed at 5 K

compression machines, see Figure 8.2. For the R404A / R23 / R14 cascade, depending on the isentropic efficiency of the compressors (0,35, 0,45 or 0,55), there are different cooling capacities. For the temperature differences and isentropic efficiency, data from standard cascade refrigeration machines was used. The refrigeration coefficient with the most realistic value of the isentropic efficiency of 0,45 was set at 100 %.

In addition to good heat exchangers with small temperature differences, the isentropic efficiency of the compression and expansion machine is crucial for the energy efficiency of a cold air

Figure 8.4
Turbo compressor-expander with electric motor on common shaft (© Mirai Intex)

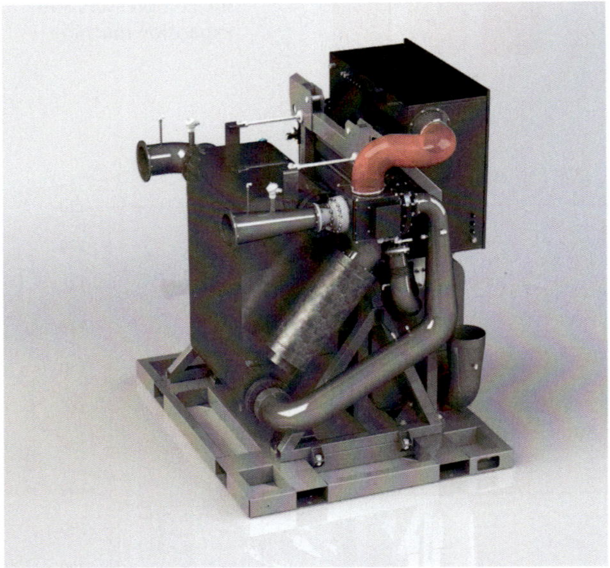

Figure 8.5
Cold air refrigeration machine for air supply to cryogenic chamber for therapeutic purposes (© Mirai Intex)

refrigeration machine. Figure 8.3 shows the influence of the isentropic efficiency of the machines on the cooling capacity of a cryogenic air cycle refrigerator.

The required turbo-expander compressor units for such low temperatures were unavailable for a long time. Now systems are available from both a Japanese and a European supplier; see Figure 8.4. The turbo machines from these two manufacturers achieve an isentropic efficiency of almost 80 % for the compressor and approximately 75 % for the expansion machine. Depending on the size, these turbomachines rotate between 40000 and 100000 rpm. The higher the refrigeration capacity, the lower the speed.

The turbomachine unit uses either magnetic or air gas bearings. Therefore, they run completely oil-free and therefore allow operation directly on a cryo chamber, see Figure 8.5. Such cold air refrigeration machine can also be designed as a closed circuit and the systems can then be made more compact due to the higher system pressures that are then possible. Alternatively, the refrigeration capacity can be multiplied by installing more units of the same size. In addition to air, dry nitrogen is a suitable working substance in such closed circuits. Such systems for freeze-drying systems can replace the current cold vapour compression systems.

In addition, there is of course also the possibility of cooling with liquid nitrogen, but this makes little sense from an energy perspective if the application does not really require −196 °C (−320.8 °F) – the normal boiling point of liquid nitrogen.

As the above explanations show, there will probably no longer be a need for the exemption in the EU F-Gas Regulation for refrigeration systems intended for applications for cooling products below −50 °C (−58 °F) in the near future.

8.4 Literature

[8-1] Göpfert T., Hesse U.: *Survey on Non-flammable Low GWP Refrigerant Mixtures based on Carbon Dioxide for Applications below 220 K.* ICR 2015, August 16–22, 2015, Yokohama, Japan, Paper ID: 143

[8-2] Kauffeld, M., Maurath, T.: *N_2O/CO_2 – Mixtures as Refrigerants for Temperatures below −50 °C.* 25th IIR International Congress of Refrigeration, Montreal, Canada, August 24–30, 2019. Paper ID: 1532, DOI: 10.18462 / iir.icr.2019.1532

[8-3] Di Nicola, G., Giuliani, G., Polonara, F., Stryjek, R.: *Solid-liquid Equilibria for the CO_2 + N_2O, CO_2 + R32, and N_2O + R32 Systems.* Fluid Phase Equilibria, 256 (1-2), 2007, pp. 86–92

[8-4] Kauffeld, M., Maurath, T., Germanus, J., Askar, E.: *N_2O/CO_2-Mixtures as Refrigerants for Temperatures below −50 °C.* International Journal of Refrigeration 117 (2020) 316–327, https://doi.org/10.1016/j.ijrefrig.2020.04.026

[8-5] Kauffeld, M., Maurath, T., Germanus, J., Askar, E.: *N_2O/CO_2-Gemische als Kältemittel für Temperaturen unter −50 °C – Teil 1 und 2,* ki Luft- und Kältetechnik. 56 (11 2020 and 12 2020) pp. 42–46 and pp. 49–53, ISSN 1865-5432

[8-6] Kauffeld, M.: *Untersuchung von Kaltluftprozessen unter besonderer Berücksichtigung kleiner Kompressions- und Expansionsmaschinen (Investigation of cold air processes with special consideration of small compression and expansion machines).* DKV-Forschungsbericht No. 39, 1993, ISBN 3-922- 429-40-8

9 Lubricants for natural refrigerants

Wolfgang Bock, Christian Puhl

9.1 Lubricants for ammonia (NH_3 – R717)

Conventional refrigeration oils are generally not miscible with ammonia (DIN 51503, Group KAA: refrigeration oils immiscible with ammonia). Some special systems that work on the principle of dry evaporation use what are known as partially miscible oils. These oils are based on polyalkylene glycol (PAG) and behave significantly differently than conventional immiscible ammonia oils based on mineral oil, alkylbenzene, polyalphaolefin (PAO) or synthetic hydrocarbons, especially with regard to the refrigerant solution. When using partially miscible PAG oils (DIN 51503, Group KAB: refrigeration machine oils miscible with ammonia), it is particularly important to ensure that there is a low water content in the refrigeration system; otherwise problems with the chemical stability of the refrigerant circuit can occur. In addition, the high dissolution of ammonia in these partially miscible oils represents a potential hazard with regard to the wear of bearings and compressor components (including cavitation through outgassing of NH_3 from the oil). In summary, it can be said that the use of partially miscible oils based on polyalkylene glycol has not yet been fully investigated and is generally regarded as critical.

In the case of conventional (immiscible) ammonia oils, the products that are based on mineral oil are of primary importance. Here, primarily naphthenic selective raffinates are used as the oil base. This group of refrigeration oils have proven themselves over decades. The products are available worldwide and are characterised by good aging and cold flow properties. They are suitable for use down to evaporation temperatures of approximately –30 °C (–22 °F). The pour point of an immiscible ammonia refrigeration oil should always be below the evaporation temperature. Otherwise, an oil with better cold flow properties should be used [9-1, 9-2, 9-3].

In addition to determining the pour point (ISO 3016), a measure of the cold viscosity of an ammonia refrigeration oil is the determination of the so-called U-tube flowability. Figure 9.1 shows the excellent cold flow properties of synthetic refrigeration oils compared to mineral oil and alkylbenzene.

The most important synthetic refrigeration oils for ammonia are lubricants based on polyalphaolefin (PAO) and alkylbenzene (AB). Both lubricant types are also – like mineral oils – not miscible with liquid ammonia. These fully synthetic refrigeration oils are significantly superior to mineral oil-based products in terms of aging stability and cold flowability. Especially at very low evaporation temperatures below –45 °C (–49 °F), PAO-based refrigeration oils are characterised by their good flowability. In applications in which plate heat exchangers are used, the selection criteria focus on good low-temperature refrigeration fluidity of the refrigeration oil. PAO refrigeration oils are the first choice for these applications [9-1].

Other outstanding product properties of PAO refrigeration oils are a minimised depositing / coking tendency, a very good viscosity-temperature relationship (high viscosity index, VI) and accordingly a reliable lubricant film thickness with excellent wear protection properties. Due to the low evaporation loss of PAO, significantly less oil is discharged from the compressor into the refrigeration circuit. This results in higher system efficiency; the oil consumption of the system is reduced, and the service life of the oil filling is therefore significantly increased. Selected

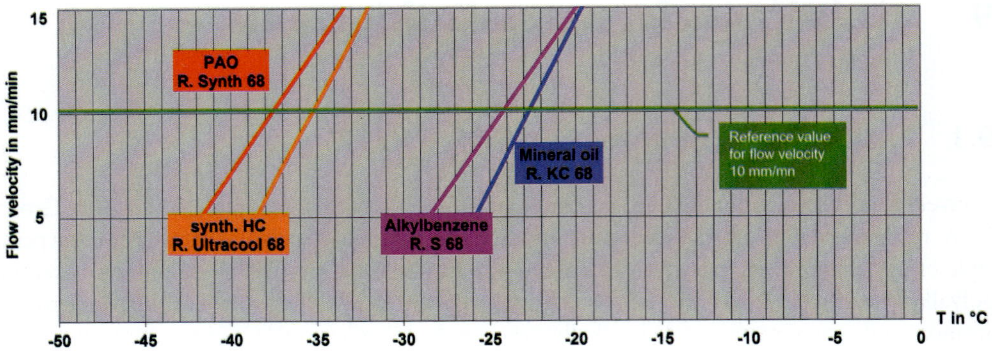

Figure 9.1 Cold flow velocity of various refrigeration oils according to temperature

market-relevant PAO refrigeration machine oils can be listed according to the American NSF H1 standard (NSF = National Sanitation Foundation: approval of so-called "food grade" lubricants). NSF H1 lubricants are products of the highest purity, which due to their composition can be used in food manufacturing processes.

The so-called hydrocrack oils or hydrogenated mineral oils occupy a special position between classic, naphthenic mineral oils and synthetic PAO ammonia refrigeration oils. They can be above the naphthenic refrigeration oils in terms of their level of performance – cold flow properties, depositing tendency, oil carry-over – but are outperformed by PAO refrigeration oils.

In R723 refrigeration applications (ammonia / dimethyl ether mixture 60 % / 40 %), usually PAO refrigeration oils are utilised. The dimethyl ether content in the refrigerant partially dissolves the refrigerant in the refrigeration oil (immiscible with pure ammonia), which in theory should facilitate the return of oil from the refrigerant circuit to the compressor. The extent to which this effect actually occurs in practice certainly has to be checked in appropriate reference systems. In conclusion, a complete assessment of the overall system including the lubricant budget cannot be made here [9-1].

9.2 Lubricants for hydrocarbons (propane – R290, propene – R1270, isobutane – R600a etc.)

There are various perspectives when looking at the lubricant functionality for hydrocarbon refrigeration applications. First of all, the general high miscibility of hydrocarbon refrigerants with commonly used refrigeration oil types is worth mentioning. Both mineral oil as well as synthetic lubricants such as alkylbenzene, PAO and polyol ester (POE) oils, form homogeneous mixtures over a wide temperature and concentration range. There is no separate oil phase here. Therefore, these refrigeration oils provide a reliable oil transport through the system and heat transfer in the heat exchangers.

The group of PAG lubricants behave differently. They have a much lower miscibility with hydrocarbons and are therefore often used for open systems (gas transport / gas compression in the chemical and petrochemical industry). They may have a pronounced miscibility gap (phase

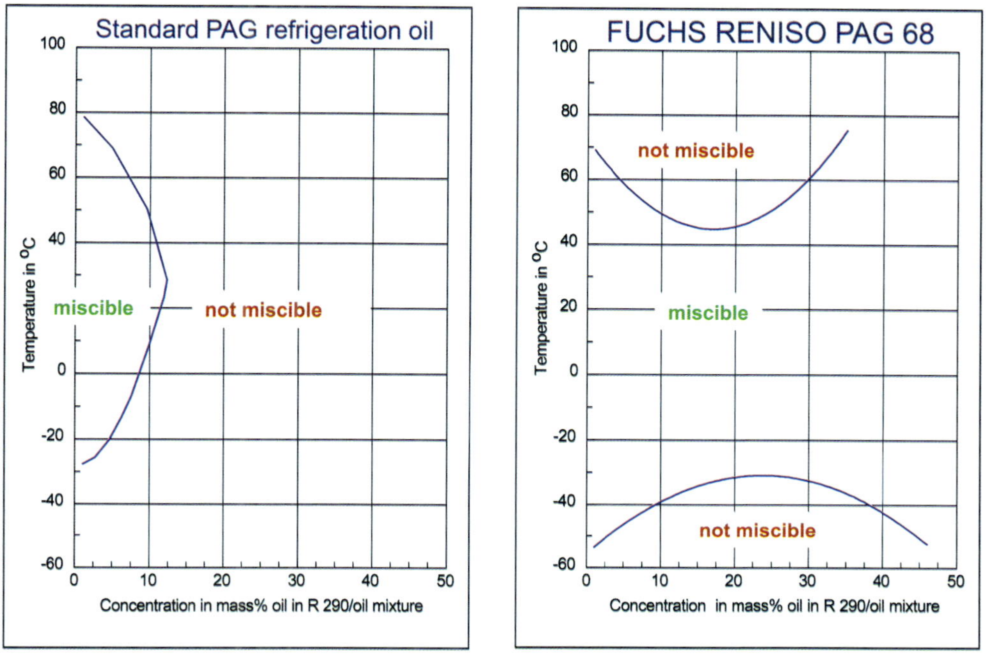

Figure 9.2 Miscibility for PAG type with propane (R290) (© Fuchs Schmierstoffe GmbH)

separation) with hydrocarbon gases depending on the chemical structure of the PAG base fluid, see Figure 9.2.

The use of PAG refrigeration machine oils is also an option but should be agreed to by the system planner or operator in order to avoid negative effects on the refrigeration system components (insufficient oil return, material compatibility).

A special issue of hydrocarbon refrigeration applications is the high solubility of the hydrocarbon in the refrigeration oil. The strong absorption of refrigerant in the oil results in a significant dilution of the lubricant. This in turn can drastically reduce the load-bearing properties of the lubricating film. Increased wear in the compressor and a shortening of the compressor service life can sometimes occur because of this dilution and the associated decrease in the lubricating film thickness in the moving parts. In addition to measures on the compressor design side (more powerful crankcase oil heaters to expel the refrigerant component from the oil), care must also be taken to select a lubricant with a higher initial viscosity for use in hydrocarbon applications. This will compensate for the dilution that occurs. The operating parameters for temperature and refrigerant pressure must be given special attention. In general, the following applies: lower temperatures and / or higher refrigerant pressure leads to greater absorption and thus to greater dilution. In addition, the use of refrigeration oils with selected lubricant additives is definitely advisable to prevent possible wear in the contact areas.

PAO, POE and PAG-based refrigeration oils have proven their worth in practice, both because of the low refrigerant solubility in the lubricant and because of the good viscosity-temperature properties (high VI). When using mineral oil or alkylbenzene products, which are also suitable in principle, it is absolutely necessary to increase the initial viscosity when selecting the appropriate refrigeration machine oil.

In the field of household refrigerators, isobutane (R600a) is primarily used as the refrigerant. Low-viscosity mineral oils (ISO VG 5 and lower) with wear protection additives (AW = anti-wear or EP = extreme pressure additives) have proven themselves as lubricants for the full hermetic compressors installed in these household appliances.

9.3 Lubricants for carbon dioxide (CO_2 – R744)

Main criteria for choosing suitable CO_2 refrigeration oils are, in addition to good miscibility properties with liquid CO_2, high thermal stability and reliable lubrication properties. Likewise, especially in CO_2 systems, the water content in the refrigeration oil has to be reduced to a minimum to ensure chemical stability and trouble-free functioning of the cooling circuit.

9.3.1 Miscibility properties with liquid CO_2

CO_2 refrigeration oils can be divided according to their miscibility behaviour with CO_2 as follows:

- practically immiscible polyalphaolefins (PAO)
- partially miscible polyalkylene glycols (PAG)
- fully miscible polyol esters (POE)

Polyalphaolefins show a very limited miscibility behaviour with the refrigerant CO_2. They are generally only miscible in an oil concentration range above approximately 70 %, which poses certain problems when using these lubricants: They can only be used if an effective oil separator is installed in the refrigeration system. In the case of polyalphaolefin, oil that enters the refrigerant circuit does not adequately return to the compressor. This will lead to problems with the system. One possible consequence is an oil migration into the circuit piping – as a result, insufficient lubrication and thus massive wear can occur in the compressor. A deterioration of the heat transfer process in the heat exchangers can also occur. Suitable oil separator types for PAO refrigeration oils in CO_2 refrigeration systems are the coalescent separators, which are clearly superior to centrifugal separators [9-4].

Specially developed polyalkylene glycols (PAG) can have better miscibility properties with CO_2 than PAO oils. In addition to the range of high oil concentrations at which these special PAGs (so-called double end-capped PAGs) can be mixed with CO_2, there is also a concentration range of around 1–3 % oil in the liquid CO_2, where the lubricant is completely dissolved in the liquid refrigerant phase. In this way, oil can be transported back to the compressor at normal oil circulation rates. In larger refrigeration systems, such as supermarket refrigeration installations with widely branched tubing systems, the use of PAG refrigeration oils can lead to difficulties. This is due to the poor miscibility behaviour of PAG together with CO_2. However, PAG refrigeration oils (containing specific additives) are very suitable for compact systems, such as car air conditioning (AC) systems.

Polyol ester oils (POE) show excellent miscibility with CO_2. There is complete miscibility between the refrigerating machine oil and the refrigerant CO_2, especially at low evaporation temperatures. For this reason, the negative effects on the heat transfer in the evaporator will be averted and the oil transport in large, branched tubing systems (e.g., in the supermarket) will also reliably take place. The use of an oil separator can be avoided, depending on the system configuration.

Polyol ester oils also show excellent stability even under the influence of the refrigerant CO_2.

9.3.2 Thermal stability

The thermal stability of the refrigeration oil is particularly important in transcritical applications. In mobile AC systems for example, gas temperatures of over 200 °C (392 °F) can easily be reached. Therefore, the refrigeration oil must be highly thermally resistant. Part of the typical test programme for the development of suitable CO_2 refrigeration oils is the storage at 220 °C (428 °F) for several weeks at 50 bar (725 psi) CO_2 pressure in suitable autoclaves (see Figure 9.3). After the storage period, the oil is analysed to determine whether the lubricant has started to decompose. A decomposition can manifest itself for example in an increase in the neutralisation number (acidification of the oil) or in a change in viscosity – or in general – in a change in the chemical structure which can be examined by means of an infrared spectrum.

When comparing the different types of refrigeration oil, a clear order can be drawn up in terms of thermal stability. Polyol ester oils (POE) have the highest thermal resistance, followed by PAG and PAO lubricants. Mineral oils on the other hand show only a limited thermal stability behaviour and are therefore not suitable for use with CO_2.

In addition to the base fluids, the additives in the CO_2 refrigeration oil also have to undergo the same high temperature stability testing.

Figure 9.3
Autoclaves for thermal stability research on CO_2 refrigeration oils (© Fuchs Schmierstoffe GmbH)

9.3.3 Lubrication properties

CO_2 as a high-pressure refrigerant has a very strong influence on the refrigeration oil viscosity. Due to the high pressures in the system, the lubricant will be well diluted. CO_2's very good solvent properties additionally increase the effect of such dilution. The consequence of CO_2 dissolving in the oil is a decrease in viscosity. This fact must be taken into account when preselecting the refrigeration oil by choosing a lubricant with a higher initial viscosity. It should also be ensured that the refrigeration oil contains anti-wear (AW) additives enhancing the wear protection in the contact area of the metal surfaces.

The influence of such highly effective AW additives can also be demonstrated in test bench trials. In special roller bearing lifetime tests, lubricants with suitable additives deliver significantly longer running times without wear scars on the rolling elements or raceways (see Figure 9.4) when compared to refrigeration oils without additives [9-1].

The working principle of AW additives is particularly important under mixed friction conditions. Mixed friction scenarios exist when the lubricating film becomes thinner and the rough tips of the metal surfaces come into direct contact, usually with a combination of a sliding and rolling action. The mixed friction scenario with materials in direct contact stands in contrast to the fluid friction or hydrodynamics regime with a separating lubricant film (= the lubrication gap), see Figure 9.5. Such mixed friction scenarios can be caused on the one hand by the high CO_2 oil dilution, or on

Figure 9.4 CO_2 high-pressure bearing test: comparison of the degree of wear (roller bearing and raceway) after 20 hours' running time (© Fuchs Schmierstoffe GmbH)

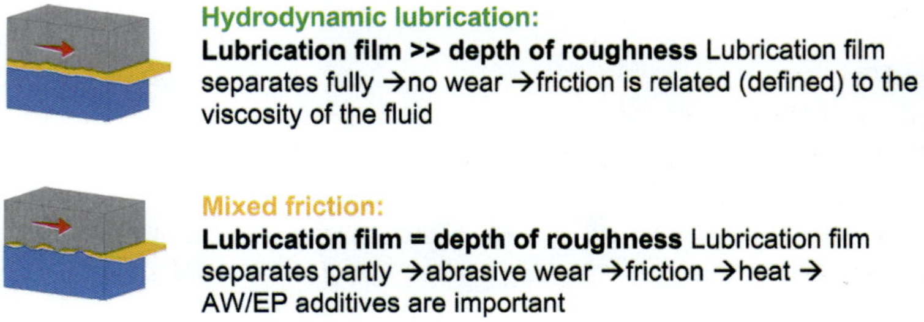

Figure 9.5
Various lubrication regimes: mixed friction (sliding / rolling) and fluid friction / hydrodynamics (© Fuchs Schmierstoffe GmbH)

the other hand, by high loads on the bearings, high temperatures in the lubricating film and the presence of low sliding speeds (when switching the compressor on and off). Under these conditions, suitable lubricant AW additives can form highly effective protective layers on the metal surface (the so-called reaction layers), which effectively protect against wear [9-1, 9-2, 9-3].

9.3.4 Water content

Reducing the water content in the refrigeration oil to a minimum is essential for trouble-free operation of the CO_2 refrigeration system. As a result of an increased water content in the refrigeration machine oil, malfunction and thus failure of the refrigeration system can occur (e.g., freezing of the expansion valve). In addition, the risk of corrosion of circuit components rises. Copper plating can also occur on highly loaded bearings and affect chemical stability of the oil-refrigerant system. The presence of increased water concentrations in the CO_2 cycle can lead to the formation of hydrates (clathrates). The formation of carbonic acid, which can attack materials in the refrigerant circuit, is also a possible consequence of a high water content in the CO_2 refrigeration oil.

Because of the high importance of a low water content in CO_2 systems, the maximum water content in refrigeration oils is limited to 150 ppm as a warning value in accordance with DIN 51503-2. From a manufacturing perspective, special care must also be taken in the production and storage of CO_2 refrigeration oils. Containers in which CO_2 refrigeration oils are delivered are always purged with dry nitrogen before filling and the oil is blanketed with dry nitrogen in order to maintain a low water content below 50 ppm (DIN 51503-1, for POE group KB: max. H_2O content 100 ppm). The penetration of moisture into the refrigeration oil poses a great danger, especially for POE and PAG lubricants, since these lubricants are highly hygroscopic (water-absorbing). Contact with the moist ambient air must be avoided with these lubricants [9-4].

9.3.5 Areas of application for CO_2 refrigeration oils

In some niche applications CO_2 refrigeration oils based on PAO are used. These are mostly large-scale refrigeration systems in which the low-temperature cascade contains CO_2 and another refrigerant (e.g., ammonia) is used for normal cooling. Highly efficient oil separators are installed in these systems to prevent the refrigeration machine oil from entering the refrigeration circuit and causing negative effects to the refrigerating capacity.

In terms of volume the largest group of CO_2 refrigeration oils are lubricants based on polyol esters (POE) with special anti-wear (AW) additives. These are used both in low-temperature cascades in supermarkets and for systems in which both the low and the medium-temperature circuits contain CO_2. For piston compressors in subcritical CO_2 applications a nominal kinematic viscosity of 55 mm^2/s (55 cSt resp. 5109500 ft^2/s) (at 40 °C / 104 °F) has generally proven to be most suitable for this group of refrigeration oils. In supercritical supermarket systems, POE lubricants higher than VG 80 are mainly used due to the higher pressures and the associated higher dilution. Higher viscosities are recommended for AC and heat pump applications. Highly viscous products such as ISO VG 170 oils are also used for screw compressors. The main area of application for CO_2 screw compressors are subcritical marine refrigeration systems. Very good practical experience has been gained with these POE-based CO_2 refrigeration oils over a period of more than fifteen years.

In contrast to this, for CO_2 in car AC applications special polyalkylene glycol (PAG)-based refrigeration oils are preferred. One of the reasons for this is the need for worldwide compatibility with the R134a/R1234yf AC oils which are also based on polyalkylene glycol.

Also, these special PAG-based CO_2 refrigeration oils have been used with success over the last years in other transcritical installations as heat pumps and stationary AC systems due to their favourable low dilution behaviour with CO_2.

9.4 Selection table – Refrigeration oil type for natural refrigerants

	POE	PAG	PAO	Alkylbenzene	Mineral Oil
NH_3	-- chemically reactive together with NH_3	only for DX systems / not generally recommended	++	+	+
Hydrocarbons (R290, R1270 etc.)	++	++	++	o	o
CO_2	++	+	-/o not miscible with CO_2	- not miscible and not sufficiently thermally stable	- not miscible and not sufficiently thermally stable

9.5 Literature

[9-1] Fuchs Schmierstoffe GmbH: *RENISO Kältemaschinenöle, (RENISO Refrigeration Oils)*, Mannheim: 2020

[9-2] BITZER Kühlmaschinenbau GmbH: *Kältemittel-Report 20 (Refrigerant Report 20)*, Sindelfingen: 2018

[9-3] Bock Kältemaschinen GmbH: *Schmierstofftabellen – Kältemaschinenöle für Kältemittelverdichter, (Lubricant tables - refrigeration oils for refrigeration compressors)*, Frickenhausen

[9-4] Research Council Refrigeration Technology: *Ölhaushalt in CO2-Kälteanlagen (Oil balance in CO_2 refrigeration plants)*, Research Report FKT 82/01, Frankfurt: March 2008

DIN 51503-1:2021-11 Lubricants – Refrigerator oils – Part 1: Minimum requirements

DIN 51503-2:2015-09 Lubricants – Refrigerator oils – Part 2: Used refrigerator oils

ISO 3016:2019-04 Petroleum and related products from natural or synthetic sources – Determination of pour point

Further reading

Bock, W., Puhl, C.: *Kältemaschinenöle (Refrigeration machine oils)*, VDE VERLAG GmbH, Berlin, Offenbach: 2010

10 Coolants

This chapter is based on the presentation by Roger Rosander, Temper Technology AB, at the Eurammon Symposium on 29 June 2018, in Schaffhausen, Switzerland, and completed by Michael Kauffeld and Michael Eckert. We thank Roger Rosander for providing us with text and graphics for this book, and Reto Centerhorn, Temper Technology AB, for valuable comments.

The limitations of fluorinated refrigerants are increasingly redirecting the focus towards so-called indirect refrigeration systems because many natural refrigerants such as ammonia, propane and isobutane are either toxic, harmful to health or flammable. Therefore, the refrigeration installation is usually set outside of buildings or public areas. The heat of the building or a cold room is then removed by the air conditioning or refrigeration system using a secondary fluid, known as a coolant, brine, or heat transfer fluid (HTF).

If the refrigerant is fed directly to the evaporators via the pipe network, this is called a direct system. However, if the refrigeration system cools down a coolant in a central evaporator, which is then led to the individual consumers via a separate pipe network, the system is an indirect system. Typical examples are the chilled-water circuits for cooling buildings. The schematic of an indirect system is shown in Figure 10.1. The intermediate coolant is usually water, a water/glycol mixture or water/other freezing point depressing substance such as salts (calcium chloride, potassium carbonate, potassium formate as well as potassium acetate and sodium chloride). For very cold applications (< –30 °C / –22 °F), other coolants based on oils or alcohols are used.

Using indirect systems, the refrigeration system can be built in a compact format and installed in a separate machine room. Due to the larger coolant pipe diameter and the additional equipment required for this (such as pumps and storage), the investment costs for indirect systems are usually higher than for direct evaporation systems. The advantages and disadvantages of the two systems must therefore be compared for each individual case in order to find the right solution for the individual application [10-1].

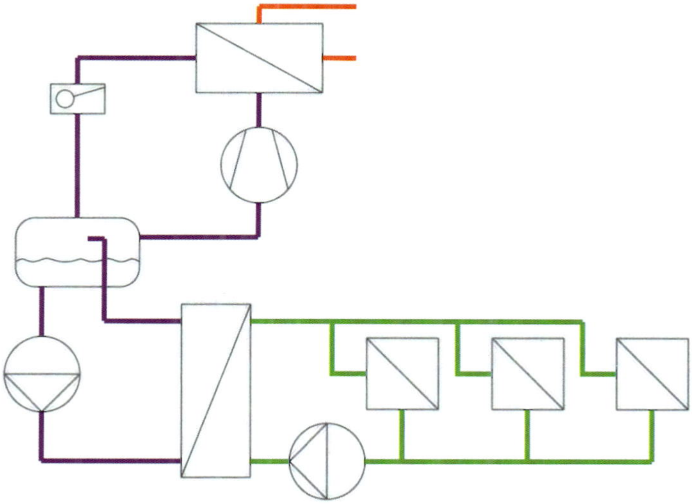

Figure 10.1
Indirect system

Advantages of indirect systems:

- No harmful or flammable refrigerants in common areas.
- If the primary refrigerant is ammonia or a hydrocarbon, workplace safety and security measures can be reduced to the actual refrigeration system located in the separate machine room.
- Commercially available pipes and low system pressures can be used for secondary coolant.
- Plastic pipes can be used; however, they must be diffusion-proof.
- Do not require certified welders.
- Not influenced by moisture or water.
- Cheaper installation of secondary coolant piping – cost effective and quick.
- Large selection of installation components available.
- Low refrigerant charge in primary system, resulting in lower refrigerant costs and less leakage potential.
- The refrigeration system is often designed using the (energy-efficient) flooded evaporation system for controlling the coolant circuit.
- The cooling machinery is located in separate, lockable machine rooms. All refrigerant-carrying system components are arranged within the same machine room.
- The actual refrigeration system can be prefabricated by suitable specialist refrigeration companies. On-site assembly is quick and efficient.
- The coolant system can be completed on site by experienced installers and hydraulic installation companies; this also applies to the extension or adaptation of the indirect coolant system.
- The workplace safety of people in the refrigerated area or in the refrigerated goods packing area is improved because only the coolant circulates in these areas.
- The cooling performance of a refrigeration system is more precisely and easily controlled with an indirect coolant.
- Indirect coolant enables more precise control of the temperature and humidity in the rooms to be cooled.
- Heat recovery on the condenser side (free heat) is easier to integrate into energy saving design (e.g. for defrost).
- Easier system expansion options.
- Coolant systems do not have to comply with pressure equipment directives (with the exception of evaporating CO_2).
- Risk of damage or loss of goods due to escaping refrigerant is largely excluded compared to direct evaporating systems. Damages or leaks in the secondary refrigerant piping network can be repaired more easily and quickly. It is noted that the cost of damage to goods can quickly exceed the additional costs of installing an indirect system.

Disadvantages of indirect systems:

- In certain cases, corrosive liquids limit choice of construction materials.
- Pressure loss in the coolant lines must be considered.
- Additional energy requirements for pumps needed to circulate coolants.
- Larger pipe diameters for coolants, necessitating cost increases.
- A further stage of heat transfer causing a possible negative impact on energy efficiency.
- The heat transfer capacity of directly evaporating heat exchangers is much higher than the output of heat exchangers that are used for secondary coolants. Therefore, heat exchangers for indirect systems have to be made larger to achieve the same cooling capacity.

Indirect systems are particularly suitable for food processing rooms and for air conditioning applications in office buildings.

The waste heat from indirect refrigeration systems can be used for defrosting, thus making electrical defrosting superfluous. Instead of heating energy costs, only the costs for operating the defrost pumps are incurred. Condensation or subcooling heat from a refrigeration system running at 30 °C (86 °F) is usually sufficient for defrosting. The ice on the air cooling heat exchangers is melted, and the defrost water can be led away, while during electrical defrosting (and especially when the temperature is too high), the ice evaporates, and the resulting steam quickly re-freezes on the evaporator tubes and fins. Defrosting with brine often has a payback period of not more than three to four years, and depending on the electricity price, it can be even a shorter time frame.

10.1 Requirements for coolants

All coolants have advantages and disadvantages, apart from perhaps water, which, however, can only be used above 0 °C (32 °F) in frost-free locations. When choosing a heat transfer fluid (HTF), the cooling needs of the application and the specific system must be taken into account. In addition, health and environmental risks, fire risk and, on the other hand, energy efficiency (thermophysical properties), corrosion (electrical conductivity), corrosion protection, installation (restrictions due to type of materials used), service life and life-cycle costs must be considered.

10.2 Types of heat transfer fluids (HTF)

Only single-phase HTFs are considered in this chapter. In particular for deep-freezing, evaporative CO_2 is also popular and widely used (see Chapters 5.5.1 and 10.4.1). In the area of medium-temperature cooling, ice slurry is an alternative to liquid HTFs (see Chapter 10.4.2).

There are different types of HTFs: water-based and organic liquids such as silicone oils or hydrocarbons. The liquids can be divided into different groups according to Table 10.1.

Table 10.1 Division of heat transfer fluids (HTF)

Group	Example
Alcohols	Ethanol
Glycols	Monoethylene glycol (MEG) Monopropylene glycol (MPG or 1,2-MPG) including PDO (1,3-Monopropylene glycol)
NH_3 / Water	Ammonia spirit
Organic fluids	e.g. Marlotherm XC
Salts / Brines	Inorganic: calcium chloride $CaCl_2$ Organic: potassium formate, potassium acetate

In the following, individual HTF liquids are examined in more detail.

Ethanol: 

- high viscosity even at high temperatures
- compatible with most materials of construction
- often no need for corrosion protection
- combustible at high ethanol concentrations, therefore normally usable down to –10/–15 °C (14 / 5 °F)
- risk of explosion when above 30 % ethanol in water, especially at elevated temperatures
- official safety approval may be required

Monoethylene glycol – EG [10–6]:

- good thermophysical properties for cooling and heat pump applications
- viewed as health risk as EG doesn't taste unpleasant, but is highly toxic for humans, the minimum lethal dose is 1–1,5 ml/kg or ≈ 100 ml (3.5 fl oz) concentrated glycol in an adult
- short-term exposure can result in irritation to eyes, skin and respiratory tract
- repeated or long-term exposure can result in kidney failure and brain injury
- EG is slightly flammable

Monopropylene glycol-1,2-propanediol – MPG

- propylene glycol has a high viscosity at low temperatures; this already increases sharply at –5 / –7 °C (23 / 19.4 °F)
- high viscosity leads to high energy consumption for the coolant pumps
- compatible with most construction materials
- propylene glycol is not toxic, the minimum lethal dose for an adult is 15 ml/kg, much higher than that of EG.

Ammonia in water – ammonium hydroxide NH_4OH

- very good thermal properties
- strong smell even with very low ammonia concentration
- very high pH (13–14), corrosive
- requires high quality material and definitely no copper
- the respective safety regulations for the use of ammonia must be observed

Hydrocarbons – for example Marlotherm XC or ethyl methylbenzene CAS No.: 98-82-8

- very low freezing point, below –90 °C (–130 °F); mainly used in industrial low-temperature refrigeration systems
- very low viscosity
- low specific heat and thermal conductivity
- compatible with most construction materials
- flammable
- fatal if swallowed
- respiratory tract irritation
- toxic to aquatic organisms

Calcium chloride $CaCl_2$

- very good thermal properties
- freezing point > –45 °C (–49 °F)
- high pH: 10–11
- cannot use stainless steel; titanium heat exchangers required
- very corrosive, dichromate inhibitors required, which have very negative health properties and environmental influences

Potassium acetate / formate salts and mixtures

- very good thermal properties
- very low freezing point < –60 °C (–76 °F)
- compatible with most construction materials
- not dangerous; acetate is a preservative E261
- pH: potassium acetate: 8–9, potassium formate 10–11
- electrical conductivity: potassium acetate: 135 mS/cm, potassium formate 250 mS/cm
- corrosive to zinc and soft solder

10.3 Liquid coolants for medium- and low-temperature applications

The individual coolants are compared below. A distinction is made between medium- and low-temperature applications.

10.3.1 Cooling of foodstuffs

Table 10.2 compares the coolants presented above with regard to pressure drop, heat transfer coefficient and pump energy requirement for normal cooling. The typical HTF inlet temperature for refrigerated cabinets or coldrooms is approximately –8 °C (17.6 °F). A suitable freezing point for liquid HTFs is then –15 °C (5 °F).

Table 10.2 shows that monoproplene glycol has a high viscosity and Marlotherm XC has a very low specific heat capacity and thermal conductivity.

Table 10.2 Material data from selected coolants for medium-temperature cooling

Coolant	Ethanol	MPG	NH_3/Water	Marlotherm XC	$CaCl_2$	Temper-15
Concentration in %	24,4	32,9	10,8	100	18	–
Density ρ in kg/m³	974	1,037	961	885	1168	1121
Dynamic viscosity η in 10^{-3} Pa s	9,38	12,95	2,51	0,91	3,70	4,16
Kinematic viscosity ν in mm²/s	9,63	12,49	2,61	1,03	3,17	3,71
Specific heat capacity c_p in J/(kg·K)	4,284	3,855	4,235	1,790	3,122	3,381
Thermal conductivity λ in W/(m·K)	0,419	0,411	0,477	0,135	0,535	0,486
Density ρ in lb/ft³	60.81	64.74	60.00	55.25	72.92	69.98
Dynamic viscosity η in 10^{-3} lb /(ft·s)	6.30	8.70	1.69	0.61	2.49	2.80
Kinematic viscosity ν in cSt	9.63	12.49	2.61	1.03	3.17	3.71
Specific heat capacity c_p in Btu/(lb·°F)	1.023	0.921	1.012	0.428	0.746	0.808
Thermal conductivity λ in Btu/(h·ft·°F)	0.242	0.237	0.276	0.078	0.309	0.281

The table is valid for a brine temperature of –8 °C (17.6 °F).

Table 10.3 Reynolds numbers of HTFs for normal cooling applications

Coolant	Reynolds number	Conditions:
Ethanol	1661	pipe diameter: 16 mm (0.63")
Monopropylene glycol (MPG)	1282	flow velocity: 1 m/s (3.28 ft/s)
NH_3	6134	freezing temp.: −15 °C (−5 °F)
Marlotherm XC	15504	operating temp.: −8 °C (−17.6 °F)
$CaCl_2$	5044	
Temper-15	4306	

Table 10.3 shows an example of the Reynolds numbers of the various HTFs for a typical 16 mm diameter pipe in a heat exchanger. It is clear that the flow is laminar with ethanol and propylene glycol, whereas it is turbulent with the other fluids. Turbulent flow is advantageous for heat transfer, see Figure 10.2.

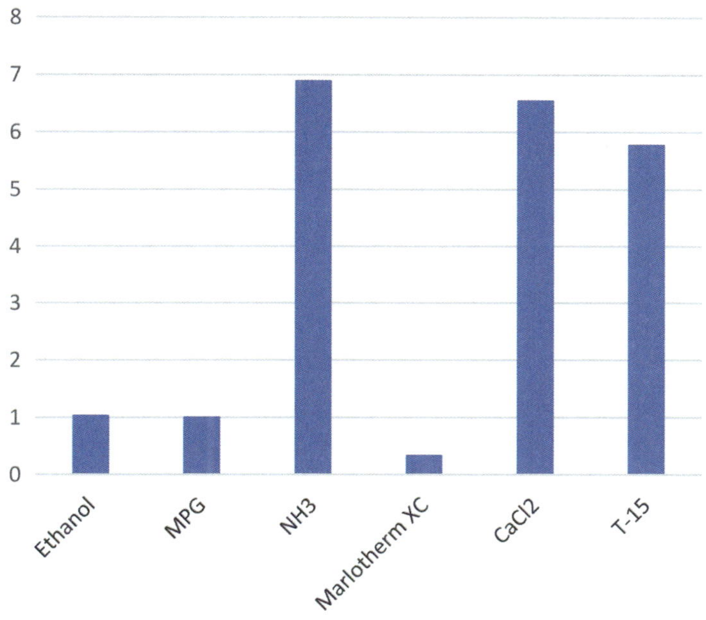

Figure 10.2 Relative heat transfer coefficient based on MPG (=1)

Despite laminar flow, ethanol and glycol have a high pressure drop, see Figure 10.3.

Due to the laminar flow, ethanol and glycol have a low heat transfer coefficient (see Figure 10.2). Ammonia, water and the salt solutions have a high heat transfer coefficient (turbulent flow). Marlotherm XC has a very low heat transfer coefficient due to its low specific heat capacity and thermal conductivity.

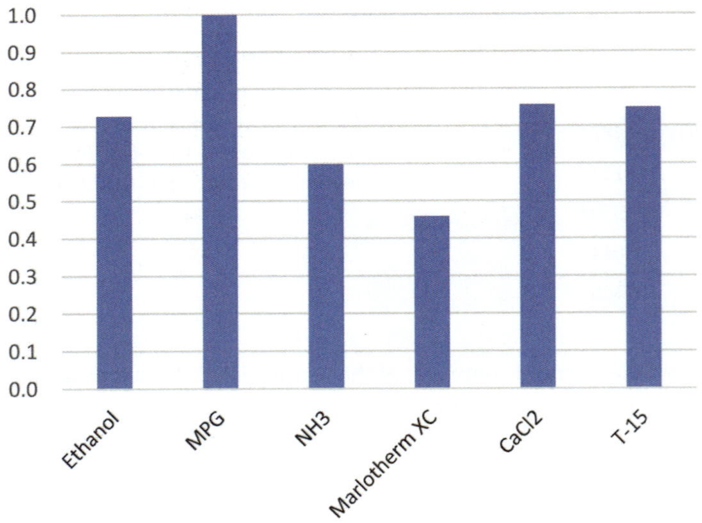

Figure 10.3
Relative pressure losses when compared to MPG (=1)

Energy requirements of the pumps

Using the thermo-physical properties of the different products, the ratio of the required pump power between two fluids can be calculated [10-2]. This value, which is called the Pump Power Demand Ratio (PPR), describes the amount of energy required to pump a liquid 1 relative to a liquid 2 so as to obtain the same heat transfer performance.

$$PPR_{12} = \left(\frac{\nu_1}{\nu_2}\right)^{1,95} \cdot \left(\frac{\rho_1}{\rho_2}\right)^{-0,15} \cdot \left(\frac{\lambda_1}{\lambda_2}\right)^{-2/3} \cdot \left(\frac{c_{p1}}{c_{p2}}\right)^{-1,05}$$

with

ν	kinematic viscosity
ρ	density
λ	thermal conductivity
c_p	specific heat capacity

Despite laminar flow, ethanol and glycols have a high energy consumption, see Figure 10.4.

Ammonia and the salt solutions have a low pumping energy requirement due to the good heat transfer properties and the low viscosity. Marlotherm XC lies in between.

10.3.2 Deep freezing food

The typical HTF inlet temperature for low-temperature units or rooms is approximately –32 °C (–25.6 °F). A suitable freezing point for liquid HTFs is then –40 °C (–40 °F).

Table 10.4 shows that MPG has a very high viscosity so that centrifugal pumps cannot be used. Although Marlotherm XC has a very low viscosity, its specific heat capacity and thermal conductivity are also very low.

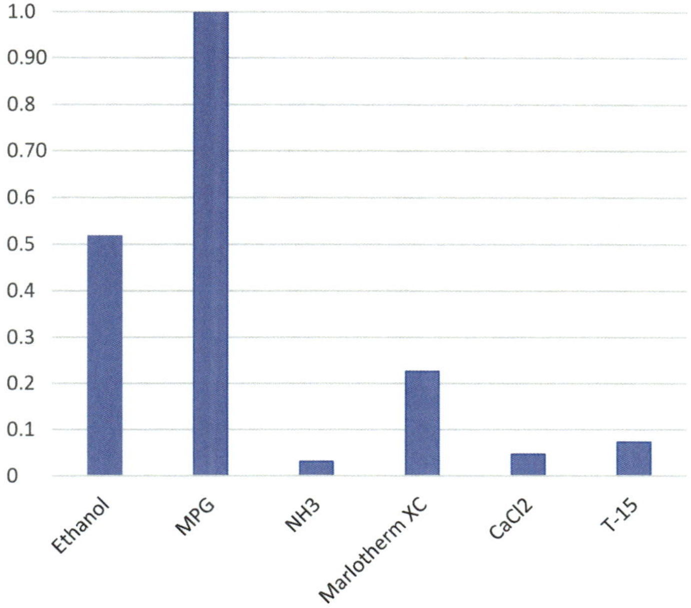

Figure 10.4
Relative pump energy requirement based on MPG (=1)

Table 10.4 Material properties from selected coolants for low-temperature cooling

Coolant	Ethanol	MPG	NH_3/Water	Marlotherm XC	$CaCl_2$	Temper-40
Concentration in %	53,1	54,0	21,0	100	28,3	–
Density ρ in kg/m³	949	1068	939	904	1258	1225
Dynamic viscosity η in 10^{-3} Pa s	42,90	364,7	7,85	1,41	17,02	27,65
Kinematic viscosity ν in mm²/s	45,22	341,5	8,33	1,56	13,24	22,57
Specific heat capacity c_p in J/(kg·K)	3367	3359	4296	1750	2682	2865
Thermal conductivity λ in W/(m·K)	0,294	0,321	0,385	0,140	0,490	0,408
Density ρ in lb/ft³	59.25	66.68	58.62	56.44	78.54	76.48
Dynamic viscosity η in 10^{-3} lb/(ft·s)	28.83	245.07	5.27	0.95	11.44	18.58
Kinematic viscosity ν in cSt	45.22	341.50	8.33	1.56	13.24	22.57
Specific heat capacity c_p in Btu/(lb·°F)	0.804	0.802	1.026	0.418	0.641	0.684
Thermal conductivity λ in Btu/(h·ft·°F)	0.170	0.185	0.222	0.081	0.283	0.236

The table is valid for a brine temperature of –32 °C (–25.6 °F).

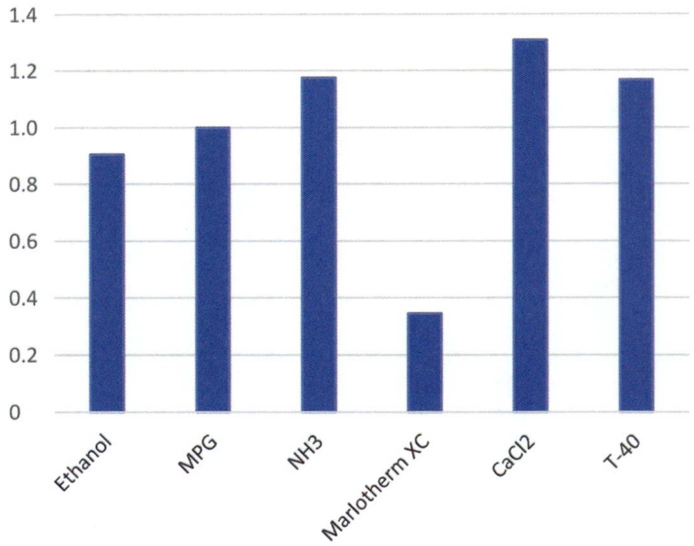

Figure 10.5 Relative heat transfer coefficient, based on MPG (=1)

Table 10.5 Reynolds numbers for coolants for low-temperature cooling applications

Coolant	Reynolds number	Conditions:
Ethanol	354	pipe diameter: 16 mm (0.63") freezing temp.: –40 °C (–40 °F) operating temp.: –32 °C (–26.6 °F)
MPG	47	
NH_3	1921	
Marlotherm XC	10286	
$CaCl_2$	1208	
Temper-15	709	

Table 10.5 shows an example of the Reynolds numbers of the various coolants for a typical 16 mm diameter pipe of a heat exchanger. It is clear that the flow is laminar with ethanol and propylene glycol, whereas it is turbulent with the other fluids. Turbulent flow is advantageous for heat transfer, see Figure 10.5.

Glycol shows a high pressure drop despite the laminar flow (see Figure 10.6). Saline solutions and ammonia have low pressure drops due to the laminar flow. Marlotherm XC (turbulent flow) has a very low viscosity.

There is almost no difference between the fluids with laminar flow. Only Marlotherm XC with its turbulent flow deviates due to the low specific heat capacity and thermal conductivity.

Figure 10.7 shows the pump power requirement ratio, on the left for normal cooling with MPG as reference fluid and on the right for low-temperature cooling (without MPG) with ethanol as reference fluid 2.

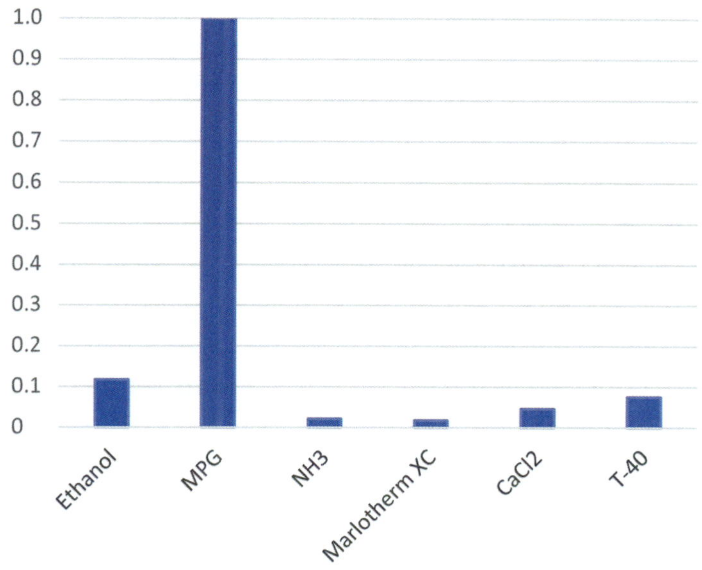

Figure 10.6
Relative pressure loss, based on MPG (=1)

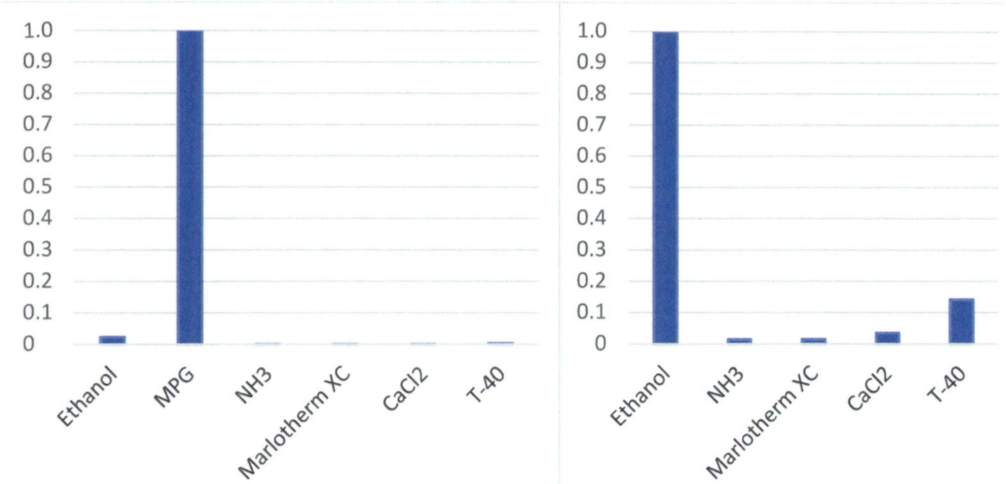

Figure 10.7 Relative pumping energy requirements; left: medium-temperature (MT) cooling w.r.t. MPG (=1); right: low-temperature (LT) cooling for better comparison without MPG, applicable to ethanol (=1).

10.4 HTF with phase change

In order to compensate for the disadvantage of the low energy content and the heat transfer properties of single-phase HTFs, which are sometimes poorer compared to evaporating refrigerants, it is possible to use HTFs with a phase change. Vaporising HTFs or those with a melting component are suitable here. Carbon dioxide has proven itself as an evaporating HTF in many hundreds of plants, particularly in the Scandinavian region. In the case of HTFs with a melting component, ice slurry is used in the area of medium-temperature cooling and partly for air conditioning. For air conditioning applications, aqueous solutions with a melting paraffin component can also be used.

10.4.1 Evaporating carbon dioxide

For *evaporating secondary fluids,* the secondary coolant is used in the same manner as is known for direct expansion refrigeration systems. The only difference is that the refrigerant which evaporates at the cooling duty point is usually not the same as that in the primary refrigeration system. So far, carbon dioxide (CO_2) has been used as an evaporating secondary refrigerant, see also chapter 5.5.1. The high standstill/idle pressures of CO_2 are of concern. But even whilst in operation, the actual pressures are significantly higher than with other HTFs or refrigerants – in operation at 0 °C (32 °F) pressure will be approximately 35 bar (507 psi). However, the pipes have very small diameters due to the very high heat of vaporisation of CO_2. As a result, the walls need not be very thick.

In addition, it must be ensured that non-vaporised liquid CO_2 runs back to the pumping station. The thermodynamic properties of CO_2 are so advantageous that such refrigeration systems do not consume more energy than comparable HFC direct expansion systems [10-3].

CO_2 as an evaporating coolant has advantages and disadvantages compared to liquid (water-based) coolants as listed in Table 10.6.

Table 10.6 Comparison of CO_2 as HTF versus water-based HTFs

Advantages	Disadvantages
Natural product Low pressure losses, small pumps, non-corrosion mediums, small temperature differences, with simultaneous use of CO_2 as coolant no temperature difference compared to HTF	Higher operating pressures, higher maintenance cost due to materials, only qualified competent pipe fitters, water sensitive, two-phase system

10.4.2 Ice Slurry

As a starting product for *melting coolants* (ice slurry), all water-based refrigerants described in chapter 10.2 can be used. So far, most practical experiences have been made using ethanol or glycol as an additive in indirect refrigeration systems. Sodium chloride (sea water) is also used especially in fisheries.

Ice slurry is a mixture of very fine ice particles in a liquid. The freezing point, e.g., for an 8 % ethanol / water mixture is –3,5 °C (25.7 °F). At this temperature the first water particles freeze. Frozen ice particles always contain pure water; therefore, this makes the remaining liquid richer in ethanol, and the freezing point drops. Again, the remaining liquid becomes richer in ethanol. With further cooling, the freezing point drops further.

If a change in temperature of the HTF (coolant) from –6 to –3,5 °C (21.2 to 25.7 °F) is allowed, the enthalpy change for the ice slurry is about twelve times greater than for a conventional liquid water-based coolant. With an ethanol content of 8 %, this temperature change in the ice slurry corresponds to a change in the ice concentration from 35 to 0 %. The ice slurry pressure drop at moderate ice concentrations, up to 25 % ice content, is more or less the same as with conventional HTFs. At the same time, however, the heat transfer coefficient is about twice as good for melting ice slurry as for conventional coolants/refrigerants [10-4]. The high heat content together with the good heat transfer values means that pipe diameters can be reduced by about 50 %, with approximately the same pressure losses. At the same time, the power requirement of the pump drops to approximately 1/8 of the size required for conventional coolants [10-5].

10.5 Corrosion protection

Most HTFs are based on aqueous solutions – in single or two-phase (ice slurry) – this can cause corrosion on metal surfaces. Corrosion requires an electrical potential difference in a metal or between two different metals, as well as an electrolyte, i.e. a liquid that can hold an electrical charge (electrons).

Corrosion caused by electrical potential differences between metals of different nobility is usually called galvanic corrosion. Corrosion can also be seen in material with only one metal owing to, for instance, concentration gradients in the electrolyte/fluid or impurities in the metal chemical structure or elsewhere; see Figure 10.8.

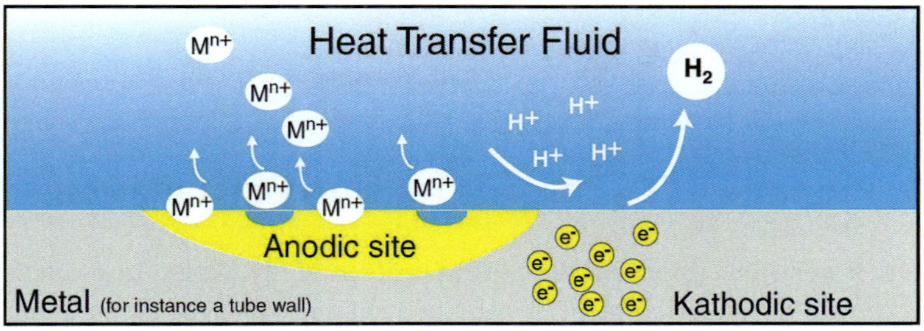

Figure 10.8
Typical electrolytic reaction

The pre-state of corrosion

The pre-state of corrosion is when the electrons in the cathodic area or the metal atoms, that have become positively charged, have not yet left the metal surface. This electrical potential diversion occurs all the time in electrical conductors such as metals. The charge displacements with the corresponding potential equalisation often take place without any corrosion occurring.

Traditional corrosion protection

Usually metals are protected by a so-called sacrificial anode or by adding a corrosion-protecting inhibitor. Traditional corrosion inhibitors form a mechanical protective film that prevents electron transport in the electrolyte and thus prevents metal ion release, see Figure 10.9. Unfortunately, the protective film not only reduces electron transport or metal ion release, but also heat transfer.

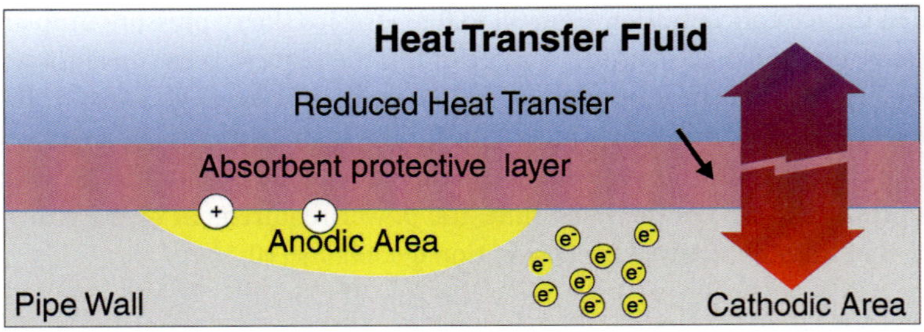

Figure 10.9
Principle of traditional corrosion protection

Adsorption inhibitors: Due to electrochemical bonds, they accumulate on the metal surface at the initial stage of corrosion and form a reliable, local protective film, see Figure 10.10. If necessary, adsorption corrosion inhibitors form protective films of a monomolecular (very low) thickness that do not (measurably) influence the heat transfer. This enables almost perfect heat transfer from the liquid to the metal or from the metal to the liquid in heat exchangers.

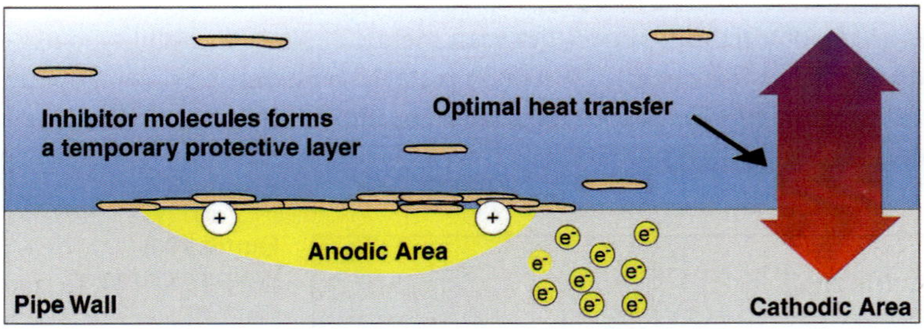

Figure 10.10
Principle of adsorption inhibitors

10.6 Case study: Propane – CO_2 cascade ice slurry installation

In 2013, 10 dysfunctional refrigeration systems in the central canteen of the Karlsruhe Student Union were replaced by a central refrigeration plant. A container set up in front of the building (see Figure 10.11), holds three propane refrigeration systems which produce ice slurry between 3 pm and 9 am. This is stored in a 40 m^3 (1412 ft^3) vertical vessel and supplies the entire cafeteria with –4 °C (24.8 °F) cold ice slurry during the day. All medium-temperature cooling points are cooled with this ice slurry. There are seven decentralised CO_2 refrigeration systems installed for the low-temperature rooms. These eject their desuperheat to the cellar air. The CO_2 is liquefied at approximately 0 °C (32 °F), i.e. 35 bar (507 psi). The heat of condensation is absorbed by the ice slurry circuit. Instead of control valves, small, decentralised, speed-controlled centrifugal pumps take over the individual supply of the cooling duty points with ice slurry. For future heat recovery, and also in order to keep the propane charge quantities small, the three independent propane refrigeration systems transfer their condenser heat to a glycol circuit, which in turn is cooled by ambient air until the heat recovery will be installed.

Overall, by lowering the condensing temperature at night and using the energy-efficient ice slurry as well as the energetically optimal CO_2 low-temperature cooling cycles, an installation energy consumption can be achieved that corresponds to that of a direct expansion HFC system. Due to the exclusive use of natural refrigerants with very low greenhouse potential, the

Figure 10.11
Machine container and ice slurry storage vessel

CO_2 balance of this plant is better than that of a comparable HFC plant and significantly better than that of the old plants.

Figure 10.11 shows the machine plant room container with control cabinet space (left) and emergency stop switch, which is separated airtight from the machine room (right). All components of the three propane refrigeration systems are housed in the right part. This space has a propane gas sensor at floor level and corresponding emergency ventilation (air outlet to the right of the window). On the right-hand side of the picture the 40 m³ ice slurry vertical storage vessel is shown.

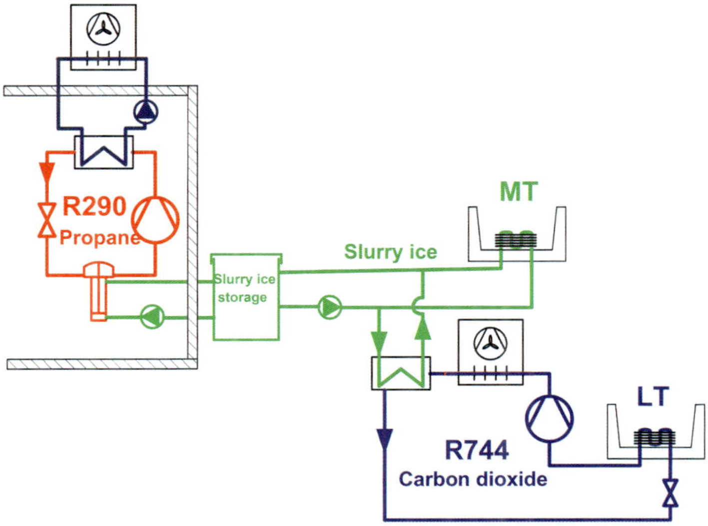

Figure 10.12
Schematic layout of propane ice slurry CO_2 refrigeration system in the Adenauerring cafeteria of the Karlsruhe Student Union.

Technical specifications:

- 6 ice makers with scrapers with a cooling capacity of 14 kW (3.98 TR) each (approximately 3,5 metric tons / 3.85 $tons_{US}$ / 3.44 $tons_{Imp}$ of ice in 24 hours each)
- 3 separate propane cooling circuits (each containing 10 kg / 22 lb of R290)
- Cooling of the condensers with ethylene glycol circuit
- 7,5 % ethanol in ice slurry fluid, ice content 5 to 25 %
- 40 m^3 (1412 ft^3) cylindrical storage (energy content at 25 %, ice content approximately 900 kWh / 3070000 Btu)
- Ice production from 3 pm to no later than 9 am
- Propane refrigeration systems switched off from 9 am to 3 pm

10.7 Integral system for cooling goods, air conditioning and heating in supermarkets

Simon Brieden, Roland Handschuh

This system, developed by Cool Expert GmbH, uses a combination of R290 (propane) for > 0 °C (32 °F) cooling and R744 (CO_2, subcritical) for deep-freezing. It is characterised by high environmental friendliness and high energy efficiency.

This system was developed as a demonstration and sample plant. In order to gain extensive experience in the use of natural refrigerants, a plant consisting of a service counter, a closed-door-type and an open-type refrigerated shelf, a deep-freeze shelf, a liquid-cooled plug-in freezer, a deep-freeze room and ceiling air-conditioning units were installed, see Figure 10.13.

Figure 10.13
Demonstration and sample market building with refrigerated units (© Cool Expert GmbH)

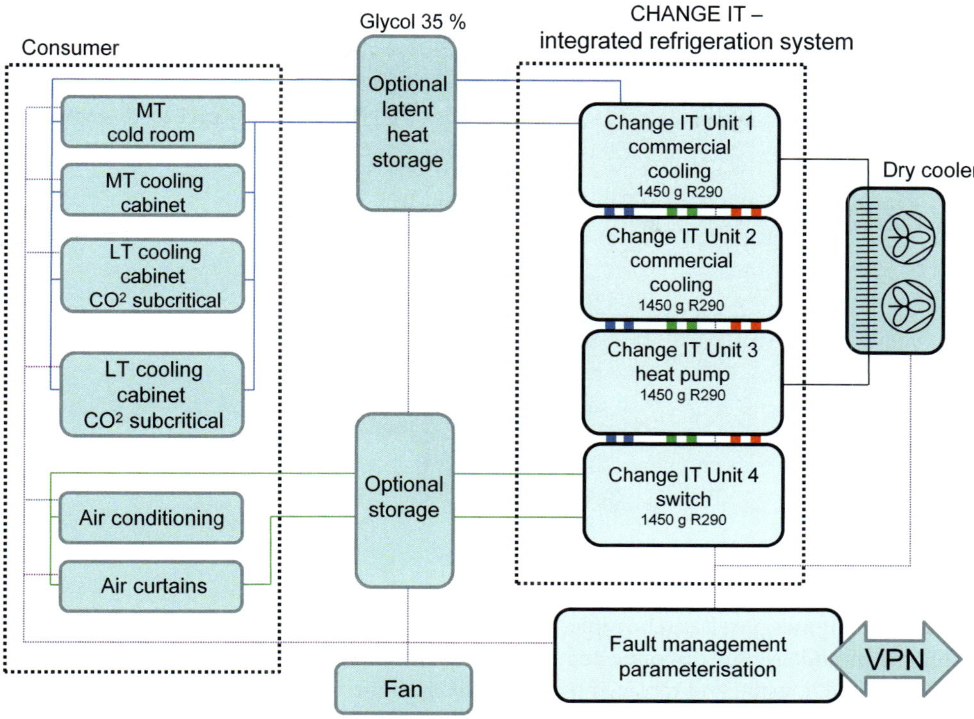

Figure 10.14 System overview integral refrigeration system CHANGE IT (© Cool Expert GmbH)

The heat is transported via a load-adapted coolant system with temperatures for commercial cold rooms between –8 and –4 °C (17.6 and 24.8 °F), for air conditioning in summer between 6 and 15 °C (42.8 and 59 °F), and for heating between 30 and 50 °C (86 and 122 °F). A control algorithm continuously calculates the actual needs of the consumers and determines the most energy-efficient supply temperature.

The entire heat management is carried out by the modular, built-in refrigeration system. In the example, this is a system with two commercial cooling modules, a heat pump module and a switch-over module, which can be used alternately, see Figure 10.14. The modules are each charged with propane with a refrigerant charge of ≤ 1450 g (3.3 lb) and provide a total cooling capacity of approximately 100 kW (28.4 TR). The commercial refrigeration modules supply the medium-temperature cooling supply points directly and the low-temperature supply points indirectly. In heating mode, the entire cooling capacity and the electrical drive capacity of the compressor can be recovered for heating the interiors. A heat exchanger collects heat from the outside air. It ejects heat when it is not needed for heating purposes and absorbs heat when the heat recovered from the commercial refrigeration system is not sufficient to heat the building.

The heat pump module absorbs heat from the outside air in winter to heat the building and emits the heat to the outside environment to cool the interiors in summer.

The switch-over module is used as a heat pump module in normal operation. In the event of a malfunction in a commercial cooling module, this is automatically replaced by the switch module. This ensures system availability for commercial refrigeration even if one module fails.

Figure 10.15
Integral cooling / heat management system including hydraulic modules (© Cool Expert GmbH)

If a module malfunctions, it can be replaced within 30 minutes without opening the refrigerant circuit. The module is only disconnected electrically and on the glycol side and can then be removed from the housing and replaced by a spare device. The module is repaired at the factory, tested for performance and is available again as an exchange module.

The system can optionally be equipped with heat or cold storage, for example to avoid power peaks via load management or to reserve heat that is currently not required for later heating processes.

The heart of this integral system is a control system that enables and evaluates communication between all components. In this way it can be ensured at all times that the entire system is running with the least energy consumption.

The cooling of the > 0 °C (32 °F) cooling duty points takes place directly via decentralised pumps, which also ensures the hydraulic balancing of the glycol circuit. If there is a need for defrosting, an indirect electrical heat exchanger (Defro Power Pack) is used.

Defrosting is initiated by switching a three-way valve. The glycol of the respective cooler is warmed up, see Figure 10.16. With this hot brine defrost, each core tube of the cooler has full hot brine flow, so that a complete defrost takes place in a short time. Another advantage is that the goods to be cooled do not have to be stored elsewhere when replacing the defrost heater. The Defro Power Pack works with maximum glycol temperatures of 15 °C (59 °F), which minimises the entry of superfluous heat into the defrosted refrigerator room. In combination with the MIC QKL e3 refrigeration controller (Figure 10.17) with integrated latent heat management and adaptive defrost detection, it is proven that up to 80 % defrost energy can be saved on the refrigeration side compared to conventional defrosting with electric heating elements.

At the low-temperature supply points, the generated heat is removed by a subcritical CO_2 installation. The heat is transported and fed to the glycol circuit via a decentralised pump; see Figure 10.18.

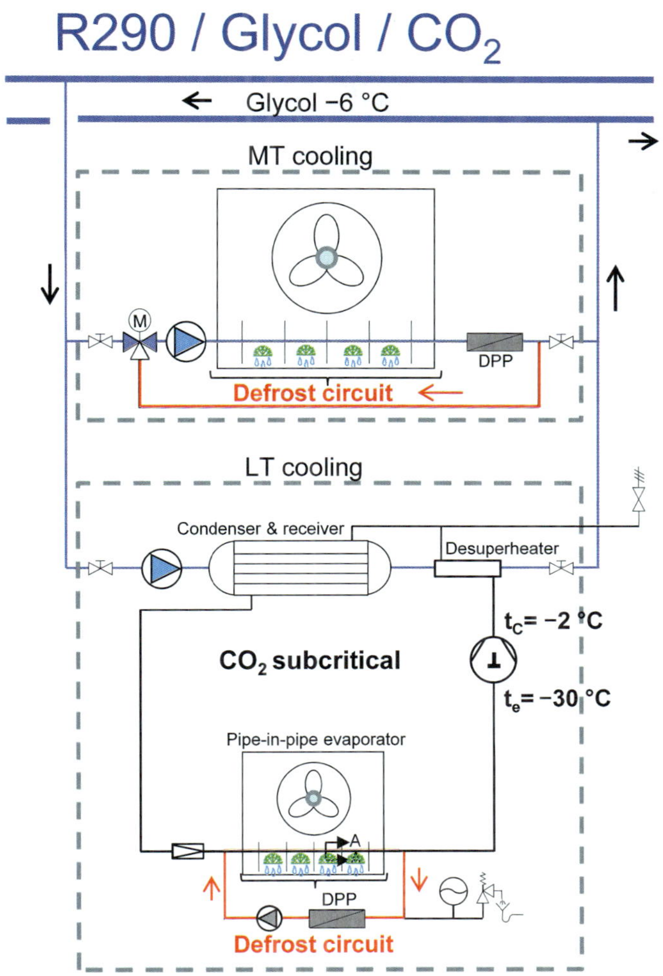

Figure 10.16
Connection of the > 0 °C (32 °F) and deep-freeze duty points to the glycol circuit with Defro Power Pack defrosting (© Cool Expert GmbH)

Figure 10.17
MIC QKL e3 refrigeration controller with adaptive defrost detection and latent heat management (© Cool Expert GmbH)

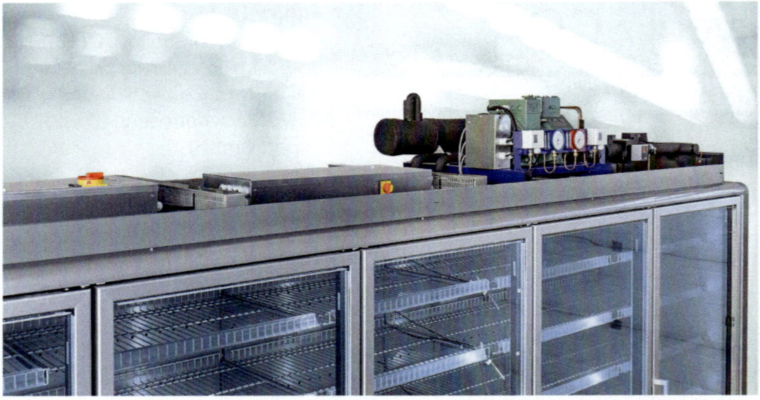

Figure 10.18 Subcritical CO_2 unit on the freezer wall shelf in the demonstration market building (© Cool Expert GmbH)

Also, the CO_2 evaporator is defrosted with Defro Power Pack. The evaporators are equipped with tube-in-tube technology. The outer pipe is filled with glycol, the refrigerant CO_2 flows in the inner pipe during the cooling process, see Figure 10.19. During the defrost, the glycol is heated in the jacket area, so that even in this case there is even temperature distribution and defrosting over the whole surface area of the evaporator coil. The CO_2 evaporation process is stabilised due to the heat store effect of the glycol. With this pipe-in-pipe system, too, compared to electric defrosting, no goods have to be removed from storage in the event of a malfunction.

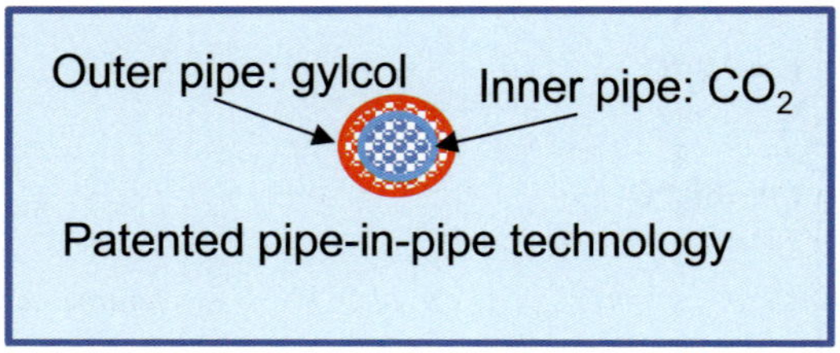

Figure 10.19 Pipe-in-pipe technology

The building is equipped with glycol air conditioning units. Depending on requirements, these units can be used for cooling or heating the building.

Warm or cold glycol is provided via the same two-pipe network to the air conditioning units. The waste heat from the existing refrigeration system is primarily used for this heating. The heat pump module is only switched on when the waste heat is no longer sufficient.

The regulation and control work hand in hand with the monitoring system, see Figure 10.20.

Via an integrated data recording PLC, all relevant data of the entire system are recorded, are transmitted via the internet, and can be viewed and analysed in the event of service needs. In addition, energy consumption measurements are carried out on all electrical consumers for energy assessments and benchmark comparisons for different systems. The heat quantities of the different glycol circuits are used to calculate the coefficient of performance for the heat pump circuit and the energy efficiency ratio for the refrigeration cycle. Energy and HACCP reports are sent regularly by email.

The fault message management enables user-defined messages that are precisely adapted to the system and configured and sent according to priorities. This ensures maximum system availability and reliability. In addition, the service company can react quickly in the event of a malfunction. This allows the service deployment to be planned in advance by evaluating the records.

Figure 10.20 Switchboard showing monitoring and fault demonstration management (© Cool Expert GmbH)

10.8 Literature

[10-1] Rosander, R.: *Heat Transfer Fluids - an overview*. Eurammon Symposium "Challenges of the F-Gas Regulation and Eco-Design Directive in 2018 and Compelling Solutions with Natural Refrigerants", June 28–29, 2018, Schaffhausen, Switzerland

[10-2] Sherwood, G.: *Secondary Heat Transfer Systems and the Application of a New Hydroflouroether*, 1995 International CFC and Halon Conference

[10-3] Rees, B., Rohrer, C., Shapiro, D., Trachta, J.: *Performance testing and comparison of liquid overfeed and cascade CO2 systems with R404A primary*. IIR 22nd Int. Congr. of Refr., Beijing, China, 21-26 August 2007, B2-123

[10-4] Kauffeld, M., Kawaji, M., Egolf, P.W. et al.: *Handbook on Ice Slurries. International Institute of Refrigeration*. 2005

[10-5] Kauffeld, M., Christensen, K.G., Lund, S., Hansen, T.: *Experience with ice slurry*. Int. Congr. of Refr., Sydney, Australia, September 19-24, 1999

[10-6] Melinder, Å.: *Properties of Secondary Working Fluids (Secondary Refrigerants or Coolants, Heat Transfer Fluids) for Indirect Systems*. 2nd edition, Publication date: 2010, ISBN: 9782913149830

Appendix

Conversion Table SI units to Imperial / US units

SI		IU	
1	mm	0.03937	in
1	m	39,37	in
1	m	3.2808	ft
1	mm^2	0.00155	in^2
1	m^2	10.764	ft^2
1	m^3	35.3147	ft^3
1	l	0.22	gal (Imp)
1	l	0.2642	gal (US)
1	m/s	3.28	ft/s
1	m^3/h	0.58857833	ft^3/m = cfm
1	m^3/h	35.3147	ft^3/h = cfh
1	g	0.03527337	oz
1	kg	2.20462	lb
1	t	0.9843	$tons_{IMP}$
1	t	1.102	$tons_{US}$
1	m^3/kg	16.02	ft^3/lb
1	dm^3/kg	0.01602	ft^3/lb
1	mg/m^3	9.9885E-07	oz/ft^3
1	mg/m^3	1.6036E-07	oz/gal (Imp.)
1	mg/m^3	1.3353E-07	oz/gal (US)
1	g/m^3	0.00099885	oz/ft^3
1	g/m^3	0.07015689	gr/gal (Imp.)
1	g/m^3	0.05841783	gr/gal (US)
1	kg/m^3	0.06243	lb/ft^3
1	kg/dm^3	62.4219725	lb/ft^3
1	kg/h	2.20462	lb/h
1	bar	14.5038	psi
1	bar	67196.8975	pdl/ft^2
1	Pa	0.67196898	pdl/ft^2

SI		IU	
1	Pa	0.00014504	psi
1	Pa s	0.02088534	lbf·s/ft²
1	Pa s	0.67196898	pdl·s/ft²
1	Pa s	0.67196895	lb/(ft·s)
1	g/mol	1	lb/lbmol
1	g/mol	0.0022046	lb/mol
0	°C	32 [· 1.8 + 32]	°F
1	kW	0.28434514	TR
1	kW	3414.13452	Btu/h
1	kWh	3412.8	Btu
1	kJ/(kg · K)	0.2388459	Btu/(lb °F)
1	W/(m · K)	0.57778921	Btu/(h·ft·°F)
1	W/(m · K)	6.93347126	Btu·in/h·ft²·°F
1	W/(m² · K)	0.17611019	Btu/(h·ft²·°F)
1	kg/kW	0.0006461	lb/Btu(IT)/h
1	kg/kW	2.32597615	lb/Btu(IT)/s
1	kJ/kg	0.42992261	Btu/lb
1	kJ/m³	0.02685	Btu/ft³
1	mm²/s	1	cSt
1	mm²/s	1.0764E-05	ft²/s
1	mm²/s	0.00155	in²/s